Givan

to assist her in carrying the torch of SPC Implementation for ASQ and the world.

John T. Burr

Elementary Statistical Quality Control

Statistical Distributions
Professor N. Balakrishnan
McMaster University

Statistical Process Improvement
Professor G. Geoffrey Vining
Virginia Polytechnic Institute

Stochastic Processes
Professor V. Lakshmikantham
Florida Institute of Technology

Survey Sampling
Professor Lynne Stokes
Southern Methodist University

Time Series
Sastry G. Pantula
North Carolina State University

1. The Generalized Jackknife Statistic, *H. L. Gray and W. R. Schucany*
2. Multivariate Analysis, *Anant M. Kshirsagar*
3. Statistics and Society, *Walter T. Federer*
4. Multivariate Analysis: A Selected and Abstracted Bibliography, 1957–1972, *Kocherlakota Subrahmaniam and Kathleen Subrahmaniam*
5. Design of Experiments: A Realistic Approach, *Virgil L. Anderson and Robert A. McLean*
6. Statistical and Mathematical Aspects of Pollution Problems, *John W. Pratt*
7. Introduction to Probability and Statistics (in two parts), Part I: Probability; Part II: Statistics, *Narayan C. Giri*
8. Statistical Theory of the Analysis of Experimental Designs, *J. Ogawa*
9. Statistical Techniques in Simulation (in two parts), *Jack P. C. Kleijnen*
10. Data Quality Control and Editing, *Joseph I. Naus*
11. Cost of Living Index Numbers: Practice, Precision, and Theory, *Kali S. Banerjee*
12. Weighing Designs: For Chemistry, Medicine, Economics, Operations Research, Statistics, *Kali S. Banerjee*
13. The Search for Oil: Some Statistical Methods and Techniques, *edited by D. B. Owen*
14. Sample Size Choice: Charts for Experiments with Linear Models, *Robert E. Odeh and Martin Fox*
15. Statistical Methods for Engineers and Scientists, *Robert M. Bethea, Benjamin S. Duran, and Thomas L. Boullion*
16. Statistical Quality Control Methods, *Irving W. Burr*

17. On the History of Statistics and Probability, *edited by D. B. Owen*
18. Econometrics, *Peter Schmidt*
19. Sufficient Statistics: Selected Contributions, *Vasant S. Huzurbazar (edited by Anant M. Kshirsagar)*
20. Handbook of Statistical Distributions, *Jagdish K. Patel, C. H. Kapadia, and D. B. Owen*
21. Case Studies in Sample Design, *A. C. Rosander*
22. Pocket Book of Statistical Tables, *compiled by R. E. Odeh, D. B. Owen, Z. W. Birnbaum, and L. Fisher*
23. The Information in Contingency Tables, *D. V. Gokhale and Solomon Kullback*
24. Statistical Analysis of Reliability and Life-Testing Models: Theory and Methods, *Lee J. Bain*
25. Elementary Statistical Quality Control, *Irving W. Burr*
26. An Introduction to Probability and Statistics Using BASIC, *Richard A. Groeneveld*
27. Basic Applied Statistics, *B. L. Raktoe and J. J. Hubert*
28. A Primer in Probability, *Kathleen Subrahmaniam*
29. Random Processes: A First Look, *R. Syski*
30. Regression Methods: A Tool for Data Analysis, *Rudolf J. Freund and Paul D. Minton*
31. Randomization Tests, *Eugene S. Edgington*
32. Tables for Normal Tolerance Limits, Sampling Plans and Screening, *Robert E. Odeh and D. B. Owen*
33. Statistical Computing, *William J. Kennedy, Jr., and James E. Gentle*
34. Regression Analysis and Its Application: A Data-Oriented Approach, *Richard F. Gunst and Robert L. Mason*
35. Scientific Strategies to Save Your Life, *I. D. J. Bross*
36. Statistics in the Pharmaceutical Industry, *edited by C. Ralph Buncher and Jia-Yeong Tsay*
37. Sampling from a Finite Population, *J. Hajek*
38. Statistical Modeling Techniques, *S. S. Shapiro and A. J. Gross*
39. Statistical Theory and Inference in Research, *T. A. Bancroft and C.-P. Han*
40. Handbook of the Normal Distribution, *Jagdish K. Patel and Campbell B. Read*
41. Recent Advances in Regression Methods, *Hrishikesh D. Vinod and Aman Ullah*
42. Acceptance Sampling in Quality Control, *Edward G. Schilling*
43. The Randomized Clinical Trial and Therapeutic Decisions, *edited by Niels Tygstrup, John M Lachin, and Erik Juhl*
44. Regression Analysis of Survival Data in Cancer Chemotherapy, *Walter H. Carter, Jr., Galen L. Wampler, and Donald M. Stablein*
45. A Course in Linear Models, *Anant M. Kshirsagar*
46. Clinical Trials: Issues and Approaches, *edited by Stanley H. Shapiro and Thomas H. Louis*
47. Statistical Analysis of DNA Sequence Data, *edited by B. S. Weir*
48. Nonlinear Regression Modeling: A Unified Practical Approach, *David A. Ratkowsky*
49. Attribute Sampling Plans, Tables of Tests and Confidence Limits for Proportions, *Robert E. Odeh and D. B. Owen*

50. Experimental Design, Statistical Models, and Genetic Statistics, *edited by Klaus Hinkelmann*
51. Statistical Methods for Cancer Studies, *edited by Richard G. Cornell*
52. Practical Statistical Sampling for Auditors, *Arthur J. Wilburn*
53. Statistical Methods for Cancer Studies, *edited by Edward J. Wegman and James G. Smith*
54. Self-Organizing Methods in Modeling: GMDH Type Algorithms, *edited by Stanley J. Farlow*
55. Applied Factorial and Fractional Designs, *Robert A. McLean and Virgil L. Anderson*
56. Design of Experiments: Ranking and Selection, *edited by Thomas J. Santner and Ajit C. Tamhane*
57. Statistical Methods for Engineers and Scientists: Second Edition, Revised and Expanded, *Robert M. Bethea, Benjamin S. Duran, and Thomas L. Boullion*
58. Ensemble Modeling: Inference from Small-Scale Properties to Large-Scale Systems, *Alan E. Gelfand and Crayton C. Walker*
59. Computer Modeling for Business and Industry, *Bruce L. Bowerman and Richard T. O'Connell*
60. Bayesian Analysis of Linear Models, *Lyle D. Broemeling*
61. Methodological Issues for Health Care Surveys, *Brenda Cox and Steven Cohen*
62. Applied Regression Analysis and Experimental Design, *Richard J. Brook and Gregory C. Arnold*
63. Statpal: A Statistical Package for Microcomputers—PC-DOS Version for the IBM PC and Compatibles, *Bruce J. Chalmer and David G. Whitmore*
64. Statpal: A Statistical Package for Microcomputers—Apple Version for the II, II+, and IIe, *David G. Whitmore and Bruce J. Chalmer*
65. Nonparametric Statistical Inference: Second Edition, Revised and Expanded, *Jean Dickinson Gibbons*
66. Design and Analysis of Experiments, *Roger G. Petersen*
67. Statistical Methods for Pharmaceutical Research Planning, *Sten W. Bergman and John C. Gittins*
68. Goodness-of-Fit Techniques, *edited by Ralph B. D'Agostino and Michael A. Stephens*
69. Statistical Methods in Discrimination Litigation, *edited by D. H. Kaye and Mikel Aickin*
70. Truncated and Censored Samples from Normal Populations, *Helmut Schneider*
71. Robust Inference, *M. L. Tiku, W. Y. Tan, and N. Balakrishnan*
72. Statistical Image Processing and Graphics, *edited by Edward J. Wegman and Douglas J. DePriest*
73. Assignment Methods in Combinatorial Data Analysis, *Lawrence J. Hubert*
74. Econometrics and Structural Change, *Lyle D. Broemeling and Hiroki Tsurumi*
75. Multivariate Interpretation of Clinical Laboratory Data, *Adelin Albert and Eugene K. Harris*
76. Statistical Tools for Simulation Practitioners, *Jack P. C. Kleijnen*
77. Randomization Tests: Second Edition, *Eugene S. Edgington*

78. A Folio of Distributions: A Collection of Theoretical Quantile-Quantile Plots, *Edward B. Fowlkes*
79. Applied Categorical Data Analysis, *Daniel H. Freeman, Jr.*
80. Seemingly Unrelated Regression Equations Models: Estimation and Inference, *Virendra K. Srivastava and David E. A. Giles*
81. Response Surfaces: Designs and Analyses, *Andre I. Khuri and John A. Cornell*
82. Nonlinear Parameter Estimation: An Integrated System in BASIC, *John C. Nash and Mary Walker-Smith*
83. Cancer Modeling, *edited by James R. Thompson and Barry W. Brown*
84. Mixture Models: Inference and Applications to Clustering, *Geoffrey J. McLachlan and Kaye E. Basford*
85. Randomized Response: Theory and Techniques, *Arijit Chaudhuri and Rahul Mukerjee*
86. Biopharmaceutical Statistics for Drug Development, *edited by Karl E. Peace*
87. Parts per Million Values for Estimating Quality Levels, *Robert E. Odeh and D. B. Owen*
88. Lognormal Distributions: Theory and Applications, *edited by Edwin L. Crow and Kunio Shimizu*
89. Properties of Estimators for the Gamma Distribution, *K. O. Bowman and L. R. Shenton*
90. Spline Smoothing and Nonparametric Regression, *Randall L. Eubank*
91. Linear Least Squares Computations, *R. W. Farebrother*
92. Exploring Statistics, *Damaraju Raghavarao*
93. Applied Time Series Analysis for Business and Economic Forecasting, *Sufi M. Nazem*
94. Bayesian Analysis of Time Series and Dynamic Models, *edited by James C. Spall*
95. The Inverse Gaussian Distribution: Theory, Methodology, and Applications, *Raj S. Chhikara and J. Leroy Folks*
96. Parameter Estimation in Reliability and Life Span Models, *A. Clifford Cohen and Betty Jones Whitten*
97. Pooled Cross-Sectional and Time Series Data Analysis, *Terry E. Dielman*
98. Random Processes: A First Look, Second Edition, Revised and Expanded, *R. Syski*
99. Generalized Poisson Distributions: Properties and Applications, *P. C. Consul*
100. Nonlinear L_p-Norm Estimation, *Rene Gonin and Arthur H. Money*
101. Model Discrimination for Nonlinear Regression Models, *Dale S. Borowiak*
102. Applied Regression Analysis in Econometrics, *Howard E. Doran*
103. Continued Fractions in Statistical Applications, *K. O. Bowman and L. R. Shenton*
104. Statistical Methodology in the Pharmaceutical Sciences, *Donald A. Berry*
105. Experimental Design in Biotechnology, *Perry D. Haaland*
106. Statistical Issues in Drug Research and Development, *edited by Karl E. Peace*
107. Handbook of Nonlinear Regression Models, *David A. Ratkowsky*

108. Robust Regression: Analysis and Applications, *edited by Kenneth D. Lawrence and Jeffrey L. Arthur*
109. Statistical Design and Analysis of Industrial Experiments, *edited by Subir Ghosh*
110. *U*-Statistics: Theory and Practice, *A. J. Lee*
111. A Primer in Probability: Second Edition, Revised and Expanded, *Kathleen Subrahmaniam*
112. Data Quality Control: Theory and Pragmatics, *edited by Gunar E. Liepins and V. R. R. Uppuluri*
113. Engineering Quality by Design: Interpreting the Taguchi Approach, *Thomas B. Barker*
114. Survivorship Analysis for Clinical Studies, *Eugene K. Harris and Adelin Albert*
115. Statistical Analysis of Reliability and Life-Testing Models: Second Edition, *Lee J. Bain and Max Engelhardt*
116. Stochastic Models of Carcinogenesis, *Wai-Yuan Tan*
117. Statistics and Society: Data Collection and Interpretation, Second Edition, Revised and Expanded, *Walter T. Federer*
118. Handbook of Sequential Analysis, *B. K. Ghosh and P. K. Sen*
119. Truncated and Censored Samples: Theory and Applications, *A. Clifford Cohen*
120. Survey Sampling Principles, *E. K. Foreman*
121. Applied Engineering Statistics, *Robert M. Bethea and R. Russell Rhinehart*
122. Sample Size Choice: Charts for Experiments with Linear Models: Second Edition, *Robert E. Odeh and Martin Fox*
123. Handbook of the Logistic Distribution, *edited by N. Balakrishnan*
124. Fundamentals of Biostatistical Inference, *Chap T. Le*
125. Correspondence Analysis Handbook, *J.-P. Benzécri*
126. Quadratic Forms in Random Variables: Theory and Applications, *A. M. Mathai and Serge B. Provost*
127. Confidence Intervals on Variance Components, *Richard K. Burdick and Franklin A. Graybill*
128. Biopharmaceutical Sequential Statistical Applications, *edited by Karl E. Peace*
129. Item Response Theory: Parameter Estimation Techniques, *Frank B. Baker*
130. Survey Sampling: Theory and Methods, *Arijit Chaudhuri and Horst Stenger*
131. Nonparametric Statistical Inference: Third Edition, Revised and Expanded, *Jean Dickinson Gibbons and Subhabrata Chakraborti*
132. Bivariate Discrete Distribution, *Subrahmaniam Kocherlakota and Kathleen Kocherlakota*
133. Design and Analysis of Bioavailability and Bioequivalence Studies, *Shein-Chung Chow and Jen-pei Liu*
134. Multiple Comparisons, Selection, and Applications in Biometry, *edited by Fred M. Hoppe*
135. Cross-Over Experiments: Design, Analysis, and Application, *David A. Ratkowsky, Marc A. Evans, and J. Richard Alldredge*
136. Introduction to Probability and Statistics: Second Edition, Revised and Expanded, *Narayan C. Giri*

137. Applied Analysis of Variance in Behavioral Science, *edited by Lynne K. Edwards*
138. Drug Safety Assessment in Clinical Trials, *edited by Gene S. Gilbert*
139. Design of Experiments: A No-Name Approach, *Thomas J. Lorenzen and Virgil L. Anderson*
140. Statistics in the Pharmaceutical Industry: Second Edition, Revised and Expanded, *edited by C. Ralph Buncher and Jia-Yeong Tsay*
141. Advanced Linear Models: Theory and Applications, *Song-Gui Wang and Shein-Chung Chow*
142. Multistage Selection and Ranking Procedures: Second-Order Asymptotics, *Nitis Mukhopadhyay and Tumulesh K. S. Solanky*
143. Statistical Design and Analysis in Pharmaceutical Science: Validation, Process Controls, and Stability, *Shein-Chung Chow and Jen-pei Liu*
144. Statistical Methods for Engineers and Scientists: Third Edition, Revised and Expanded, *Robert M. Bethea, Benjamin S. Duran, and Thomas L. Boullion*
145. Growth Curves, *Anant M. Kshirsagar and William Boyce Smith*
146. Statistical Bases of Reference Values in Laboratory Medicine, *Eugene K. Harris and James C. Boyd*
147. Randomization Tests: Third Edition, Revised and Expanded, *Eugene S. Edgington*
148. Practical Sampling Techniques: Second Edition, Revised and Expanded, *Ranjan K. Som*
149. Multivariate Statistical Analysis, *Narayan C. Giri*
150. Handbook of the Normal Distribution: Second Edition, Revised and Expanded, *Jagdish K. Patel and Campbell B. Read*
151. Bayesian Biostatistics, *edited by Donald A. Berry and Dalene K. Stangl*
152. Response Surfaces: Designs and Analyses, Second Edition, Revised and Expanded, *André I. Khuri and John A. Cornell*
153. Statistics of Quality, *edited by Subir Ghosh, William R. Schucany, and William B. Smith*
154. Linear and Nonlinear Models for the Analysis of Repeated Measurements, *Edward F. Vonesh and Vernon M. Chinchilli*
155. Handbook of Applied Economic Statistics, *Aman Ullah and David E. A. Giles*
156. Improving Efficiency by Shrinkage: The James-Stein and Ridge Regression Estimators, *Marvin H. J. Gruber*
157. Nonparametric Regression and Spline Smoothing: Second Edition, *Randall L. Eubank*
158. Asymptotics, Nonparametrics, and Time Series, *edited by Subir Ghosh*
159. Multivariate Analysis, Design of Experiments, and Survey Sampling, *edited by Subir Ghosh*
160. Statistical Process Monitoring and Control, *edited by Sung H. Park and G. Geoffrey Vining*
161. Statistics for the 21st Century: Methodologies for Applications of the Future, *edited by C. R. Rao and Gábor J. Székely*
162. Probability and Statistical Inference, *Nitis Mukhopadhyay*

163. Handbook of Stochastic Analysis and Applications, *edited by D. Kannan and V. Lakshmikantham*
164. Testing for Normality, *Henry C. Thode, Jr.*
165. Handbook of Applied Econometrics and Statistical Inference, *edited by Aman Ullah, Alan T. K. Wan, and Anoop Chaturvedi*
166. Visualizing Statistical Models and Concepts, *R. W. Farebrother and Michael Schyns*
167. Financial and Actuarial Statistics, *Dale Borowiak*
168. Nonparametric Statistical Inference, Fourth Edition, Revised and Expanded, *edited by Jean Dickinson Gibbons and Subhabrata Chakraborti*
169. Computer-Aided Econometrics, *edited by David EA. Giles*
170. The EM Algorithm and Related Statistical Models, *edited by Michiko Watanabe and Kazunori Yamaguchi*
171. Multivariate Statistical Analysis, Second Edition, Revised and Expanded, *Narayan C. Giri*
172. Computational Methods in Statistics and Econometrics, *Hisashi Tanizaki*
173. Applied Sequential Methodologies: Real-World Examples with Data Analysis, *edited by Nitis Mukhopadhyay, Sujay Datta, and Saibal Chattopadhyay*
174. Handbook of Beta Distribution and Its Applications, *edited by Richard Guarino and Saralees Nadarajah*
175. Survey Sampling: Theory and Methods, Second Edition, *Arijit Chaudhuri and Horst Stenger*
176. Item Response Theory: Parameter Estimation Techniques, Second Edition, *edited by Frank B. Baker and Seock-Ho Kim*
177. Statistical Methods in Computer Security, *William W. S. Chen*
178. Elementary Statistical Quality Control, Second Edition, *John T. Burr*
179. Data Analysis of Asymmetric Structures, *edited by Takayuki Saito and Hiroshi Yadohisa*

Additional Volumes in Preparation

Elementary Statistical Quality Control

Second Edition

John T. Burr
Rochester Quality Associates

Marcel Dekker New York

Although great care has been taken to provide accurate and current information, neither the author(s) nor the publisher, nor anyone else associated with this publication, shall be liable for any loss, damage, or liability directly or indirectly caused or alleged to be caused by this book. The material contained herein is not intended to provide specific advice or recommendations for any specific situation.

Trademark notice: Product or corporate names may be trademarks or registered trademarks and are used only for identification and explanation without intent to infringe.

Library of Congress Cataloging-in-Publication Data
A catalog record for this book is available from the Library of Congress.

ISBN: 0-8247-9052-9

This book is printed on acid-free paper.

Headquarters
Marcel Dekker, 270 Madison Avenue, New York, NY 10016, U.S.A.
tel: 212-696-9000; fax: 212-685-4540

Distribution and Customer Service
Marcel Dekker, Cimarron Road, Monticello, New York 12701, U.S.A.
tel: 800-228-1160; fax: 845-796-1772

World Wide Web
http://www.dekker.com

The publisher offers discounts on this book when ordered in bulk quantities. For more information, write to Special Sales/Professional Marketing at the headquarters address above.

Current printing (last digit):

10 9 8 7 6 5 4 3 2 1

PRINTED in the UNITED STATES of AMERICA

In Memoriam to the First Author
Irving W. Burr

Preface to the Second Edition

During the last few years of his life Irving sought to bring this book on elementary quality control up to date. While the examples, the philosophy, and the techniques are timeless, the notation, the language, and the applications change. This is what he sought to update and what I have undertaken to accomplish. Having spent 24 years in a chemical analytical industrial environment using statistics as well as teaching their use, I took an early retirement and have spent the last 17 years teaching and consulting in the quality field. These last years have been a busy time riding the crest of the quality wave in the United States but at last I find the time to complete the work that my father started and so dearly loved.

The intent is to keep intact as much of Irving's original work as possible, adding where appropriate discussions on process capability, establishment of realistic specifications, cautions on the blind use of acceptance sampling and the introduction of additional methods of statistical control charting. And in doing this try to emulate Irving's incredible talent of simplifying the difficult and illustrating the concepts with real examples from either industrial or service industries.

Irving's observations on the need and importance of statistical quality control in the 1940s and 1950s as written above are a just as important now as then. They may be even more important now that the US is entering into the era of the global market, competing with established powerhouses of production and quality such as Germany and Japan and emerging superpowers such as China, the European Economic Community, and the Asian Economic Community. Quality has become and will continue to be not an attribute of a product or service but a requirement for continued acceptability by the customer. Quality must be built into the product or service and this is accomplished through the use of what this book is all about, statistical quality control.

All of the management models such as TQC (Total Quality Control), TQM (Total Quality Management), MBNQA (Malcolm Baldrige National Quality Award), or any other of the many others "developed" by consultants to incorporate quality into the way we do business rely on the use of the techniques described and illustrated in this book. Without these timeless techniques we could not manage our processes, our companies or our businesses efficiently or effectively. I am reminded of the 1900s tale of the maker of horseshoe nails. The 40-head stamping machine had been generating 2.5% curved (nonconforming) nails for several years. Three persons were assigned to sort the nails into two piles, good and curved. (Horses and blacksmiths did not like the latter for obvious reasons.) A bright young engineer spotted the relationship in the numbers. He tested the product from each die and found that for three years one die had produced nothing but curved horseshoe nails! The die was blanked, productivity reduced by 2.5% and quality was improved to 100%. Three people were put to more useful and productive work. We must focus on the process rather than the product.

This book is intended for use by people who work with processes and who wish to make decisions based on data and the conclusions drawn from the analysis of data rather than being based on guess work, supposition, or intuition. And strange as it may seem in a book on statistical methods, it is my experience that 85% of statistical quality control or statistical process control is in the use of the eyeball and brain of the person looking at the data rather than the use of statistics. The latter is required to refine the decisions and to give guidelines to the process operator. The early chapters continue to be in the book since knowledge of variability and how it is distributed are necessary to the understanding of the process.

John T. Burr

Preface to the First Edition

The objective of this book is to present the basic methods of statistical quality control in as simple, natural, and straightforward a manner as possible. No previous knowledge of statistics has been assumed; only elementary algebra has been used. The aim is thus to make the book accessible to the maximum number of people. It can be used for industrial courses and self-study, as well as in trade schools, junior colleges, and even high schools. Those who master this book enhance their employability in industry, not only in quality control and inspection departments, but also in many other positions.

Methods of statistical quality control began to be developed in the 1920s, especially in the Bell Telephone Laboratories. A big impetus in their usage was given by production for World War II, aided by a series of intensive 8-day courses in the subject. As an outgrowth, the American Society for Quality Control was formed in 1946 and currently numbers some 20,000 members.

Applications of statistical quality control have been profitably made in virtually all segments of industry and in procurement. The reason is not difficult to find, that is, variation in even the most refined production is universally present. Thus, control of processes and decision making on incoming lots are based on data subject to variation, and therefore it is natural to use the very methods developed for handling problems involving variation, namely statistical methods.

After an introduction to fundamental statistics and probability, a major portion of the book is spent on control charts for process control, and on acceptance sampling for decision making on lots or processes. Following this material is a chapter on statistical tolerancing for assemblies, and brief chapters on studying relationships by correlation and on concepts of reliability. All data in examples and problems are from actual cases, except for data obtained by experiments with known populations. The reader should not be concerned if he or she is not familiar with a product or production process mentioned. The author may not be either. If not familiar, just call the product or piece a "widget" and pay attention to the numerical data.

Statistical quality control methods are of increasing importance in industry because of (1) ever-refining requirements, making measurement and control of variation more important, (2) the need for obtaining the best possible performance from production processes, (3) the necessity of saving material and avoiding spoilage, (4) the need for economy and cost cutting in production, (5) heightened competition both at home and abroad, (6) the demands of warrantees and consumerism, (7) increasing use of national and international standards, (8) the increasing needs for reliability, and (9) the need to make sure that production methods for quality are sound, for maximum customer satisfaction, and at the extreme, for defense in court in liability cases.

The present book is more elementary in character than the author's *Statistical Quality Control Methods* (Dekker, New York, 1976). The latter book assumes a bit of statistical background and some elementary calculus, and includes derivations and proofs, as well as some more specialized techniques. The present book aims toward a wider readership.

Answers to the old-numbered problems are provided for the reader's benefit, especially for self-study.

As in the other book, this book is the outgrowth of a long association with industrial people, both as students and as those with whom I consulted or whose plants I visited; and also with the many professors and industrialists with whom I have taught.

Irving W. Burr

Contents

Elementary Statistical Quality Control

1

Why Statistics?

Throughout industry measuring, inspection and testing is being done every minute by people and by automatic devices. The results are in the form of numbers, e.g., measurements or a count of the number on nonconforming pieces in a sample. These are numbers or data. They are obtained in order to take action on a manufacturing process for improving and rectifying it or for a decision on incoming products. An ever present character of such data or results is that they vary: from time to time, piece to piece, sample to sample. And this is true even if the production process is held as constant as is humanly possible. The same can be said for administrative or service functions where the product is a service rather than a manufactured unit. The service delivered, e.g., maintenance, consultation, testing or project management, can be highly variable. It also can be evaluated against some defined criteria, standard or with respect to the delivery of similar or equivalent services. Wherever we have variation, we have a statistical problem, whether we know it or not. Thus in such problems, we ought to be using the very methods designed for analysis of data, i.e., statistical methods. They can be used to minimize the chances of wrong decisions. Fortunately, there are many statistical techniques which can be readily learned by anyone with even quite limited background. They will greatly aid in process improvement and control and in decision making on lots of products or in the delivery of services.

1.1. Statistical Quality Control

Control of quality of output is very old, certainly going back to the building of the pyramids in Egypt, and even to the making of arrowheads in stone age times. But statistical quality control, i.e., the use of statistical methods in quality control, is a relatively recent development which began at the Bell Telephone Laboratories in the 1920s.

The objectives of statistical quality control are to collect data in a sound, unbiased manner and to analyze the results appropriately so as to obtain satisfactory, dependable and economic quality. A mirror-like finish on all pieces, within a total tolerance of 0.0002 in. may not be necessary for the use at hand so we say "satisfactory". "Dependable" quality is obtained when the manufacturing or service process are adequately controlled and we can rely upon them to produce the desired quality consistently. Moreover the production of quality must be done "economically" so as to hold costs down, to be competitive and able to sell at a profit. Statistical quality control methods enable us to obtain maximum benefit out of production and inspection data and at lowest cost.

1.2. Data: Statistics for Action

The more closely we try to work, the more trouble we have with variation. This pencil with which the author is writing is one-quarter of an inch in diameter, to the nearest quarter of an inch. Everyone measuring its diameter in any place or angle would obtain the same result: 1/4 in. The same is true for measurements to the nearest 0.1 in.: all measurements would be 0.3 in. But probably if measured to the nearest

0.01 in., and certainly if to the nearest 0.001 in., measurements at different places or angular directions or even at the same place and direction, would differ. What this illustrates is that the closer we try to work, the more we are bothered by variation. Fortunately the diameter of the pencil is not a very critical dimension. Other characteristics are of greater importance, e.g., gluing or hardness of the lead.

The very fact that industry has continually tried to do a better job and to work to ever closer tolerances has increased the need to handle varying results statistically. Statistical quality control methods can be used to avoid or decrease the likelihood of error. For example, on a production process, we can adjust its average level (reset it) when we should leave it alone or we can fail to adjust when, in fact, we should adjust it. On an incoming lot, we can err by rejecting a lot which is satisfactory or we can accept a lot which is unsatisfactory. Statistical quality control methods enable us to determine the risks of such wrong decisions and to control these risks within economic limits.

1.3. PATTERNS OF VARIATION

Numerical results or data tend to form characteristic patterns or forms of variation. For example, if we have a fairly large collection of measurements and make up numerical classes and then find how many measurements there are in each class, we will find that some classes in the middle have many measurements, while those classes relatively far out in either direction have few measurements. Such a "tabulation" gives us a picture of how the data run, i.e., a pattern of the variation. Such patterns are determined by the manufacturing process and measuring technique from which the data are obtained. The pattern or "distribution" for the process may or may not represent satisfactory quality performance. For example, virtually all of the measurements may lie between the specification limits or a sizable proportion of the measurements may lie beyond one specification limit or outside both limits. Then we have the question of what action to take.

Patterns or distributions of data take various shapes. Through the years, standard distributions or "models" of variation have been developed to enable us to analyze data. In this book, we shall be especially interested in three such models or statistical laws. We shall find them of great use in applications and decision making. Fortunately they can be readily understood and applied without extensive use of mathematics.

1.4. WIDE APPLICABILITY

Statistical methods can be used wherever we have variation. Variation of production exists everywhere in industry and some statistical technique can help us in process improvement, control experimentation, and decision making on product. The relatively few techniques to be presented in this book can be applied advantageously to most industrial production, administrative and service jobs. The authors do not know of any industry where the methods cannot be applied. Do not let anyone say to you, "I can see how they can be applied in Henry's operation; but our processes are different." This particularly applies to administration or services where the statement is, "Statistical quality control is good for the shop floor but not here in the office." In this book, we are not interested in the differences between processes which anyone can see. We are interested in what is common to processes whether they be manufacturing or administrative. And we are looking at the single common element of all processes, variation.

The statistical quality control methods to be presented here are applicable with the use of a little imagination beyond the usual manufacturing environment that makes products. Thus the methods are applicable to the service industries such as insurance, banking, food service, health care and consumer sales as well as administrative services, e.g., sales, marketing, R&D, design, purchasing, accounting and secretarial functions. In fact, wherever we can have a series of results which vary and a conceptual collection or "population" of the results, we can use the methods to improve the process.

1.5. SUMMARY

We can obtain satisfactory, dependable and economic quality from our processes and those of our suppliers by the use of statistical quality control methods. They will enable us to gain economical control of processes by finding and eliminating the causes of poor quality. And they will provide the tools for sound decisions. So on with the show!

2

Characteristics of Data and How to Describe Them

This chapter presents methods of describing and summarizing data so as to picture their distribution, to analyze them and to facilitate the making of decisions. We are here thinking of the description of a sample of data, i.e., observed data from some process or lot.

2.1. TWO BASIC CHARACTERISTICS OF DATA

There are two main characteristics of numerical data which we always consider in describing our data. These are (1) the average or level of the numbers and (2) the variability amongst the numbers. Let us consider an example on the outside diameter of a brass "low-speed plug". Three of some 20 samples, each of five, were the following in 0.0001 in. units:

Sample 1	1921	1919	1924	1924	1925
Sample 2	1922	1923	1921	1923	1924
Sample 3	1923	1926	1926	1929	1929

By inspection of the numbers, we can see that the first two samples seem to be averaging or centered at about the same level while the third sample averages much higher. This is the first characteristic. But looking again at the first two samples which averaged the same, we see that the variability among the numbers in sample 1 is much greater than that for sample 2. In fact the total extent of variation which we call the range, is 6 = 1925 − 1919 for the first sample, but is only 3 = 1924 − 1921 for the second sample. Now consider samples 1 and 3. For each the range is 6, but the average for sample 3 is far above that for sample 1. These two distinct characteristics of data appear perhaps even more clearly in the accompanying graphical picture using dots on scales.

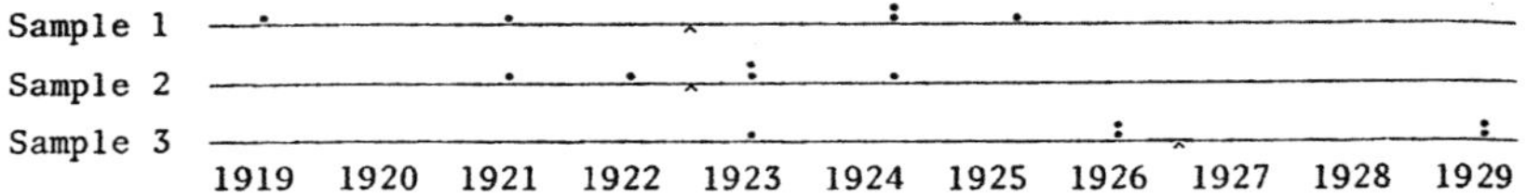

The small arrow point below each line indicates the "arithmetic mean", average or balance point for the sample. The first two are at 1922.6, while the third is at 1926.6.

In order to describe or summarize a sample of numbers, we need to provide two statistical measures, an average and a measure of the variability (range). It is not enough to give only the average nor only the range. For example, the average annual temperatures of Chicago, Illinois, and Bermuda are very nearly the same; it is the variability of the temperatures that gets us in trouble in Chicago.

2.2. MEASURING AVERAGE LEVEL AND VARIABILITY

Averages. The objective of an average is to measure the central or balance point around which a set of numbers seems to cluster. An average is also said to locate the central tendency. The one average with which almost everyone is familiar is called the arithmetic mean when we might be considering several different averages and their properties.

But we shall in this book call the arithmetic mean, the average.

We have then (recognizing that there exist other averages) the following definition:

Definition 2.1. The average of a set of n numbers, say x_1, $x_2, \ldots, x_n$, is the sum of the numbers divided by n. (The three dots in the definition stand for all x's between x_2 and x_n.) It is often clearer to use symbols instead of words as long as the meaning of the symbols is clear. Let us let x with a bar over it represent the average of the x's. This is called x-bar. Then we have:

$$\bar{x} = \frac{\text{sum of } x\text{'s}}{\text{number of } x\text{'s}} = \frac{x_1 + x_2 + \cdots + x_n}{n} \tag{2.1}$$

A most convenient symbol to use in formulas is the capital Greek letter sigma, i.e., Σ. It is used to mean the sum of whatever is written after it. Thus, Σx means "sum of the x's" or ΣR means "sum of the R's". Then, if we use this useful symbol, (2.1) can be written:

$$\bar{x} = \frac{\Sigma x}{n} \tag{2.2}$$

which is briefer and just as clear.

Let us now use the definition for the three samples of diameters. For sample 1, we have

$$\begin{aligned}\bar{x} &= \frac{1921 + 1919 + 1924 + 1924 + 1925}{5} \\ &= \frac{9613}{5} = 1922.6 \equiv 0.19226\,\text{in.}\end{aligned}$$

For the other samples, we have

$$\bar{x} = \frac{9613}{5} = 1922.6$$

$$\bar{x} = \frac{9633}{5} = 1926.6$$

These three averages, x-bars, locate the arrowpoints on the graphical picture and describe the level at which the respective samples of diameters tend to center.

Variability measures. As with averages, there are many measures of variability which could be discussed; but in this book we shall only need to include two such measures. We have already introduced one, the "range".

Definition 2.2. The range, R, of a set of numbers is the difference between the highest number and the lowest. The formula is:

$$\text{range} = R = \text{maximum } x - \text{minimum } x \tag{2.3}$$

The range therefore measures the total amount of variation within the sample of n numbers. For the three samples, we have:

$$R_1 = 1925 - 1919 = 6$$
$$R_2 = 1924 - 1921 = 3$$
$$R_3 = 1929 - 1923 = 6$$

and so, as we have seen, the first and third samples have twice as much total variation as does the second sample. The range is a simple measure of variability, is readily understood, is easily obtained and is widely used for small samples in statistical quality control.

There is another measure of variability, however, which is of much importance; in fact its concept is basic in all statistics, not just in statistical quality control. This is the so-called standard deviation. Let us develop it from basic needs.

When we have a sample of data, it seems to make much sense to try to measure how far away from x-bar the numbers, x, lie on the average, i.e., how big a deviation from x-bar is only to be expected? Some deviations of x's from x-bar will of course be greater than this typical deviation, while others will be smaller. Consider sample 1:

Data (x)	Deviation ($x - x$-bar)	Deviations squared ($x - x$-bar$)^2$
$x_1 = 1921$	$x_1 - x$-bar $= -1.6$	$(x_1 - x\text{-bar})^2 = (-1.6)^2 = 2.56$
$x_2 = 1919$	$x_2 - x$-bar $= -3.6$	$(x_2 - x\text{-bar})^2 = (-3.6)^2 = 12.96$
$x_3 = 1924$	$x_3 - x$-bar $= +1.4$	$(x_3 - x\text{-bar})^2 = (+1.4)^2 = 1.96$
$x_4 = 1924$	$x_4 - x$-bar $= +1.4$	$(x_4 - x\text{-bar})^2 = (+1.4)^2 = 1.96$
$x_5 = 1925$	$x_5 - x$-bar $= +2.4$	$(x_5 - x\text{-bar})^2 = (+2.4)^2 = 5.76$
x-bar $= 1922.6$	$\Sigma(x - x\text{-bar}) = 0$	$\Sigma(x - x\text{-bar})^2 = 25.20$

The first number, $x_1 = 1921$, "deviates" 1.6 units from the average, x-bar, being below it, hence the minus sign is added. This is obtained by $x_1 - x$-bar $= 1921 - 1922.6 = -1.6$. Similarly, the fifth deviation is $x_5 - x$-bar $= 1925 - 1922.6 = +2.4$, x_5 being above x-bar. Now what we wish is some sort of average of these five deviations, $x - x$-bar. If we try the same game of averaging them as we did in finding x-bar, we run "aground" because the sum of the deviations here is zero. In fact the sum of the deviations is always zero, because of the +'s and −'s adding out to zero. Thus, some other average of the deviations is in order. To get around the trouble with the signs of the deviations, we square them. Now it would seem logical to divide this sum of the squared deviations, $\Sigma(x - x\text{-bar})^2$ by 5 and then take the square root; so as to get back to the original unit (0.0001 in.). In fact division by n has been often practiced in the past. But modern usage calls for division of $\Sigma(x - x\text{-bar})^2 = 25.20$ by $n - 1 = 5 - 1 = 4$ and taking the square root of the result. Hence, the standard deviation is:

$$s = \sqrt{\frac{25.20}{4}} = \sqrt{6.30} = 2.51 \equiv 0.000251\,\text{in.}$$

Note that of our five deviations, $x - x$-bar, only one -3.6 is larger in size than $s = 2.51$, while one other $+2.4$ is nearly the same size. We choose now to define the standard deviation, s, by an appropriate formula which is seen to be inherent in the preceding development.

Definition 2.3. The standard deviation, s, of a set of n numbers, $x_1, x_2, \ldots, x_n$, is given by:

$$s = \sqrt{\frac{\Sigma(x - \bar{x})^2}{n - 1}} \tag{2.4}$$

In review, for comparison let us find the standard deviation for sample 2:

(x)	$(x - x\text{-bar})$	$(x - x\text{-bar})^2$
1922	−0.6	0.36
1923	+0.4	0.16
1921	−1.6	2.56
1923	+0.4	0.16
1924	+1.4	1.96
9613	0	5.20

$$x\text{-bar} = 1922.6 \quad s = \sqrt{\frac{5.20}{4}} = \sqrt{1.30} = 1.14 \equiv 0.000114\,\text{in.}$$

So here $s = 1.14$, or a bit less than half of the s for sample 1. This time two of the five deviations are larger in size than s and three are similar.

We now need to make a few comments. The first is that the technique for finding the standard deviation, s, may well seem a bit (or even quite) complicated to the reader. And certain it is, that it is more time consuming than finding the range. Indeed the movement toward the use and application of control charting of measurements did not really begin to get off the ground until it was found that for small samples the range was just about as good to use as the standard deviation, and thus much time could be saved in estimating the amount of variation in the sample. If a computer or an appropriate calculator is used it is just as easy to find s as to find R. Then too the concept of the standard deviation is enormously important. Finally, early workers in statistical quality control were so proud at having mastered the standard deviation that its Greek letter symbol, the small sigma, σ, was incorporated in the official seal of the American Society for Quality Control (now the American Society for Quality). Some one once asked why the Greek letter was chosen and the reply could have been, "It's all Greek to me!".

2.3. CONDENSING DATA INTO A FREQUENCY TABLE

When you have a relatively large sample of data, such as we see in Table 2.1, it is desirable to condense them into a so-called frequency table. About all you get from looking over such a long list of numbers, e.g., Table 2.1, is eye strain. A frequency table will help us picture the data.

Definition 2.4. A frequency table gives numerical classes which cover the range of data and lists the frequency of cases with each class.

Let us therefore consider the needed steps in making a frequency table. The first step is to choose limits for the classes. There would usually be 8–12 such classes which must cover the entire range of the data. "Class limits" give the precision of the measurements or data. In Table 2.1, the charge weights were to the nearest gram; so the class limits are to the gram. Each possible measurement must have one and only one class into which it can go. The extreme charge weights are 424 and 511 g.; therefore the range, R, is $511 - 424 = 87$ g. If we include 10 possible charge weights per class, we shall have 9 or 10 classes. Convenient classes are 420–429, 430–439 and so on. (The beginner might think that such classes have only nine possible measurements each but look closely and you will see 10). Note that every

Table 2.1 Charge Weights for 99 Insecticide Dispensers in Grams (Specifications 454 ± 27 g)

476	478	473	459	485	454	456	454	451	452
458	473	465	492	482	467	469	461	452	465
459	485	447	460	450	463	488	455	478	464
441	456	458	439	448	459	462	495	500	443
453	457	458	470	450	478	471	457	456	460
457	434	424	428	438	460	444	450	463	467
476	485	474	471	469	487	476	473	452	449
449	477	511	495	508	458	437	452	447	427
443	457	485	491	463	466	459	471	472	472
481	443	460	462	479	461	476	478	454	

possible charge weight will have one and only one class to which it belongs since a recorded weight such as 429.6 will never occur. (We are not making measurements to the 0.1 g.)

The next step is to list all the classes in a column and take each charge weight in turn and tally, just like the tallying of votes. Thus, 476 is tallied in the class 470–479. Then we take the next, 478, and make a second tally in this class and so on. This gives us Table 2.2. after we count the tallies in each class. Looking at the class frequencies, we see that they rise steadily from 3 to a maximum of 28 and then fall fairly steadily to a minimum of 1 at the 510–519 class. The midvalue column gives the midmost charge weight in each class. This is often called the "class mark". For this we note that a weight measured as a 420 g might come from any true weight 419.5–420.5 g. and likewise one measured as 429 g might be from a true weight 428.5–429.5 g. So the extreme possible true weights in the 420–429 class are 419.5 and 429.5. These are called "class boundaries". The midvalue of this interval is 424.5 as recorded in the third column. Similarly, the other midvalues are as shown in Table 2.2.

By looking at the table, we see that the point of greatest concentration is about 454.5 g., i.e., about the middle of the

Table 2.2 Frequency Table for Charge Weights of Insecticide Dispenser, from Data of Table 2.1 (Specifications 454 ± 27 g)

Class in grams	Frequency, f	Midvalue, x
420–429	3	424.5
430–439	4	434.5
440–449	10	444.5
450–459	28	454.5
460–469	19	464.5
470–479	20	474.5
480–489	8	484.5
490–499	4	494.5
500–509	2	504.5
509–519	1	514.5
Total frequency	99	

specification limits. However, we note that considerably more than half of the remaining charge weights (namely, 54) lie above the 450–459 class. This means that the process average was above the specified "nominal" weight of 454 g. This overfill above the required average proved to be costing the company about $14,000 per month (in 1970 $'s). Also note that since the specification limits were 427 and 481 g, there were a few of these 99 charge weights out of specification. By actual count in Table 2.1, one weight was below 427 g and 14 above 481 g. In the total production of which the 99 charge weights were but a sample, there would, of course, be far more outside the limits. By using statistical control methods, the overfill cost was soon cut to $2000 per month as well as more safely meeting the specification limits.

Two kinds of frequency graphs which we show in Figure 2.1 are of further help in visualizing the frequency distribution or manner in which the frequencies in the classes are distributed. The frequency polygon plots the frequencies, f, against the midvalues. Note that the broken line graph is connected to the midvalues having zero frequency (414.5 and 524.5) which lie next to those with some frequency. The histogram has blocks with bases between the class boundaries and heights representing frequencies. They emphasize the class intervals. Either type of frequency graph may be used according to individual preference.

A frequency table, or even better, one of the two frequency graphs helps us visualize the pattern of variation for the numbers, their average and point of greatest concentration and their typical variability and total spread. Also we may find out something about the curve shape especially if we have a substantial sample, e.g., at least 100 observations. That is, the frequencies may decrease to zero much more slowly on one side of the highest frequencies than they do on the other, e.g., eccentricities (distances between two center points) or percent impurity can never be negative but a few relatively large values may occur. In this case starting at a zero measurement, the highest frequency may be quickly reached but then the frequencies decrease more slowly and may "tail out" quite a distance.

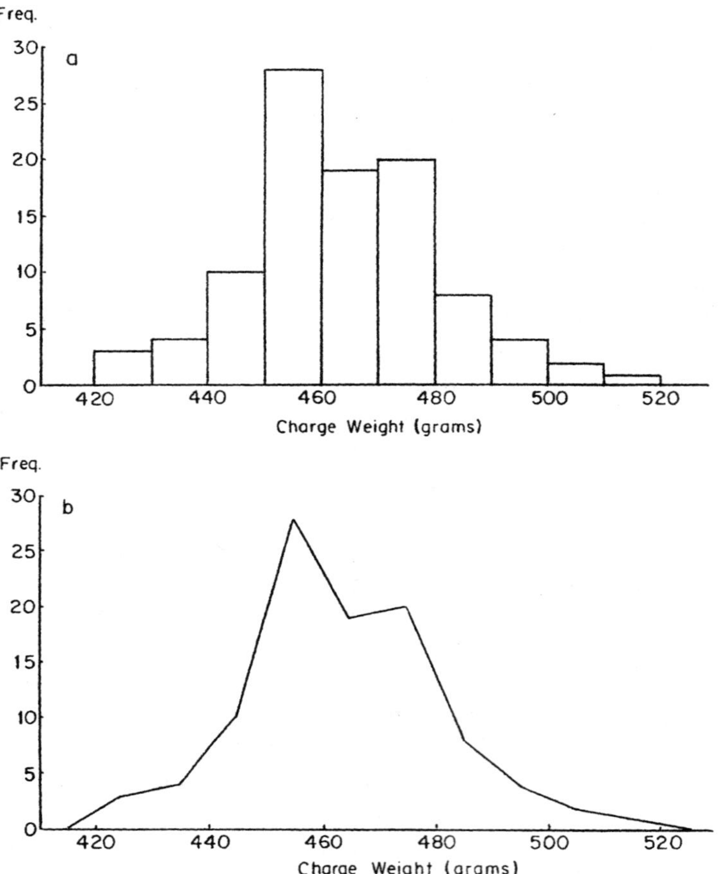

Figure 2.1 (a) Histogram and (b) frequency polygon for charge weights of an insecticide dispenser. Specifications $427 + 27$ g. Data in Table 2.2.

A frequency table and/or graph may reveal irregularities, e.g., the frequencies may suddenly stop and become zero in one or both "tails" giving a "bob-tailed" appearance. This could be the result of 100% sorting to eliminate extreme values in one or both directions. Such an abrupt stop might thus be at a specification limit. Or there may be alternating higher and lower frequencies, e.g., if classes are in 0.0005 in.

for some nominal dimension, there might be many more cases measured as whole ten-thousandths of an inch than at half ten-thousandths.

In any case a frequency graph can be of much aid in picturing the pattern of variation to an associate, an executive or in a report, e.g., in pressing out railroad car wheels from heated steel cylindrical blocks, a high rate of rejection (about 35%) was experienced. Upon weighing a substantial sample of blocks, an average of 650 lb and range of 70 lb showed up. A frequency graph was shown to the person who cut off the blocks. He was amazed and said he thought he could do better. He was able to cut the range to 20 lb and the rejection rate of wheels dropped to 5–10%.

2.4. SAMPLE DATA VS. POPULATION

Up to now we have been talking about samples of industrial product and the associated data taken by some method of measurement. The product might be an auto component and we take a sample from the shipment, making a careful inspection of each to seek out any "nonconformities". These nonconformities may be anything from a critical nature which could make the component dangerous, down to the mildest kind of imperfection. Each component upon inspection yields 0, 1, 2, ... nonconformities, i.e., a whole number or integer. Such "counted" data are called discrete data.

Definition 2.5. Discrete data are the result of counting nonconformities or nonconforming pieces in a sample so that only whole numbers or integers may occur.

Now suppose that our sample of the auto component came from a large lot of components. We could, given enough time, inspect the entire lot for nonconformities. This would yield a "population distribution" of counts of nonconformities on the component for the lot. Now we would hope that the "sample distribution" of counts would be similar to that of the population or lot. In order to make it likely, we should have drawn the sample of components in an unbiased manner.

One unbiased method is to draw the sample from the lot in such a way that all components in the lot have the same chance to be chosen for the sample. This is called "random sampling".

Definition 2.6. We say that we are doing random sampling when each item in the lot has the same chance to be chosen for the sample as has every other item in the lot.

Unless we take a sample using a method which will give an unbiased representation of the population, our sample distribution may be widely different from the population distribution, e.g., a sample of 100 piston ring castings was taken from a lot of 3000 and inspected. It yielded 25 nonconforming castings out of the 100. It was decided to sort the remaining 2900 in the lot so as to salvage the conforming ones. Only four more nonconforming castings were found in the 2900! Question: Was the sample of 100 chosen at random from the 3000?

In the example just given the "population" was the lot of which 29 were nonconforming. The so-called fraction nonconforming in the lot was 29/3000 or about 0.01 (1%). Meanwhile the sample fraction nonconforming was 25/100 or 0.25 (25%). The sample fraction was wildly different from that of the population. (The industrial explanation was that the castings were poured in stacks of molds and one stack must have been poured with cool iron, the sample having happened to be chosen from the part of the box containing these nonconforming ones.)

We have just seen two types of discrete data, namely, counts of the number of nonconformities on the items in a sample or a lot and secondly, counts of the number of pieces in a sample or lot which are found to be nonconforming. The presence of one or more nonconformities on a piece may or may not make it a "nonconforming piece" depending upon the seriousness of the nonconformities. This must be spelled out in the definitions of what constitutes a nonconformity and a nonconforming unit. The other type of data is called measurement data.

Definition 2.7. Measurement data are obtained whenever we take an item of product and compare it against a standard unit of the measurement. This, in general, involves reading a scale of some kind and the resulting measurement is so many whole units and some decimal fraction; e.g., 1.0626 in. or 31.28 cm^3. Indefinitely many decimal places are theoretically possible but in practice the number of places is limited by the measuring technique. Such data are also called continuous data.

For measurement data, the populations are perhaps more complicated than for discrete data. The reason is that each item of product does not have just one measurement; instead repeated measurements of the item could yield varying results. One example of a measurement population is the distribution of measurements of the diameter of a shaft obtained by very precise repeated measurements in just one place on the shaft. The variation in this distribution reflects the "repeatability" error in the measuring technique. The population size is not finite; since the number of repeated measurements we could take is unlimited.

On the other hand, if we take a lot of shafts and measure the diameter of each just once, we have a distribution of measurements. But the variation in these observed measurements reflects not only shaft-to-shaft variation but also variation from place to place within a single shaft and also measurement variation (or errors) for repeated measurements on one place on one shaft (repeatability). Because of these different sources of variation, we can never actually pick out the population distribution for the lot.

The appropriate statistical models for discrete and for measurement data are quite different and we must therefore be careful as to which type of data we have so as to use the appropriate model, as we shall see.

2.5. INTERPRETATION OF *X*-BAR AND *S*

These notations are for the average and the standard deviation for a sample of data. We think of x-bar as a typical value of the x's in a sample or as a point around which the numbers, x, tend to cluster. It is a one-number description for the level of the numbers. We may also say that it is a balance point for the x's; since, as we saw in Section 2.2, the (algebraic) sum of the deviations, $x - x$-bar, is zero. See the examples in that section.

As we have seen in Section 2.2, the standard deviation, s, is a typical deviation of an observation, x, from the average, x-bar. It is an "only-to-be-expected" deviation, descriptive of the variation within the sample. Having by now discussed frequency distributions, we are in a position to add another interpretation. This is for substantial samples of, say, at least 50 observations which are well behaved data with frequencies increasing to a maximum and then the frequencies decreasing. Most data from homogeneous conditions will show such a tendency. We can then say that:

---

1. Between x-bar $- s$ and x-bar $+ s$ will lie about 68% of the cases.
2. Between x-bar $- 2s$ and x-bar $+ 2s$ will lie about 95% of the cases.
3. Between x-bar $- 3s$ and x-bar $+ 3s$ will lie about 99% of the cases.

These proportions come from the so-called normal distribution, but they also apply to data which are reasonably similar to "well behaved" data. Looking at the little table just given, we can say that a deviation of $2s$ or greater in size will occur only about one time in 20 and a deviation of $3s$ or more in size will be quite rare.

2.6. EFFICIENT CALCULATION OF *X*-BAR AND *S*: CODING

We now consider ways in which the average and standard deviation may be more easily calculated. Of course the use of a computer or a calculator with a statistics function could essentially eliminate the need to hand-calculation of the mean or standard deviation for a set of data. However, it is still worthwhile to learn the simplifications which can be made.

And in doing so, the person may well become better acquainted with the characteristics of the two measures. Also the use of coding the data to simplify the calculations, is used commonly in measurement, for example, the gauge that may be set to tell the number of 0.001 in. a piece is above a lower specification of 0.620 in. Thus, a piece measuring 0.624 in. will be read out as a +4 and one of 0.618 in. as a −2. And then too there may not be a computer or calculator handy.

Let us consider sample number 1 of Section 2.1. The x's were 1921, 1919, 1924, 1924, and 1925. Suppose that we subtract the lowest value (1919) from each and call the difference y. Then the five y values are 2, 0, 5, 5, and 6. Is it not quite apparent that the y's vary just the same amount as do the x's? But if that is not clear, let us find the mean and standard deviation:

$y = x - 1919$	$y - y$-bar	$(y - y$-bar$)$
2	−1.6	2.56
0	−3.6	12.96
5	+1.4	1.96
5	+1.4	1.96
6	+2.4	5.76
18	0	25.20

$$\bar{y} = \frac{18}{5} = 3.6$$

$$s_y = \sqrt{\frac{25.2}{5-1}} = \sqrt{6.30} = 2.51$$

and hence $s_y = s_x$, the subscript showing of what variable we have found the standard deviation. Moreover, we can find x-bar and y-bar by merely adding back on 19,119 which we subtracted from each x to find y. Thus x-bar $= y$-bar $+ 1919 = 3.6 + 1919 = 1922.6$.

There was no magic in 1919. We could as well have subtracted say 1923 which is a "guess" at x-bar. Then the y's would have been:

$y = x - 1923$	$y - y$-bar	$(y - y\text{-bar})$
−2	−1.6	2.56
−4	−3.6	12.96
+1	+1.4	1.96
+1	+1.4	1.96
+2	+2.4	5.76
−2	0	25.20

$$\bar{y} = \frac{-2}{5} = -0.4$$

$$s_y = \sqrt{\frac{25.2}{5-1}} = \sqrt{6.30} = 2.51$$

$$\bar{x} = -0.4 + 1923 = 1922.6$$

We mention that in finding the deviations for these y's one must be quite careful with signs. Thus, for the first deviation $y - y\text{-bar} = -2 - (-0.4) = -2 + 0.4 = -1.6$ and for the last $y - y\text{-bar} = +2 - (-0.4) = 2 + 0.4 = 2.4$. We can check on our arithmetic by summing these differences for they must sum to zero. Note that coding close to the expected average will result in the potential difficulty with signs but has the advantage of smaller numbers in the calculation of the x-bar and s. Coding to the lowest value of the data eliminates this problem but results in larger numbers. Coding can also be done to nominal or the middle value of the specification range or at some other logical number, e.g., 1900 for this case.

So far we have not seen much gain, except that for x-bar we have avoided the addition of all of the 191x values in the x's. There is a shortcut we can use for finding s_y from the y's however. Since the y's are such simple numbers, they are easily squared. We may then use the following formula:

$$s_y = \sqrt{\frac{n\Sigma y^2 - (\Sigma y)^2}{n(n-1)}} \tag{2.5}$$

which may be proved from (2.4). Our work then looks like the following:

y	y^2
2	4
0	0
5	25
5	25
6	36
18	90

$$s_y = \sqrt{\frac{5(90) - 18^2}{5(4)}} = \sqrt{\frac{450 - 324}{20}} = \sqrt{\frac{126}{20}} = \sqrt{6.3} = 2.51$$

This approach is a considerable help, especially when y-bar is a messy decimal instead of a number as simple as 3.6 or -0.4. It means we do not have to round off any numbers until we do the final division.

Finding s, however, for small samples is not often done (unless a computer or calculator is used), the range being used instead. But when a large sample is being analyzed, as in a frequency table, we have a large gain in the calculation time required as in Table 2.3.

Let us consider again the data on charge weights as given in Table 2.2. See Table 2.3 which gives the midvalues, x, in the first column and the corresponding frequencies, f, in the second column. For x-bar we want the average of 99 charge weights and could treat the three cases in the 420–429 class as though all were at the midvalue 424.5 and so on. Then we would proceed as follows:

$$\bar{x} = \frac{\Sigma fx}{\Sigma f} = \frac{3(424.5) + 4(434.5) + 10(444.5) + \cdots + 1(514.5)}{3 + 4 + 10 + \cdots + 1}$$

$$\bar{x} = \frac{45{,}875.5}{99} = 463.39$$

This is not too troublesome for x-bar but finding s for the x's along this line would give much trouble. Using this approach for s in (2.4), we would now need to find

Table 2.3 Frequency Table for Charge Weights of Insecticide Dispenser, from Table 2.2, and Calculations by Coding for $\bar{x}$ and s

Midvalue, x	Frequency, f	Coded variable v	vf	v^2f
424.5	3	−4	−12	48
434.5	4	−3	−12	36
444.5	10	−2	−20	40
454.5	28	−1	−28	28
464.5	19	0	0	0
474.5	20	+1	+20	20
484.5	8	+2	+16	32
494.5	4	+3	+12	36
504.5	2	+4	+8	32
514.5	1	+5	+5	25
	$\Sigma f = 99$		$\Sigma vf = -11$	$\Sigma v^2 f = 297$

$\bar{v} = \frac{\Sigma vf}{\Sigma f} = \frac{-11}{99} = -0.11$

$s_v = \sqrt{\frac{n\Sigma v^2 f - (\Sigma vf)^2}{n(n-1)}} = \sqrt{\frac{99(297)-(-11)^2}{99(98)}} = \sqrt{\frac{29{,}282}{9702}} = \sqrt{3.018} = 1.737$

$x_0 = 464.5 \quad d = 10$

$\bar{x} = 464.5 + 10(-0.11) = 463.4\,\text{g}$

$s_x = 10(1.737) = 17.37\,\text{g}$

$$\Sigma f(x - x\text{-bar})^2 = 3(424.5 - 463.39)^2 + 4(434.5 - 463.39)^2 + \cdots + 1(514.5 - 463.39)^2$$

which is basically $\Sigma(x - x\text{-bar})^2$ taking the frequencies into account.

So instead we "code" the x's into, say v values. To do this we choose a midvalue somewhere in the middle. See Table 2.3. It is fun to try to guess the one nearest x-bar. Here, we chose 464.5. Then for the "coded" variable, v, we subtract 464.5 from each midvalue, x, in turn and also divide by the difference, 10, between consecutive midvalues (i.e., the class interval). Thus, for $x = 474.5$, $v = (474.5 - 464.5)/10 = +1$; for $x = 494.5$, $v = (494.5 - 464.5)/10 = +3$; and for $x = 424.5$, $v = (424.5 - 464.5)/10 = -4$ and so on. This gives the v column in the table. Then we say there are three v's of -4, four v's of

-3, and so on. To find the sum of all 99 v's, we multiply each v by its frequency:

$$\Sigma vf = (-4)3 + (-3)4 + (-2)10 + \cdots + (+5)1 = -11$$

Likewise for the sum of 99 v^2's we have:

$$\Sigma v^2 f = (-4)^2 3 + (-3)^2 4 + (-2)^2 10 + \cdots + (+5)^2 1 = 297$$

(The $v^2 f$ column is perhaps most easily found by multiplying the two previous columns because v times $vf = v^2 f$, for example, $(-4) \cdot [(-4)3] = (-4)\,(-12) = 48$.)

Then we have for the coded variable, v:

$$\bar{v} = \frac{\Sigma vf}{f} = \frac{-11}{99} = -0.11$$

Using Σvf as analogous to Σy in (2.5), $\Sigma v^2 f$ like Σy^2 in (2.5) and Σf line n in (2.5) we have:

$$s_v = \sqrt{\frac{99(297) - (-11)^2}{99(98)}} = 1.737$$

as shown at the bottom of Table 2.3.

Then to find x-bar and s_x, we merely decode as shown there also. We first subtracted 464.5 from each and next divided by the class interval, 10 g. To decode we first multiply v-bar by 10 g then add 464.5 back on, as shown. To find s_x from s_v, we merely multiply by 10 g (464.5 g does not enter in) as shown. The 10 g gives back the unit grams in each case.

Let us now summarize this calculation by coding as follows, using d for the class interval and x_0 for the guessed average which we subtract from each midvalue:

$$v = \frac{x - x_0}{d} \equiv x_0 - dv \tag{2.6}$$

$$\bar{v} = \frac{\Sigma vf}{f} = \frac{\Sigma vf}{n} \tag{2.7}$$

$$s_v = \sqrt{\frac{n\Sigma v^2 f - (\Sigma vf)^2}{n(n-1)}} \tag{2.8}$$

$$\bar{x} = x_0 + d\bar{v} \tag{2.9}$$

$$s_x = ds_v \tag{2.10}$$

Note that if we have small samples and treat each frequency, f, as 1 and also do not divide by any d to find the y's, we have $v = y$ and (2.7) is y-bar and (2.8) becomes (2.5).

2.7. CURVE SHAPE

Sometimes a frequency distribution shows frequencies decreasing from the maximum frequency much more rapidly in one direction than in the other, i.e., the distribution is not symmetrical around the point of greatest concentration. There are available objective measures of the extent of this lack of symmetry. But we shall not consider them here. For information on measures of curve shape, the reader is referred to Burr (1976).

In some frequency distributions there may be more than one maximum frequency, i.e., the frequencies of the classes on either side of a particular class are less. This can occur if the sample size is small; however, when the sample size is greater than 40 or 50, it might be very useful to investigate the cause of multiple peaks in the frequency distribution.

2.8. POPULATION VS. SAMPLE CHARACTERISTICS

The arithmetic mean or average of a population distribution and also the standard deviation are analogous to those of a sample. In order to distinguish between sample and population characteristics, however, we use different symbols.

We follow the standard practice. In particular, we shall use the following:

	Sample	Population
Average	$\bar{x}$ (x-bar)	μ = mu (the Greek letter m)
Standard deviation	s	σ = sigma (the Greek letter s)
Sample size	n	N

These notations are nearly universal in applied statistics.

There is one difference in the calculation of s for a sample and σ for a finite population. Let us suppose that we have a lot of 400 plummets, whose weights are the quantities in question. Let each plummet be very accurately weighed, giving 400 weights in, say, grams: $x_1, x_2, \ldots, x_{400}$. Then we have for the mean or average of this finite population (note that we are not interested in all plummets or all plummets made by a single process, we are only interested in these 400 plummets).

$$\mu = \frac{\Sigma x}{N} = \frac{\Sigma x}{400}$$

where N is the population size as noted above. This analogous to (2.2). For the standard deviation of the finite population:

$$\sigma = \sqrt{\frac{\Sigma(x - \mu)^2}{N}} = \sqrt{\frac{\Sigma(x - \mu)^2}{400}}$$

where the sum of the squared deviations from the population mean, μ, is divided by N, not $N - 1$, as it would be if the 400 weights were a sample of 400 plummets from a large lot of plummets or from a manufacturing process. When the population is of unlimited or infinite size, then other methods of calculation are used which we shall not describe here. But in any case, for a population arising under some set of homogeneous conditions of production and measurement, there will at least theoretically be an average, μ, and a standard deviation, σ. And we can regard our sample x-bar and s as estimates of them if the sample is chosen in an unbiased manner.

2.9. PROBABILITY AS AREA, THE NORMAL CURVE

In Section 2.5 we saw that the "normal" distribution is symetrical about the x-bar and in "well behaved" data from a sufficiently large sample, $n \geq 30$, over 99% of the data should lie within $\pm 3\sigma$. This should hold true also for all of the population values if the sample is truly representative of the population and the data are measured, i.e., continuous. Again, these data may be from the measurement of temperature, length, circularity, weight, purity, stress level, and flow rate for example rather than counted or discrete data.

If we look at the frequency polygon in Figure 2.1, we can get a sense of the shape of the distribution of the data in the sample and also a guess at the probable shape of the distribution of the population. Since it is reasonably "well behaved", i.e., symmetrical with no apparent evidence of multiple peaks, we call this a "normal" distribution. The measurements obtained on most features of a process will exhibit this type normal distribution as long as there are no external effects acting on the process during the period of measurement, e.g., change in the process average due to a change in temperature or tool wear. The former could cause an extra peak while the latter could cause skewness.

Using Figure 2.1, we could now smooth out the frequency polygon by drawing a smooth curve through the midvalues. This curve will look quite like a normal curve or distribution. Now there is a normal curve for every population ever sampled, they may have an infinite number of averages, μ, and an infinite number of sigma's. That leaves us ∞^2 normal curves and that is an awful lot of them. How can we make all these curves have the same x-bar? We could subtract the mean from each value in the distribution.

$$X - \mu$$

Now all the μ's are ZERO; but the width of the distributions vary all over the map from very tiny to very wide. How do we now make all these distributions have the same width? The one commonality of these distributions is that they all have a width which is described by $\pm 3\sigma$; so let us divide the deviations from the mean by the standard deviation of its respective distribution:

$$\frac{X-\mu}{\sigma}$$

For some strange reason we have called this value for the "standardized" normal distribution by the title of Z. As can be seen all the infinite number of normal curves can be reduced to one through the use of this calculation whether they come from the population or a sample. The subscript, x, denotes that this is the standard deviation of the individual values in the sample or population.

$$Z=\frac{X-\mu}{\sigma_x} \quad \text{for a population} \tag{2.11}$$

$$Z=\frac{X-\overline{X}}{s_x} \quad \text{for a sample} \tag{2.12}$$

We can solve both these equations for X with these results:

$$X=\mu-Z\sigma_x \tag{2.13}$$

$$X=X\text{-bar}-Zs_x \tag{2.14}$$

These two equations give us the number of standard deviations, Z, that X is away from the mean. Let us use Figure 2.1 for an example. The lower specification limit is 400, i.e., $427-27$ and the upper specification limit is $427+27=454$. First, we calculate the Z for the lower limit using (2.12) and the average and standard deviation from Table 2.2:

$$Z=\frac{X-\overline{X}}{s_x}=\frac{400-463.4}{17.37}=-3.65$$

It can be seen that there is very little likelihood that any charge weight would be less than the lower specification limit since less than 1% of the values would be outside of $\pm 3\,s$ much less below $-3.65\,s$. Likewise we can calculate the Z

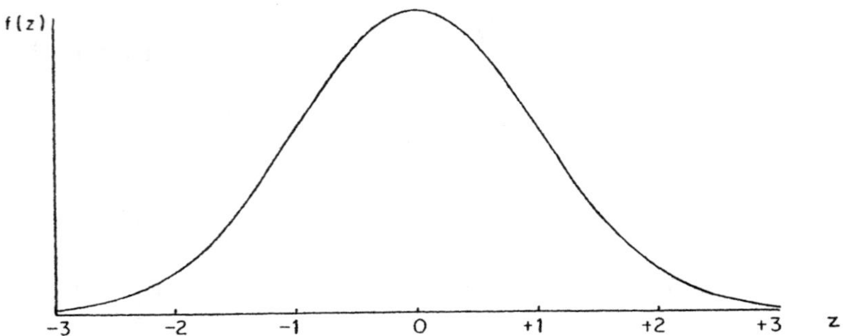

Figure 2.2 The normal curve for the standardized variable $Z = (\mathrm{x} - \mu)/\sigma$.

for the upper specification limit:

$$Z = \frac{X - \overline{X}}{s_x} = \frac{457 - 463.4}{17.37} = -0.37$$

This obviously means that more than half of the charge weights are overfilled and beyond the upper specification! Let us see this graphically in Figure 2.2.

In our example of the charge weights, we used (2.12) and the "x" is that we are asking the question about, in this case, is the upper specification limit. Since $z = -0.37$ it is to the left of the center line, i.e., the mean, of the standardized normal curve. Over half of the area under the curve is to the right of -0.37. In other words, over half of the product has charge weights greater than the upper specification limit.

Tables of this standardized normal distribution were developed over 100 years ago by Gauss without the aid of a calculator or a computer! Table A is the result of those laborious calculations. In the caption of Table A we see that the numbers in the table represent the cumulative probability that the value, z, would have, i.e., the area under the standardized normal curve to the left of that z. Looking down the left-hand column we find -0.3 near the bottom. Looking across this row we come to the column headed by 0.07. The intersection of this row and column has the number 0.3557. This is the probability of having a $z = -0.37$ or lower. Again this is the area under the curve to the left of $z = -0.37$. What then is above this z value? Since the total area under the curve must be 1.0000, we can subtract the 0.3557 from 1.0000 and see that the probability of having a charge weight greater than the upper specification is $1.0000 - 0.3557 = 0.6443$.

Let us use another example. For a normal distribution of tensile strengths with $\mu = 68{,}000$ and $\sigma = 5000$ psi, find the probability of a strength below 55,000 psi. First find z by using (2.11):

$$Z = \frac{X - \mu}{\sigma_x} = \frac{55,000 - 68,000}{5,000} = -2.60$$

From Table A, we find

$$P(x \leq 55,000) = 0.0047$$

Thus, only about 1 in 213 will lie below 55,000 (see Figure 2.3).

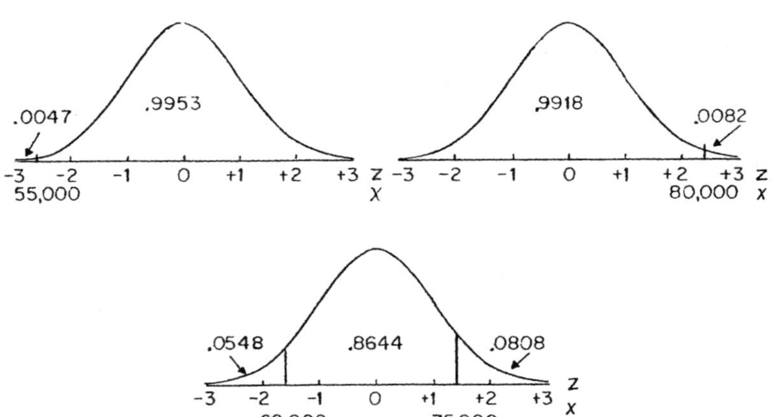

Figure 2.3 Three figures of the normal curve to illustrate the probabilities $x \leq 55{,}000$, $x > 80{,}000$ and $60{,}000 < x \leq 75{,}000$ psi, when $\mu = 68{,}000$ psi and $\sigma = 5000$ psi.

Next, we can find the probability of a strength above 80,000 psi. First, find z:

$$Z = \frac{X - \mu}{\sigma_x} = \frac{80,000 - 68,000}{5,000} = +2.40$$

Table A gives:

$$P(x \leq 80,000) = 0.9918$$

so the complementary probability, $P(x > 80{,}000) = 1 - 0.9918 = 0.0082$ which is again quite small. Next, we find the probability of a strength lying between 60,000 and 75,000. Convert these strengths to standard z's:

$$Z_L = \frac{X - \mu}{\sigma_x} = \frac{60,000 - 68,000}{5,000} = -1.60$$

$$P(x \leq 60,000) = 0.0548$$

$$Z_H = \frac{X - \mu}{\sigma_x} = \frac{75,000 - 68,000}{5,000} = +1.40$$

$$P(x \leq 75,000) = 0.9192$$

Now the probability of a strength between the two is the difference (see Figure 2.3). Thus,

$$\begin{aligned} P(60,000 < x \leq 75,000) &= P(x \leq 75,000) - P(x \leq 60,000) \\ &= 0.9192 - 0.0548 = 0.8644 \end{aligned}$$

Because the strengths which the event $60{,}000 < x \leq 75{,}000$ lacks from those of the event $x \leq 75{,}000$ are precisely those contained in the event $x \leq 60{,}000$.

Proceeding similarly to a third example, we find the important probabilities:

$$\begin{aligned} P(\mu + 1\sigma) &= P(\mu - 1\sigma < x \leq \mu + 1\sigma) = P(-1 < z \leq +1) \\ &= 0.8413 - 0.1587 = 0.6826 \end{aligned}$$

$$P(\mu + 2\sigma) = P(\mu - 2\sigma < x \leq \mu + 2\sigma) = P(-2 < z \leq +2)$$
$$= 0.9772 - 0.0228 = 0.9544$$

$$P(\mu + 3\sigma) = P(\mu - 3\sigma < x \leq \mu + 3\sigma) = P(-3 < z \leq +3)$$
$$= 0.9987 - 0.0013 = 0.9974$$

These are the exact probabilities but we can somewhat simplify them to be approximately 68% for $\pm 1\sigma$; 95% for $\pm 2\sigma$, and 99% for $\pm 3\sigma$, i.e., about two-thirds of the cases will lie within $\pm 1\sigma$, 19 out of 20 will lie within $\pm 2\sigma$ and very nearly all will lie within 3σ of the mean. Now these are for the normal distribution but it takes quite a bit of non-normality before they are much in error.

Properties of the standard normal distribution for z:

1. The total area lying between the curve and the horizontal z axis is 1, representing a total probability of 1.
2. The height of the curve represents relative frequency, also called probability density.
3. The relative frequency is greatest at $z = 0$ (or $x = \mu$).
4. Relative frequencies steadily decrease as z moves away from 0 in either direction (or x away from μ).
5. The relative frequency rapidly approaches 0 in both directions.
6. The curve is symmetrical around $z = 0$. Thus, for example, the relative frequency is the same at $z = -0.8$ as at $z = +0.8$.
7. Desired probabilities for ranges of z may be found from published tables of normal curve areas.
8. The curve has its concave side down between $z = -1$, $z = +1$ and its concave side up outside these limits, this being the point of inflection on each side of the curve.
9. The equation for the curve is not often needed because of available tables; but it is:

   $f(x) = e^{-z^2/2}/\sqrt{2\pi}$ where $e = 2.7183\ldots$ and $\pi = 3.1415\ldots.$

If we let h be the maximum height (at $z = 0$ or $x = \mu$) we have:

Ordinate	h	0.969 h	0.883 h	0.755 h	0.607 h
z	0	± 0.25	± 0.50	± 0.75	± 1.00
x	μ	$\mu \pm 0.25\sigma$	$\mu \pm 0.50\sigma$	$\mu \pm 0.75\sigma$	$\mu \pm 1.00\sigma$
Ordinate	h	0.458 h	0.325 h	0.216 h	0.135 h
z	0	± 1.25	± 1.50	± 1.75	± 2.00
x	μ	$\mu \pm 1.25\sigma$	$\mu \pm 1.50\sigma$	$\mu \pm 1.75\sigma$	$\mu \pm 2.00\sigma$
Ordinate	h	0.080 h	0.044 h	0.023 h	0.011 h
z	0	± 2.25	± 2.50	± 2.75	± 3.00
x	μ	$\mu \pm 2.25\sigma$	$\mu \pm 2.50\sigma$	$\mu \pm 2.75\sigma$	$\mu \pm 3.00\sigma$

These tabled ordinates help one draw a normal curve to any desired scale.

2.10. SUMMARY

In this chapter we have discussed one average, the arithmetic mean, for finding a typical or representative level in a sample of numbers. Although this is but one of several types of averages, it is by far the most important one for our purposes. For measures of variability, we covered the range and the standard deviation. The former is commonly used for small samples up to about 10 and the latter is used fairly exclusively for larger samples. These three measures, x-bar, R and s are objective descriptions of the characteristics of a sample of data. For a population, the characteristics, mean or μ and standard deviation or σ, correspond to x-bar and s for a sample. In addition to using x-bar and s to describe a sample, it is often desirable to construct a frequency table and draw a frequency graph, if the sample is of at least, say, 25 numbers. In addition, we have also discussed shortcuts to the calculation of x-bar and s, both for ungrouped data and for data in a frequency table. Also a distinction was made between two general types of data: (1) discrete or counted data and (2) continuous or measured data. Somewhat different approaches are used in the analysis of the two general types of data. And finally we looked at the most generally applicable probability distribution, the normal curve, and its properties.

2.11. PROBLEMS

Find for each of the following small samples, x-bar, R, and s:

2.1 Hardness Rockwell 15 T: 72, 72, 66, 67, 68, 71, specifications being 66–72 on carburetor tubes.

2.2 Tube diameters: 0.2508, 0.2510, 0.2506, 0.2509, 0.2506 in.

2.3 Dimension on a rheostat knob: 142, 142, 143, 140, 135 in 0.001 in.

2.4 Dimension on an igniter housing: 0.534, 0.532, 0.531, 0.531, 0.533 in.

2.5 Eccentricity or distance between center of cone and center of triangular base on needle valves: 50, 35, 36, in 0.0001 in.

2.6 Number of nonconformities on subassemblies: 2, 8, 3, 3, 7, 1.

2.7 Density of glass: 2.5037, 2.5032, 2.5042 g/cm^3.

2.8 Percent of silicon in steel castings: 0.94, 0.89, 0.98, 0.87.

For the following two samples of raw data, choose appropriate numerical classes and tabulate the data into a frequency table. Draw an appropriate frequency graph. Comment on the data.

2.9 Data on eccentricity of needle valves, conical point to base, in 0.0001 in. units. Maximum specification limit 0.0100 in. Suggest 0–9, 10–9, and so on for classes.

32	30	30	37	18	37	50	35	36	57	24	75	49	6	24
67	25	25	52	56	53	18	39	47	40	51	51	31	61	28
15	10	35	27	49	19	51	34	40	19	32	10	39	16	50
15	30	50	32	46	29	39	19	34	42	40	30	70	16	57
12	19	23	34	14	40	58	36	41	7	11	8	40	12	12
66	58	19	29	37	74	20	9	15	30	38	88	83	57	90

2.10 Chemical analyses for manganese in 80 heats of 1045 steel. Data in 0.01% units. Suggest 66–68, 69–71, and so on for classes. Specifications: 70–90.

74	79	77	81	72	66	75	80	76	86	84	70	80	62	74	71
68	79	81	76	79	79	84	78	74	88	71	80	79	74	76	75
81	80	80	78	76	81	70	76	79	80	79	84	75	75	76	83
88	83	79	91	73	78	82	74	81	75	76	72	83	97	76	90
79	75	74	73	93	92	70	75	86	87	79	69	79	77	76	82

Find x-bar and s for the following:

Find x-bar and s for the following:

2.11 The frequency distribution obtained in Problem 2.9.

2.12 The frequency distribution obtained in Problem 2.10

2.13 Density of glass in g/cm^3

Density	Frequency	Density	Frequency
2.5012	2	2.5052	19
2.5022	6	2.5062	10
2.5032	25	2.5072	4
2.5042	33		Total: 99

2.14 Over-all height of bomb base

Height (in.)	Frequency	Height (in.)	Frequency
0.830	1	0.834	13
0.831	3	0.835	11
0.832	11	0.836	6
0.833	14	0.837	6
			Total: 65

2.15 Spring tension in pounds

Tension (lb)	Frequency	Tension (lb)	Frequency
50.0	2	53.0	32
50.5	2	53.5	12
51.0	4	54.0	18
51.5	6	54.5	2
52.0	9	55.0	1
52.5	12		Total: 100

2.16 For problem 2.1, calculate the % out of specification that one could expect to find in the production of carburetor tubes assuming that this sample has been representative of production.

2.17 For problem 2.9, determine the % of needle valves exceeding the maximum specification limit in the production lot assuming that this sample represents the lot.

2.18 For problem 2.10, determine the % of all heats that are beyond the specification limits assuming that this sample represents the population.

2.19 For problem 2.13, calculate the % out of specification when the limits are $2.5050 \pm 0.0030\,g/cm^3$.

2.20 For problem 2.14, calculate the % out of specification when the limits are $0.830 - 0$ and $0.830 + 0.010$ in.

2.21 For problem 2.15, calculate the % of springs in the lot when the specification calls for a minimum tension of 50 pounds.

REFERENCE

IW Burr. Statistical Quality Control Method. New York: Marcel Dekker, 1976.

3

Simple Probability and Probability Distributions

Some basic knowledge of probability is needed in statistical quality control. But the amount needed is not great. In this chapter, we give just some basic concepts of probability which are needed in the interpretation of statistical techniques and how they are used. Later in this chapter, we give the derivation and formulas for the distribution models which are used in control charts for counted, i.e., discrete, data and for continuous, i.e., measured, data. The reader will find the subject of probability of interest in every day life too. We will give a few examples of this.

3.1. LIKELIHOOD OF AN EVENT OCCURRING

Probability is concerned with the likelihood of an event occurring. The event is perhaps quite likely to occur on a given trial or opportunity or it may be most unlikely. Here at the author's home on the coast of Washington state, it is quite likely that some rain will fall on any given day; since this was true on about four-fifths of the days last year. So we would call the probability of rain falling on any given day equal to 0.80.

The scale for the probability of an event is 0 to 1. If an event *cannot occur* on a trial, then the probability of its occurrence is 0. Or if an event is *certain* to occur on a trial, the probability is 1. If in the long run an event will occur half of the time then the probability is 0.5 and so on. We emphasize that a "trial" or experiment must be clearly defined. We also have to clearly define the "event".

As an example, suppose that a "trial" is drawing a piece at random from some production line and the "event" in question is that the piece drawn is a nonconforming one. Let us suppose that the probability of a nonconforming unit is 0.05. This means that 5% of the time when we draw a random piece, it is a nonconforming unit. The complementary event is that the piece drawn is a conforming one. What would you say is the probability of a piece drawn being a conforming one? The answer is 0.95.

$$P(\text{nonconforming unit}) = 0.05$$
$$P(\text{conforming unit}) = 0.95$$
$$P(\text{conforming or nonconforming}) = 1.00$$

We say that 95% of the time, in the long run, the piece drawn will prove to be a conforming one.

Similarly a golf ball might be either a conforming one or a "second" or a reject. It will always be one of the three cases. Suppose that the three respective probabilities are 0.97, 0.02 and 0.01. Their sum must be 1.00 because each ball produced must be in *one and only one* of the three cases.

3.2. OCCURRENCE RATIO

Suppose that we have a production process for which the probability of a nonconforming piece is constantly 0.08. Now what happens to the observed proportion of nonconforming pieces as we continue to sample, i.e., to the occurrence ratio? Let us give a bit of notation to help the discussion

$$p_o = \text{The constant probability of a nonconforming piece} \quad (3.1)$$

where the letter p stands for probability and the subscript, o, means that it is a population probability

$$d = \text{The number of nonconforming pieces observed} \quad (3.2)$$

$$n = \text{The number of pieces inspected or tested (sample size)} \quad (3.3)$$

$$p = d/n = \text{Sample proportion of nonconforming pieces} \quad (3.4)$$

With these notations available to us, let us ask the question again. How does $p = d/n$ behave as we sample more and more, i.e., increase n? Would we not expect that the observed occurrence ratio $p = d/n$ would tend to approach $p_o = 0.08$? Let us learn by an experiment how the approach works. We start with five samples each of size 10.

Sample		Totals		Occurrence ratio
n	d	$\sum n$	$\sum d$	$p = \sum d / \sum n$
10	0	10	0	0.0000
10	1	20	1	0.0500
10	1	30	2	0.0667
10	1	40	3	0.0750
10	0	50	3	0.0600
50	5	100	8	0.0800
50	4	150	12	0.0800
50	6	200	18	0.0900
50	4	250	22	0.0880
50	8	300	30	0.1000
50	1	350	31	0.0886
50	3	400	34	0.0850
50	5	450	39	0.0867
50	3	500	42	0.0840

(*Continued*)

Sample		Totals		Occurrence ratio
n	d	$\sum n$	$\sum d$	$p = \sum d/\sum n$
50	5	550	47	0.0855
50	5	600	52	0.0867
50	5	650	57	0.0877
50	4	700	61	0.0871
50	3	750	64	0.0853
50	6	800	70	0.0875
50	4	850	74	0.0871
50	7	900	81	0.0900
50	4	950	85	0.0895
50	4	1000	89	0.0890

The first two columns are for the current sample of 10 (and later 50). The third and fourth columns are for the total sample size and the cumulative total number of nonconforming pieces. The fifth column is based on the third and fourth columns and gives the current overall proportion nonconforming occurrence ratio $= \sum d/\sum n$.

Now how does the proportion nonconforming behave? It only "tends" to approach $p_o = 0.08$. Sometimes it gets closer and sometimes it backs away from p_o. Before the total sample size was 100, the occurrence ratio or fraction nonconforming was below 0.08. At 100 and 150, it was exactly 0.08. For higher cumulative sample sizes, the occurrence ratio was greater than 0.08. We hasten to point out that if we were to do this experiment all over again, we would get different, even inverse, results.

But in this experiment what will happen to next 1000? Gamblers like to say that the "law of averages" would indicate that the next 1000 will have a fraction less than 0.08 because the first 1000 was at 0.09 (we have to make up for the discrepancy between what we found in the first 1000 and the "true" or population fraction). The law of averages says no such thing. It does say that as the number of trials or sample size increases, the probability of the occurrence ratio lying in any range around p_o, e.g., 0.75 to 0.85 increases and approaches certainty. Misinformation about this law of averages has cost gamblers untold amounts of money.

Principle. Take any limits around the true constant probability, p_o. How close the estimate is depends upon the sample size, n, the value of p_o and upon luck or chance. But we can learn to "play the odds" well and will do so.

3.3. EXAMPLE 1 AND PROBABILITY LAWS

In this example, and the *two following ones*, we shall be illustrating some laws of probability which the reader will find make very good sense.

Consider again the production line producing pieces with a constant probability 0.08 of the piece being a nonconforming unit and such that each piece is independent of the others produced. What is meant by the latter is that the probability of the next piece being a nonconforming unit is 0.08 and of being a good one is 0.92 regardless of what the preceding pieces have been like. (This property is called *independence* and is typical of a production process which is "in control", as we have been seeing).

For a single draw of one unit, there are just two possible outcomes or events. The unit can either be conforming or nonconforming. The respective probabilities are obviously 0.92 and 0.08.

Now let us suppose that we draw a sample of two units and inspect them. What now are the possible outcomes? There are three: namely the sample may contain 0, 1 or 2 nonconforming units. Let us find the probabilities of these outcomes or events.

For the probability of no nonconforming units in the sample of two, we must have conforming units on both draws. Now 92% of the times or trials the first unit will be a conforming unit (in the long run). Of all these times in which we obtain a conforming unit on the first draw, it will be followed by a conforming unit on the second draw 92% of the time. Thus the probability of having two nonconforming units in a sample of two is $(0.92)\cdot(0.92) = 0.8464$.

For the probability of our sample containing two nonconforming units, we proceed in the same way. In 8% of the samples, the first unit sample will be nonconforming. Now in how many of these samples will we follow with a second nonconforming unit? This is clearly 8%. Thus the probability of the sample containing two nonconforming units is $(0.08)\cdot(0.08) = 0.0064$. Next we ask for the probability of the sample containing one nonconforming unit (and one conforming unit, of course). Proceeding in a similar fashion, we will in the long run start with a conforming unit (C) 92% of the time on the first draw. Then we must have a nonconforming unit (N) on the next draw in order to fulfill the condition of the question. The probability of this is 0.08. Thus a sample of two in the order of C, N will have a probability of $(0.92)\cdot(0.08) = 0.0736$. But the sample could also have the nonconforming unit first and the conforming unit second. This probability of N, C is $(0.08)\cdot(0.92) = 0.0736$. And so we find the probability of the sample containing exactly one onforming and one nonconforming unit (in either order) $(0.92)\cdot(0.08) + (0.08)\cdot(0.92) = 2(0.0736) = 0.1472$. But we could have arrived at this result in another way. Since we must have 0, 1 or 2 nonconforming units in the sample of two, we must have (letting P stand for the probability of the event listed in the parenthesis following it):

$$P(0 \text{ nonconforming}) + P(1 \text{ nonconforming})$$

$$+ P(2 \text{ nonconforming}) = 1$$

$$\text{or } P(d = 0) + P(d = 1) + P(d = 2) = 1$$

$$\text{and } P(d = 1) = 1 - [P(d = 0) + P(d = 2)]$$

$$= 1 - 0.8464 + 0.0064 = 0.1472$$

In finding these probabilities, we have assumed that the probabilities of the outcomes on the second draw were independent of what happened on the first draw, i.e., whether the first piece was a conforming or nonconforming unit did not influence the probability of the second draw. We give the definition:

Definition 3.1. Two events A and B are independent if the occurrence or nonoccurrence of A does not affect the probability of B's occurrence.

If the outcome of a nonconforming unit on a draw from a process and of a conforming unit has no relation to the outcome on the preceding or following draw, then the results are random and we say that the process is "in control".

Definition 3.2. Two or more events are mutually exclusive if only one of them can possible occur on a trial.

Thus in the last example, the three possibilities were for the sample to contain exactly 0, 1 or 2 conforming pieces. These events were therefore mutually exclusive. In such a case, the probability of some one of the events to occur on a trial is the sum of the separate probabilities of occurrence of the mutually exclusive events. In this example, the three events exhaust all the possibilities and so one of the events was certain to occur and the sum of the three probabilities was 1.

Law 3.1. If two events, A and B, are mutually exclusive then we have:

$$P(A \text{ or } B, \text{mutually exclusive events}) = P(A) + P(B) \tag{3.5}$$

We have seen this law in operation in the drawing of a sample of two pieces from a process. Also in the drawing of a sample of three, we used an obvious extension of the law.

Definition 3.3. If one of two events A and A' is certain to occur on a trial but both cannot simultaneously occur, then A and A' are called complementary events (A and A' are mutually exclusive). Thus on a single draw from the process, the events of it being a conforming piece and of its being a nonconforming piece were complementary events. The respective probabilities were 0.92 and 0.08.

Law 3.2. For any event, A, and its complementary event, A', we have

$$P(A) + P(A') = 1 \tag{3.6}$$

Another example is P(1 nonconforming unit) + P(0 or 2 nonconforming units) = 1 when there are only two units in the sample.

3.4. THE BINOMIAL DISTRIBUTION

Let us now consider Example 1 extended. Let us work out the probabilities for a sample of three taken from a process with nonconforming pieces produced randomly with constant probability, $p_o = 0.08$. There are now four ways for the sample to occur, namely 0, 1, 2 or 3 nonconforming pieces in the sample. Letting d stand for the number of nonconforming pieces in the sample, we have

$$P(d = 0) = 0.92^3 = 0.779$$

$$\text{and } P(d = 3) = 0.08^3 = 0.001$$

Now consider the probability of $d = 1$. There must be one nonconforming piece and two conforming pieces in the sample but they can occur in any order. For the order N, C, C, we have the probability 0.08(0.92)(0.92). For C, N, C and C, C, N, we have

$$P(d = 1) = 3(0.08)(0.92)^2 = 0.203$$

For the sample to contain two nonconforming pieces and one conforming piece, the probability for each of the three possible orders is $(0.08)^2(0.92)$ and we have

$$P(d = 2) = 3(0.08)^2(0.92) = 0.018$$

Note that the sum of the four probabilities is 1.001 (the extra 0.001 is due to rounding error). The sum of the four probabilities has to equal 1 because we have covered all the possibilities. These four probabilities are exactly the four terms in the binomial expansion of $(0.92 + 0.08)^3$. This is the reason the theoretical distribution we have described is called the "binomial distribution".

Now lest the reader is lost or worried let us say that it is not necessary for a quality control worker to calculate the terms of a binomial distribution very often or perhaps ever. The reasons are twofold: there are excellent books of tables available (see the references at the end of this chapter); and secondly this distribution is very easy to program into a computer. Most statistical software packages now include the calculation of the binomial probabilities. All said and done it is still worthwhile to see the binomial distribution because it is the basic one for counted or discrete data. We often assume this distribution as our model in many statistical quality control applications. In general the terms of a binomial expansion give the probabilities of the sample containing 0, 1, 2, ..., $r, \ldots, n$ nonconforming pieces. We take the binomial as follows:

$$\begin{aligned}[(1-p_o)+p_o]^2 &= (1-p_o)^n + [n/1](1-p_o)^{n-1} \\ &\times p_o + [n(n-1)/1 \cdot 2](1-p_o)^{n-2}p_o^2+ \\ &+ [n(n-1)(n-2)/1 \cdot 2 \cdot 3](1-p_o)^{n-3}p_o^3+ \\ &+ \cdots + [n(n-1)\cdots(n-r+1)/1 \cdot 2 \cdots r](1-p_o)^{n-r} \\ &\times p_o^r + \cdots + (1-p_o)^n \qquad (3.7)\end{aligned}$$

The terms written out in the binomial above give the respective probabilities $P(d=0)$, $P(d=1)$, $P(d=2)$, $P(d=r)$, and $P(d=n)$. These are the terms of the population distribution.

Sometimes we need to know the population average number of nonconforming pieces in samples of n pieces or units from a process that is in control

$$\mu_d = np_o \quad \text{and} \quad \sigma_d = \sqrt{np_o(1-p_o)} \qquad (3.8)$$

Let us take a simple case of $n=4$ and use the formula (3.7). We let $p_o=0.10$ and then have:

$$\begin{aligned}p(d=0) &= (1-0.1)^4 = 0.6561 \\ p(d=1) &= 4(0.9)^3(0.1) = 0.2916 \\ p(d=2) &= [4(3)/2](0.9)^2(0.1)^2 = 0.0486 \\ p(d=3) &= [4(3)(2)/2(3)](0.9)(0.1)^3 = 0.0036 \\ p(d=4) &= (0.1)^4 = 0.0001\end{aligned}$$

Once again we note that the sum of the five probabilities is 1, which is a good check on the arithmetic.

In these equations are calculations that designate the numbers of different orders that the nonconforming and conforming pieces can be taken to satisfy the conditions of the sample. The calculation of this order is called a combination, i.e., n things taken r at a time. The notation for this is $n\text{C}r$ or $(^n{}_r)$. Most hand calculators have an "$n\text{C}r$" key. For example if $d = 2$ and $n = 4$, we would calculate ${}_4\text{C}_2 = 6$. Where does this come from? Well we have a defining equation:

$$n\text{C}r = n!/[(n - r)!r!] \tag{3.9}$$

The $n!$ is called "n factorial" and is equal to $n(n-1)(n-2)\cdots(n-r+1)$. For $n = 4$, $n! = 4! = 4(3)(2)(1)$, i.e., it is the product of 4 times 3 times 2 times $1 = 24$. Using (3.9) ${}_4\text{C}_2 = 4!$ divided by $(4-2)!$ and $2! = 24/(2{\cdot}1)(2{\cdot}1) = 24/4 = 6$. Likewise ${}_4\text{C}_3 = 24/1!(3!) = 24/6 = 4$. Sometimes in the calculation of combinations, we might run into the case where we have $0!$, e.g., ${}_4\text{C}_4 = 4!/0!(4!)$. Zero factorial, $0!$, is defined as being equal to 1. This saves us from dividing numbers by zero. One can easily see that ${}_4\text{C}_4$ and ${}_4\text{C}_0$ are both equal to one. To generalize this ${}_n\text{C}_n$ and ${}_n\text{C}_0$ are both equal to 1.

For the sake of completeness, let us digress for a moment and talk about "permutation" here. A permutation is like a combination except that the order in which the items are taken or arranged is important. In the case above, we do not care whether the one nonconforming unit was taken first or second or third or fourth in the consideration of the probability of obtaining a single nonconforming unit in a sample of four. Thus N, C, C, C and C, N, C, C and C, C, N, C and C, C, C, N are all equivalent, i.e., there is one nonconforming unit in the sample.

On the other hand, in permutations, order is important, for example if we consider the five vowels, a, e, i, o, and u, we can select any two. In this example, "ei" is very different than "ie" and "ae" is different than "ea". Order in this case is important and must be taken into consideration in the count of the number of ways things can happen. In the case of the vowels, we can have:

ae, ai, ao, au, ei, eo, eu, io, iu, ou
but we also have
ea, ia, oa, ua, ie, oe, ue, oi, ui, uo

There are 20 ways to take (or order) two vowels from the five. This permutation is larger than the corresponding combination, ${}_5C_2 = 10$. The formula for the calculation of permutations is

$$nPr = n!/(n-r)! \tag{3.10}$$

Before continuing, we tabulate our results of our example of the binomial distribution and then picture them with a histogram.

Binomial Distribution
($p_o = .1, n = 4$)

d	P(d)	p = d/n
0	.6561	.00
1	.2916	.25
2	.0486	.50
3	.0036	.75
4	.0001	1.00
	1.0000	

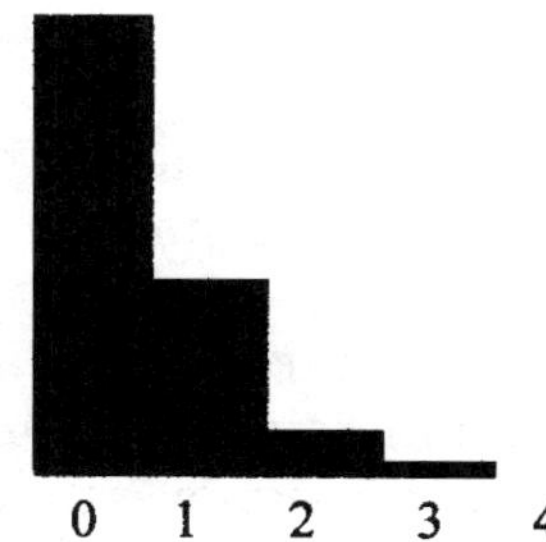

Such a distribution is sometimes described as "*J*-shaped" since as *d* increases, *P*(*d*) continually decreases.

The general case for the binomial distribution is

$$P(d) = {}_nC_r(p_o)^d(1-p_o)^{n-d} = {}_nC_r(p_o)^d(q_o)^{n-d} \tag{3.11}$$

Fortunately we do not have to make use of this formula very often, using instead either a computer or tables of binomial probabilities.

3.4.1. Binomial Tables

Instead of working out any desired cases of (3.7), it is best to try to find any needed probabilities in published tables. See US Department of the Army (1950), Harvard Univ. (1955),

Robertson (1960), Romig (1947) and Weintraub (1963) at the end of this chapter. The table entries may be in the form of particular probabilities such as $p_o = 0.06$, $n = 50$, $P\ (d = 2) = 0.22625$. For some table, i.e., Harvard Univ. (1955) and Weintraub (1963), we may have to subtract cumulative entries. This will be demonstrated in the discussion of the Poisson tables later in this chapter.

Using the example cited earlier where $n = 4$ and $p_o = 0.10$, we can verify that $\mu_d = np_o = 4(0.1) = 0.4$ by doing some arithmetic. We can take the weighted average of the five possible outcomes, i.e., $\sum d[P(d)] = 0(0.6561) + 1(0.2916) + 2(0.0486) + 3(0.0036) + 4(0.0001) = 0.4000$. Given then this estimate of μ_d (if p_o had not been known and we had obtained these probabilities from a table using a p calculated from a sample) we can now calculate the standard deviation of p using (3.8). Thus $\sigma_d = \surd[4(0.1)(0.9)] = 1.2$. This is sort of the average amount by which the various d counts differ from the average of 0.4. In Section 2.5, it was stated that very few cases will lie outside of the limits, x-bar $\pm 3s$, for a sample. The same is true for the limits of $\mu \pm 3\sigma$ for a population; but in this case, we speak of a small probability of a case on or outside of the "three sigma range". Let us see how it works out for the binomial distribution which we have been studying. The three sigma range is $0.4000 \pm 3(1.2) = -3.2$ and 4.0. Notice that we cannot have a negative number of nonconforming units; therefore much of the left side of the population distribution does not exist (making it, as we said earlier, a J-distribution). In that sense, the standard deviation being for symmetrical probability distributions does not really apply. Yet it does do a pretty good job of estimating the probabilities of the larger d's, e.g., $P(4)$ in our example $= 0.0001$.

Example. We now give six examples of binomial distributions including the sample size, n, the expected or average number of nonconforming cases in the sample, μ_d, and the only-to-be-expected deviation from μ_d, namely, σ_d. The probabilities are given to three decimal places in Table 3.1. The six histograms are shown in Figure 3.1.

Table 3.1 Examples of Binomial Distributions for Various Process Fractions Defective and Sample Sizes

	(a)	(b)	(c)	(d)	(e)	(f)
p_o	0.30	0.06	0.01	0.08	0.05	1/6
n	10	50	300	20	100	6
$E(d)=p_o$	3	3	3	1.6	5	1
σ_d	1.45	1.68	1.72	1.21	2.18	0.91
d	$P(d)$	$P(d)$	$P(d)$	$P(d)$	$P(d)$	$P(d)$
0	0.028	0.045	0.049	0.189	0.006	0.335
1	0.121	0.145	0.149	0.328	0.031	0.402
2	0.233	0.226	0.224	0.271	0.081	0.201
3	0.267	0.231	0.225	0.141	0.140	0.054
4	0.200	0.173	0.169	0.052	0.178	0.008
5	0.103	0.102	0.101	0.015	0.180	0.001
6	0.037	0.049	0.050	0.003	0.150	
7	0.009	0.020	0.021	0.001	0.106	
8	0.002	0.007	0.008		0.065	
9		0.002	0.003		0.035	
10		0.001	0.001		0.017	
11					0.007	
12					0.003	
13					0.001	

The first three cases have the same expected or average number of nonconforming cases, $\mu_d = 3$. But σ_d varies due to the sample size, n, varying. The distributions do vary a bit. If we form "three sigma limits" for example (b) we have $3 \pm 3(1.68) = 3 \pm 5.04 = -2.04$ and 8.04. But d can never be negative; so there are no possibilities of cases below 0 or, for that matter -2.04. On the other hand, there are two cases in which the 8.04 can be exceeded, $d = 9$ and $d = 10$. Looking at Table 3.1, we can see that the combined probability of these occurring is 0.003. In the first five examples, the probability of exceeding the three sigma limit is very small, i.e., 0.009 or smaller. If we were to form "one sigma limits", we would expect that there will be a substantial probability of an observed number of nonconforming cases to lie within such limits. For example (b) the "one sigma limits" are

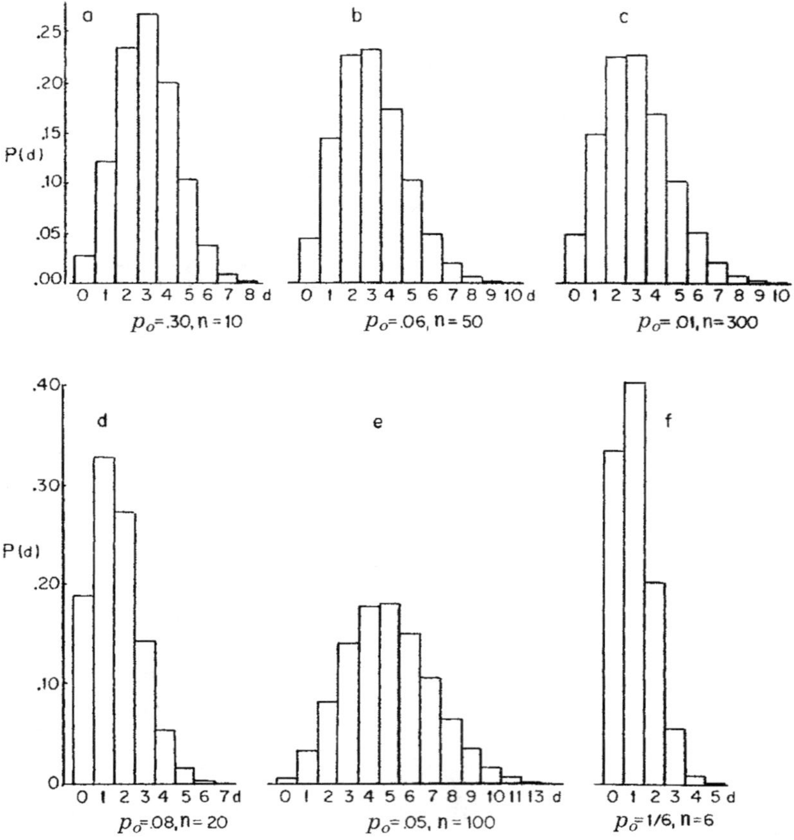

Figure 3.1 Six examples of binomial distributions from Table 3.1. Examples (a), (b), and (c) all have the same mean $np_o = 3$; but respective σ_d values are 1.45, 1.68, and 1.72. For (d) $np_o = 1.6$, $\sigma_d = 1.21$; for (e) $np_o = 5$, $\sigma_d = 2.18$; for (f) $np_o = 1$, $\sigma_d = 0.91$.

$3 \pm 1.68 = 1.32$ and 4.68 and the probability of an observation within these limits is the sum of $d = 2$, 3 and 4 probabilities, i.e., $0.226 + 0.231 + 0.173 = 0.630$. So observations within "one sigma limits" are not at all rare. The probability of an observation or count outside of "two sigma limits" can be calculated to be 0.030 for example (b). Thus observations of counts, d, of nonconforming units in samples of size, n, lying outside of "two sigma limits" for d will be rather uncommon.

Example (f) can be experimentally tested by tossing six honest dice and counting the number of dice that show some one face chosen in advance, e.g., six.

In Fig. 3.1, the first three histograms show reasonably good "bell-shaped" distributions and they are quite similar because the average number of nonconforming pieces is $\mu_d = 3$. Example (e) shows an even more symmetrical distribution due, in part, to the higher μ_d of 6. The other two examples (d) and (f) are markedly less symmetrical and have smaller means, μ_d.

3.4.2. Conditions for a Binomial Distribution

1. Constant probability, p_o, of a nonconforming piece on each draw of a piece from the population or process.
2. Results on drawings are independent of preceding drawings.
3. Sample of n draws of a piece is taken.
4. The number of nonconforming pieces in the n draws is counted.

Of course the binomial distribution is of much wider application. All that is required is that there be two and only two possible outcomes of a draw. For example, we might have green and not green (yellow and red) traffic lights, paid or unpaid accounts receivable, correctly filled out form or not, success or failure, boy or girl, win or lose, good color or substandard color. When the conditions for a binomial distribution are met in drawing from a process, we say that the process is stable or "in control", even though it may not be as good as we want or need it to be. The conditions may also be very nearly fulfilled if we take random samples from a large lot; for then the probability for a nonconforming piece on each draw will be nearly constant even though it will vary slightly. For example, if we had 10,000 records to be sampled of which 100 had errors, and we decide to take a sample of size, 20, it can be easily seen that the probability of each subsequent draw does not change appreciably. Say we get a record without an error on the first draw, $P(g) = 9900/10{,}000 = 0.990000$. Now that we have taken a good one out, the probability on the next draw drops slightly because there are fewer good ones to select, i.e., $P(g$ on second draw$) = 9899/9999 = 0.989999$ and so on. We will see later what can be done when the population or lot is small. We will have to use a different probability distribution, one that takes the probability changes into account.

3.5. THE POISSON DISTRIBUTION FOR NONCONFORMITIES

We now take up the standard model for the distribution of nonconformities on a sample. Examples are counts of the number of (1) typographical errors on a printed page, (2) breakdowns of insulation in 1000 m of insulated wire, (3) leaks in 20 radiators, (4) gas holes in castings, (5) nonconformities found in the inspection of a subassembly, (6) pinholes in a test area of a painted surface, and (7) incorrect entries on a purchase order.

If the nonconformities occur independently and not in "bunches", the Poisson distribution may well apply. However, the number of possible nonconformities should greatly exceed the average number. Also the area of opportunity for nonconformities in each sample should be the same. Then, if we know the average number of nonconformities per sample, c_o, we can find by the Poisson distribution, the probability of a sample containing no nonconformities, of having exactly one, exactly two, and so on. These are, of course, called $P(0)$, $P(1)$, $P(2), \ldots$

3.5.1. Use of Tables

Later on we shall give a formula for Poisson distribution probabilities. Usually we do not use the formula although it is a very easy calculation with most hand calculators. Rather, the use of Table B in the Appendix, further simplifies the task. In Table B we first find the row determined by c_o, the average number of nonconformities per sample. Then for any number of nonconformities, c, being considered, we find the probability of c or fewer nonconformities. For example, if the theoretical process average is $c_o = 3$ nonconformities per sample and we are interested in the probability of exactly 5 nonconformities occurring in the sample, we find in the column headed by $c = 5$ a probability of 0.916 for row, $c_o = 3$. (Note that in the table, the decimal point is omitted.) We also see that this table is of "cumulative probabilities", i.e., any given probability is for that value of c and less. Thus

$$P(5 \text{ or less nonconformities, given } c_o = 3) = 0.916$$

and

$$P(4 \text{ or less nonconformities, given } c_o = 3) = 0.815$$

These mean that we expect to find five or fewer nonconformities in 916 samples out of a 1000 samples and we also would expect to find four or fewer nonconformities in 815 samples out of 1000 samples. Now by subtracting 0.815 from 0.916 we get 0.101 which is the probability of exactly five nonconformities in a sample. This is because the event "5 or less nonconformities" or 0, 1, 2, 3, 4, 5 nonconformities contains the event 5 as well as the event of "4 or less nonconformities" or 0, 1, 2, 3, 4. About 101 samples out of 1000 samples will contain exactly $c = 5$ nonconformities. In this way, we can use the cumulative probability Table B to find the individual probabilities. Thus for $c = 3$, we have

$$P(0) = 0.050 \qquad P(0) = 0.050$$
$$P(1 \text{ or fewer nonconformities}) = 0.199 \qquad P(1) = 0.149$$
$$P(2 \text{ or fewer nonconformities}) = 0.423 \qquad P(2) = 0.224$$
$$P(3 \text{ or fewer nonconformities}) = 0.647 \qquad P(3) = 0.224$$
$$P(4 \text{ or fewer nonconformities}) = 0.815 \qquad P(4) = 0.168$$
$$P(5 \text{ or fewer nonconformities}) = 0.916 \qquad P(5) = 0.101$$
$$P(6 \text{ or fewer nonconformities}) = 0.966 \qquad P(6) = 0.050$$
$$P(7 \text{ or fewer nonconformities}) = 0.988 \qquad P(7) = 0.022$$

$P(8 \text{ or fewer nonconformities}) = 0.996 \qquad P(8) = 0.008$

$P(9 \text{ or fewer nonconformities}) = 0.999 \qquad P(9) = 0.003$

$P(10 \text{ or fewer nonconformities}) = 1.000 \qquad P(10) = 0.001$

Some published tables, e.g., Molina (1947), give individual terms like those in the second column, as well as cumulative probabilities. Kitagawa (1952) gives a very complete set of individual probabilities up to $c = 10$.

One other possibility we sometimes encounter is the need for a probability such as P(3, 4 or 5 nonconformities). For this, we could add $P(3) + P(4) + P(5) = 0.493$ or, more directly, we could use

$$P(3, 4 \text{ or } 5 \text{ nonconformities}) = P(5 \text{ or fewer}) - P(2 \text{ or fewer})$$
$$= 0.916 - 0.0423 = 0.493$$

Note that we did not subtract P(3 or fewer). If we had this event would have left only 4 or 5 and would have given only P(4 or 5). It merely requires a little care to decide which cumulative probabilities to subtract for any desired probability.

Examples. We now give some examples of the Poisson distribution in order to show some typical shapes. These are all taken from Table B by subtraction as was shown for the case $c_o = 3$. See Table 3.2. Histograms are given in Figure 3.2, picturing these distributions.

The distributions for an average of $c_o = 0.25$ are strongly J-shaped with over three-fourths of the probability at $c = 0$. Thus 77.9% of the samples can be expected to yield no nonconformities at all. For $c_o=1$, the average being one nonconformity per piece or sample, no nonconformities and one nonconformity are equally common, and together make up nearly three-fourths of the cases. However, there is scattering and even six nonconformities can occur in about one sample out of 1000. For $c_o = 1.6$, the most probable value is now one nonconformity and cases up to seven can occur. For $c_o = 3$ and 5, we find a tie for the most probable number of nonconformities at 2 and 3, and at 4 and 5, respectively, with

Table 3.2. Examples of Poisson Distributions for Various Process Averages of Defects, c'

	(a)	(b)	(c)	(d)	(e)
c'	0.25	1.0	1.6	3	5
$\sigma_c = \sqrt{c'}$	0.50	1.00	1.26	1.73	2.24
c	$P(c)$	$P(c)$	$P(c)$	$P(c)$	$P(c)$
0	0.779	0.368	0.202	0.050	0.007
1	0.195	0.368	0.323	0.149	0.033
2	0.024	0.184	0.258	0.224	0.085
3	0.002	0.061	0.138	0.224	0.140
4		0.015	0.055	0.168	0.175
5		0.003	0.018	0.101	0.176
6		0.001	0.005	0.050	0.146
7			0.001	0.022	0.105
8				0.008	0.065
9				0.003	0.036
10				0.001	0.018
11					0.009
12					0.003
13					0.001
14					0.001

considerable, tailing out to high numbers of nonconformities. These distributions are a bit more symmetrical than those of the binomial shown.

Population characteristics of the Poisson distribution. The theoretical process average number of nonconformities per sample is c_o, as we have been emphasizing. We now give the standard deviation for the count, c, of nonconformities. It can be shown to take the very simple form:

$$\sigma_c = \sqrt{c_o} = \text{population standard deviation of counts} \tag{3.12}$$

For the examples given in Table 3.2, these σ_c values are given just below the σ_o values. As usual they are an amount of

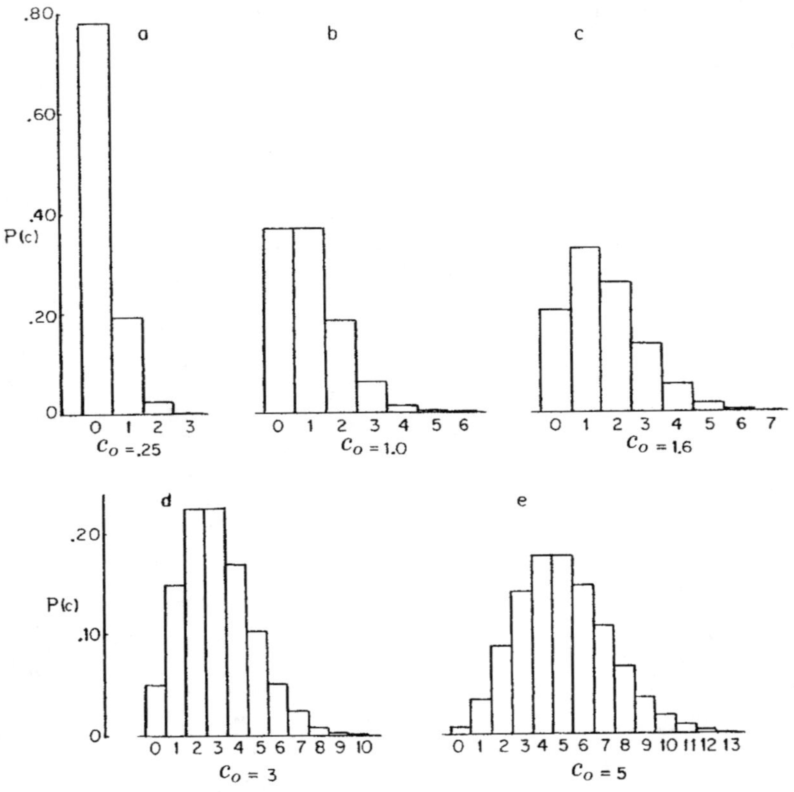

Figure 3.2. Five examples of the Poisson distribution from Table 3.2 Distribution means are listed by the histograms, namely, the c_o values. The standard deviations ($\sqrt{c_o}$) are, respectively, 0.50, 1.00, 1.26, 1.73, and 2.24.

departure from σ_o which is only to be expected. As with the binomial distribution, there will be very few occasions when the observed count, c, will be above $c_o + 3\sigma_c$. For example when $c_o = 5$, $\sigma_c = 2.24$ and $c_o + 3\sigma_c = 5 + 3(2.24) = 11.72$ and the probability of 12 or more is 0.005.

All Poisson distributions are unsymmetrical below and above c_o. But as c_o increases, the Poisson distribution becomes more symmetrical. The maximum probability always occurs at c_o, or within one count of it.

An example of the Poisson distribution which is of practical interest to us is that of occurrences of a certain type, e.g., number of fatal accidents in a particular city per week or the number of lost-time accidents in a company per month. Let us say that over the past 15 years, the average number of lost-time accidents in our company per month has been 0.3 (we have not been doing very well). Thus $c_o = 0.3$ and we would like to know the probability that on any given month, we would have no lost-time accidents, $c = 0$. Looking this up in Table B gives us $P(0) = 0.741$. Likewise $P(1) = P(\leq 1) - P(0) = 0.963 - 0.741 = 0.222$. Some texts, particularly those in reliability or safety engineering, use the notation of λ for c_o.

3.5.2. The Poisson Distribution as an Approximation to the Binomial

The Poisson distribution is very useful for approximating a binomial distribution. We shall especially make use of this fact in the chapter on acceptance sampling when we discuss decision making for acceptance or rejection of incoming lots of material.

To see the approach of probabilities by the binomial to those of the Poisson distribution, look again at examples (a)–(c) of Table 3.1. In each of the three, the average number of nonconforming units, i.e., np_o, is 3. In (a), we have $n = 10$, $p_o = 0.30$, so that we expect three nonconforming units among the 10 and can easily have as many as 5 or 6 in the 10. In (b), with $n = 50$, $p_o = 0.06$ there is the same expected number of 3; but this is in a sample of 50 not 10. Again five or six nonconforming units may readily occur, but among the 50. The probabilities for each count of nonconforming units, d, are rather similar in examples (a) and (b); but there is rather more variability in distribution (b), i.e., $\sigma_d = 1.68$ compared to 1.45 for (a). Now in example (c) with $n = 300$, $p_o = 0.01$, the expected count of nonconforming units, d, is once more 3. Now σ_d is a bit larger yet. Compare the probabilities for cases (b) and (c). Are they not quite noticeably similar? We could go further and work out the probabilities for $n = 3000$, $p_o = 0.001$ with np_o again 3. But even in case (c), we are already near the

Poisson limit. From Table 3.2(c) and Table 3.2(*d*), we have the following probabilities:

Count d or c	0	1	2	3	4	5	6	7	8	9	10
(*c*) Binomial $P(d)$	0.049	0.149	0.224	0.225	0.169	0.101	0.050	0.021	0.008	0.003	0.001
(*d*) Poisson $P(d)$	0.050	0.149	0.224	0.224	0.168	0.101	0.050	0.022	0.008	0.003	0.001

These two sets of probabilities are strikingly similar. Thus we see that if we consider a series of binomial distributions each of which has the same expectation or average count, np_o, but in which n is increased continually while p_o is correspondingly decreased (thus preserving np_o at some constant, say c_o) then the Poisson distribution with expectation c_o is approached as a limit.

n	50	300	3000	10,000	
p_o	0.06	0.01	0.001	0.0003	Poisson
np_o	3	3	3	3	$c_o = 3$

Now how good is the approximation? It begins to be of "practical use" for $n \geq 10$ and $p_o \leq 0.10$ and the approximation becomes "quite satisfactory" for $n \geq 20$ and $p_o \leq 0.05$.

But why bother with an approximation; why not use the formally correct binomial distribution when it is correct? The reason is that the Poisson distribution requires only one constant or parameter to determine it completely, namely c_o. On the other hand, the binomial distribution requires that we know two constants, n and p_o. Thus for Poisson probabilities, we enter a table such as Table B in the Appendix with c_o and a desired value of count, c (this being a double-entry table). But a binomial table, e.g., Table C in the Appendix or those in the references, must be entered with n and p_o as well as a desired count of d (this being a triple-entry table). For substantial coverage of cases particularly for all possible values of p_o, the binomial tables require vastly more entries than does a Poisson table. A fairly complete set of binomial tables could require a whole book and two books if you want to get the cumulative probabilities.

As an example, suppose for a binomial distribution $n = 100$, $p_o = 0.04$ and we desire the probability of $d = 3$. With such a sizeable n and small p_o, we can have much confidence in the approximation. Using the Poisson distribution in Table B for $np_o = 100(0.04) = 4$, we find

$$P(3 \text{ or less, given } c_o = 4) = 0.433$$

From an appropriate binomial table, we find the probability

$$P(3 \text{ or less, given } n = 100, p_o = 0.04) = 0.429$$

Or if we desire the probability of exactly four nonconforming units, we can use the Poisson in Table B with $c_o = 4$ finding

$$\begin{aligned} P(4) &= P(4 \text{ or less}) - P(3 \text{ or less}) \\ &= 0.433 - 0.238 \\ &= 0.195 \end{aligned}$$

Whereas a binomial table yields

$$P(4) = 0.197$$

Clearly these approximations by the Poisson are entirely satisfactory.

3.5.3. Conditions for a Poisson Distribution

1. Samples provide equal areas of opportunity for nonconformities (same size of unit, subassembly, length, area or quantity).
2. Nonconformities occur randomly and independently of each other.
3. The average number, c_o, remains constant.
4. The possible number of nonconformities, c, is far greater than the average, c_o.

These conditions may well occur in a production process. The quality control person can usually arrange for condition 1 to be true. Condition 4 also is commonly true. Condition 2 is, however, very possibly not true, i.e., nonconformities may occur in "bunches" (for example, dust particles on areas of paper or inspection errors on gas holes in castings). In process control using control charts for nonconformities, we actually make a test as to whether, for a process the average, c_o, is remaining constant as we shall see in Chapter 5.

Formula for P(c) for the Poisson distribution: Now for the sake of completeness, we give the formula for $P(c)$ even though we shall commonly use tables such as Table B

$$P(c \text{ nonconformities, given average } c_o) = \frac{e^{-c_o}(c_o)^c}{c!} \tag{3.13}$$

where $e =$ the natural base for logarithms, namely $e = 2.71828$. This equation can be calculated using a handheld calculator, programmable calculator or computer software. In the former case, most calculators have an "e^x" key (it is often the second function over the "ln" or natural log key). However, if you are interested in a cumulative probability you would have to calculate the probability for each c and add them up. The tables make it much easier!

3.6. HYPERGEOMETRIC DISTRIBUTION FOR NONCONFORMING UNITS

This distribution carries quite a fearsome name and possesses quite a fearsome formula but it can be very easy to use. The situation to which this distribution applies is really quite simple. It is used when we inspect a sample of items for some characteristic, e.g., nonconformity, with or without air conditioning, an ace from a deck of cards or a satisfied customer. The sample is chosen randomly from a population or lot without replacement. Thus if we draw our sample of items one at a time from the population or lot, each draw of one item will affect the chance of the next draw.

Let us work out the general formula after going over another example. Suppose we are given a lot of $N = 10$ pieces of which $D = 3$ are nonconforming (note the capital letters indicate population parameters). A random drawing of four pieces from the lot (population) are to be taken. What is the probability that none will be nonconforming, i.e., $n = 4$ and $d = 0$? One way to approach this is to look at the probability of each draw. The probability of drawing a conforming piece on the first draw is 7 out of 10, 6 out of 9 on the second draw, 5 out of 8 on the third and 4 out of 7 on the last draw thus:

$$P(0) = \frac{7}{10} \cdot \frac{6}{9} \cdot \frac{5}{8} \cdot \frac{4}{7} = 0.1667$$

Note that in this case the probability of drawing a conforming piece changes from 0.7 to 0.57 from the first to the last draw. This is due to the very small population.

For the probability of drawing one nonconforming piece is

$$P(1) = \left[\frac{7}{10} \cdot \frac{6}{9} \cdot \frac{5}{8}\right] \cdot \frac{3}{7} \cdot {}_4C_1 = 0.5000$$

The combination, ${}_4C_1$, is there because we could have taken the nonconforming piece first, second, third or fourth. Likewise the probability of obtaining two and three nonconforming pieces is (Since there only three nonconforming pieces in the lot we cannot have four in the sample!)

$$P(2) = \left[\frac{7}{10} \cdot \frac{6}{9} \cdot\right] \left[\frac{3}{8} \cdot \frac{2}{7}\right] {}_4C_2 = 0.3000$$

$$P(3) = \left[\frac{7}{10}\right] \left[\frac{3}{9} \cdot \frac{2}{8} \cdot \frac{1}{7}\right] {}_4C_3 = 0.0333$$

Since these probabilities add up to 1, we have identified all the possible cases for drawing four items from our lot of 10 of which 3 are nonconforming. If the lot were larger and the sample larger, this type of arithmetic would be burdensome; so let us try a different approach using combinations.

Let us answer the questions:

1. How many ways can we draw d nonconforming pieces from a population having D nonconforming pieces in it?

2. How many ways can we draw $(n-d)$ conforming pieces from a population having $N-D$ conforming pieces in it.
3. How many ways can we draw n pieces from the population of N? Now we multiply the results of (1) and (2) and divide by (3).

In our example above, we calculate

$$P(0) = \frac{{}_3C_0 \cdot {}_7C_4}{{}_{10}C_4} = \frac{1 \cdot 35}{210} = 0.1667$$

For a probability of a sample of size 4, we want no nonconformities out of the three in the population and we need four conforming pieces from the seven in the population. Finally there are 10 pieces in the population and we need a sample of 4. Likewise the probabilities of 1, 2, 3 and 4 nonconforming pieces in the sample are calculated:

$$P(1) = \frac{{}_3C_1 \cdot {}_7C_3}{{}_{10}C_4} = \frac{3 \cdot 35}{210} = 0.5000$$

$$P(2) = \frac{{}_3C_2 \cdot {}_7C_2}{{}_{10}C_4} = \frac{3 \cdot 21}{210} = 0.3000$$

$$P(3) = \frac{{}_3C_3 \cdot {}_7C_1}{{}_{10}C_4} = \frac{1 \cdot 7}{210} = 0.0333$$

And now we can generalize the formula

$$P(d) = \frac{{}_DC_d \cdot {}_{(N-D)}C_{(n-d)}}{{}_NC_n} \tag{3.14}$$

This can be a confusing equation to use when we are confronted with a word problem. Let us look at an example: if we were to go to a showing of 20 houses of which 12 have air conditioning what is the probability that if we look at five houses four will have air conditioning. For some people who like to work with formulas, they will find it relatively easy to assign the various subscripts to a problem like this; however, for many let us see if we can make some sense out of this problem. First, how many houses with air conditioning do we need in the sample? Answer, 4; how many houses with air are in the population? Answer, 12; therefore the first combination in the numerator is ${}_{12}C_4$. Secondly, how many houses without air do we need in the sample? Answer, 1; and likewise how many houses without air are in the population? Answer 8; thus the second combination in the numerator is ${}_{8}C_1$. Lastly how many different samples of 5 can we get in a population of 20? Answer, ${}_{20}C_5$ and this goes in the denominator. Putting them all together, we get

$$P(4) = \frac{{}_{12}C_4 \cdot {}_{8}C_1}{{}_{20}C_5} = \frac{495 \cdot 8}{15504} = 0.2554$$

The simple diagrammatic representation of this method was published in Burr (1991):

	With A/C	w/o A/C
Population	12	8
Sample	4	1

Totals Population 20, Sample 5

What if we were to work the problem from the point of view of finding only one house without air conditioning in the sample of five? It is easy to see that the only thing that happens is that we get both ${}_{8}C_1$ and ${}_{12}C_4$ in the numerator as before and ${}_{20}C_5$ in the denominator. The answer is the same!

Even with a calculator, the arithmetic in the calculation of the hypergeometric probabilities can be tedious; so when must we use this distribution and when can we approximate the probability using the binomial tables or even the Poisson tables? Since the hypergeometric distribution is used when a sample is taken without replacement, we must have a small sample from a fairly large population so that the probability of taking a unit from the population does not affect the probability of taking the next unit. This is usually considered to be when $n \leq 0.05N$, i.e., the population is 20 times larger than the sample. The hypergeometric is usually for sampling small finite populations such as a shipment of 17 drums of chemical,

9 subassemblies, 32 letters produced this day or 59 insurance policies sold this week. Even samples of 3, 4 or 5 exceed the 5% (of the population) limit for the sample. Let us try an example: Since many calculators have finite memory we will restrict our population to 150. Okay, we have a population of 150 subassemblies of which we will test 5. We will accept the whole lot if we find no nonconforming subassemblies in the sample. Unbeknownst to us the lot is 6% nonconforming. What is the probability of accepting the lot? Applying the simplified combination method, we find

$$P(0) = \frac{{}_9C_0 \cdot {}_{141}C_5}{{}_{150}C_5} = \frac{1 \cdot 432295143}{591600030} = 0.7307$$

Now the binomial probability using (3.11) is

$$P(0) = {}_5C_0(0.05)^0(0.95)^5 = 0.7738$$

There is an obvious difference of slightly over 4% in the estimate but for many cases this is good enough.

If we use Table B, we can also obtain a probability from the Poisson distribution for $np_o = c_o = 5(0.06) = 0.3$.

$$P(0) = 0.741$$

A more complete treatment of each of the three distributions and their interrelationships and approximations are given in Burr (1976). In Burr (1991), we have developed a graphical representation of the hypergeometric probability calculation.

3.7. SUMMARY

Some knowledge of probability is basic in interpreting results of quality control analysis on industrial or administrative processes and on incoming products or services as well as on the results of research. We have defined a probability in two ways: (1) as a fixed likelihood for an event to occur, lying somewhere between 0 and 1 inclusive, and determining the long-run ratio of occurrences of the even in question to the number of trials; and (2) in terms of equally likely outcomes being the number of ways the event can happen divided by the number of ways it can occur or fail to occur. We have observed that the occurrence ratio d/n tends to approach the true probability, p_o, as the number of trials increases. But this is not a fixed or invariant mathematical approach; we can, in fact expect that after n trials we may well be off from p_o by the amount

$$\frac{\sigma_d}{n} = \sqrt{\frac{np_o(1-p_o)}{n}}$$

and may even be two or three times as far away as this.

We also introduced the three most useful and applicable attribute distributions, the binomial, the Poisson, and the hypergeometric. The binomial is used when the sample is small compared to the population, the item is only one thing or another and the sample size is usually less than 50 or so. The Poisson is used when there can be more than one occurrence on a unit of the characteristic in question, the population is very large and the sample size is large, ≥ 100 or so. The hypergeometric is used when the sample is 5% or more of the population and the units taken from the population are not replaced, the sample is small (usually less than 20) and represents the exact probability for the attribute.

3.8. PROBLEMS

3.1 Toss a coin 100 times, keeping track of the number of heads thrown. After each 10 tosses, calculate the occurrence ratio of heads up to that point. Describe the approach to the assumed $p_o = 0.50$ for a balanced coin. About how far from the expected 50 heads (in 100 tosses) might we expect our observed count of heads to lie, i.e., $1\sigma_d$?

3.2 Proceed as in Problem 3.1, but roll a die and count the number of times that 6 shows. Using $p_o = 1/6$ as the expected probability.

3.3 Given a process with a fraction nonconforming $p_o = 0.02$ and taking a random sample of $n = 2$ pieces, find *P*(2 good), *P*(1 good) and *P*(0 good).

3.4 Same as Problem 3.3 but with p_o 0.01

3.5 Given as in Example 1 that $p_o = 0.08$ for the fraction nonconforming and we take $n = 3$, find P(3 good), P(2 good), P(1 good) and P(0 good).

3.6 From a lot of $N = 8$ gauges, of which $D = 1$ is slightly off, find the distribution of d for results from a random sample of $n = 2$. Show graphically.

3.7 Same as Problem 3.6, but with $n = 3$. Show graphically.

3.8 Same as Problem 3.6, but with $D = 2$ and $n = 2$. Show graphically.

3.9 Same as Problem 3.8, but with $n = 3$. Show graphically.

3.10 If the fraction nonconforming for a process is $p_o = 0.10$ on minor nonconformities and random sample of 400 is taken, how many can be expected to be nonconforming pieces. How far from this expectation would be a reasonable deviation (σ_d)? What is about the very worst discrepancy we could experience ($3\sigma_d$)?

3.11 Find the value of ${}_{10}C_4$, ${}_{10}C_6$, ${}_{10}P_4$.

3.12 From a standard deck of 52 cards, four are drawn without replacement.

a. How many different possible hands are there?
b. In how many of these will all four be of the same one rank?
c. In how many will there be three cards of the same rank and one of some other rank?
d. What then is the probability of the kind of hand in (b) and (c)?

3.13 Past data have shown an average of $c_o = 1.2$ typographical error per magazine page of print. (a) Find the probability of one or fewer typographical errors on a page. (b) Find the probability of three or more errors on a page.

3.14 Past data have shown an average of $c_o = 3.6$ leak points per radiator on the initial check test. (a) Find the probability of three or fewer leaks on a radiator. (b) Of five or more.

3.15 At final inspection of trucks, an average of $c_o = 2.0$ nonconformities is found. (a) Find the probability of two or fewer nonconformities. (b) Of exactly two nonconformities. (c) Of two or more.

3.16 Over a test area of painted surface, an average of $c_o = 5.4$ pinholes is standard for a process. (a) Find the probability of two or fewer pinholes. (b) Of over seven pinholes.

3.17 An executive picks up a sample of five brass bushings from a tote pan containing 1000 of them, of which 100 have at least slight burrs. (a) What is the probability that none in the sample have any burrs? (b) Which distribution is formally correct to use? (c) Which is the simplest distribution that you can justifiably use for the calculations?

3.18 An automatic screw machine is producing spacers with $p_o = 0.01$ for off-diameter pieces. A sample of $n = 10$ is taken randomly. (a) What is the probability of no off-diameter pieces in the sample? (b) Which distribution is formally correct? (c) Can you approximate with the Poisson distribution?

3.19 A process has produced a lot of 1000 temperature controls for automatic hot water tanks. The fraction nonconforming has been running at $p_o = 0.0020$. (a) In a sample of 100, approximate the probabilities of none, one and two nonconforming controls. (b) Which distribution is formally correct? (c) Which is the simplest distribution that you can justifiably use for the calculations?

For the following problems, find σ_c and find the probability of $c \geq c_o \pm 3\sigma_c$.

3.20 Problem 3.14.
3.21 Problem 3.15.
3.22 Problem 3.16.
3.23 Problem 3.17.

3.24 If $p_o = 0.20$, $n = 4$, find the probability of none, one, two, three, and four nonconforming units in the sample.

3.25 If $p_o = 0.10$, $n = 5$ find the probability for each of 0, 1, . . . ,5 nonconformining units in the sample.

3.26 If from a lot of seven clocks of which two are nonconforming, a random sample of two is drawn, find $P(0)$, $P(1)$ and $P(2)$.

3.27 If from a lot of nine gauges, three of which are inaccurate, a random sample of three is drawn, find $P(0)$, $P(1)$, $P(2)$ and $P(3)$.

REFERENCES

IW Burr. Statistical Quality Control Methods. NY: Marcel Dekker, 1976.

JT Burr. How can you calculate small sample probabilities without formulas. In: Quality Progress. Sep. 1991, p 51.

TC Fry. Probability and Its Engineering Uses. Princeton, NJ: Van Nostrand-Reingold, 1928.

Harvard Univ., Computing Lab. Tables of the Cumulative Binomial Probability Distribution. Cambridge, MA: Harvard Univ. Press, 1955.

T Kitagawa. Tables of Poisson Distribution. Tokyo: Balfukan, 1952.

GJ Lieberman, DB Owen. Tables of the Hypergeometric Probability Distribution CA: Stanford, CA, 1961.

EC Molina. Poisson's Exponential Binomial Limit. Princeton, NJ: Van Nostrand-Reingold, 1947.

WH Robertson. Tables of the Binomial Distribution Function for Small Values of p. Washington, DC: Tech. Services, Department of Commerce, 1960.

HG Romig. Fifty to 100 Binomial Tables. New York: Wiley, 1947.

US Department of the Army. Tables of the Binomial Probability Distribution, National Institute of Standards and Technology,

Appl. Math. Ser. 6. Washington, DC: US Government Printing Office, 1950.

S Weintraub. Cumulative Binomial Probability Distribution for Small Values of p. London: Free Press of Glencoe, Collier-Macmillan, 1963.

4

Control Charts in General

We now begin the study of a most powerful and versatile set of tools for the analysis and improvement of a production process. These tools are called "control charts". They were first invented by Dr. Walter A. Shewhart of the Bell Telephone Laboratories in 1924, and developed by him and his associates in the 1920s. He published a complete exposition of the theory, practical application, and economics of control charts in 1931 (Shewhart, 1931). Seldom has a whole field of knowledge been so well explored and its applications so well pointed out in the first publication in a field. So basic and applicable are control charts that new uses are continually being found in all *sorts* of products and industries. And yet control charting methods are simple enough to be learned and applied by one with even very modest mathematical background.

4.1. RUNNING RECORD CHARTS OF PERFORMANCE

It is a common practice of top executives, superintendents, foremen, directors of quality, chief inspectors, engineers, quality supervisors, and research workers to keep track of performance by means or plotted figures measuring such performances. This is done by periodically plotting an appropriate descriptive number of the current results. Such a number could be a measurement or some sort of count (attribute data). Although we shall be emphasizing various data indicating quality of performance of production processes, running record graphs (trend charts) can be made on any numbers of interest. Nonmanufacturing examples are gross production, absenteeism, unit costs, late shipments, hospital calls, and number of customer returns.

In Table 4.1, we give quality data on three products. On wire diameter, we list the average, x-bar, and the range, R, for samples of $n = 7$. All such diameters are supposed to lie between specification limits for individual x's of 0.096–0.102 in. The second product is a series of large aircraft, inspected at the end of production for all types of nonconformities, one of which was called "plug holes". Naturally, if production were perfect there would be no plug holes (nor any other nonconformities). But in practice they do occur, forming a distribution. The third product was packing nuts for a farm implement where considerable trouble was encountered. An intensive program was begun in October, and the present samples were taken the following January. Some substantial improvement had already occurred.

Running record graphs of the four quality characteristics are shown in Figure 4.1. Which of these four charts shows the best control, i.e., which chart shows the most nearly random pattern such as might be obtained by drawing chips randomly from a bowl? Which chart shows the poorest control, i.e., clearest evidence of some changes in production conditions, effecting changes in quality? Try to guess.

4.2. PERFORMANCE VARIES

One thing perfectly obvious in all four graphs in Figure 4.1 is *variation*. Performance does vary from sample to sample, time

Table 4.1 Running Record for Four Quality Characteristics on Three Kinds of Product

	Diameter of uncoated wire		Nonconformity "plug holes" final inspection on planes nonconformities	Packing nuts for tractor, nonconforming units in samples of 50, January	
Sample number	Average *x*-bar	Range *R*	*c*	Nonconf'g *d*	Fraction nonconf'g *p*
1	0.0985	0.0030	2	12	0.24
2	0.0980	0.0025	3	10	0.20
3	0.0970	0.0010	5	5	0.10
4	0.0985	0.0030	6	4	0.08
5	0.0973	0.0025	5	0	0.00
6	0.0972	0.0005	6	3	0.06
7	0.0980	0.0040	5	1	0.02
8	0.0965	0.0010	4	0	0.00
9	0.0970	0.0010	6	3	0.06
10	0.0975	0.0010	9	8	0.16
11	0.0980	0.0025	9	7	0.14
12	0.0980	0.0015	6	0	0.00
13	0.0975	0.0010	5	0	0.00
14	0.0980	0.0040	9	14	0.28
15	0.0984	0.0030	9	13	0.26
16	0.0985	0.0030	8	14	0.28
17	0.0980	0.0040	7	4	0.08
18	0.0970	0.0010	6	5	0.10
19	0.0968	0.0010	11	0	0.00
20	0.0975	0.0010	12	10	0.20
21	0.0972	0.0010	12		
22	0.0970	0.0010	8		
23	0.0980	0.0010	8		
24	0.0978	0.0010	7		
25	0.0985	0.0010	5		
26	0.0970	0.0020			
27	0.0985	0.0010			
28	0.0974	0.0020			
29	0.0972	0.0015			

to time. Some variation in production processes is only natural and must be lived with; such variation is unavoidable no matter how much effort is expended on the process. This

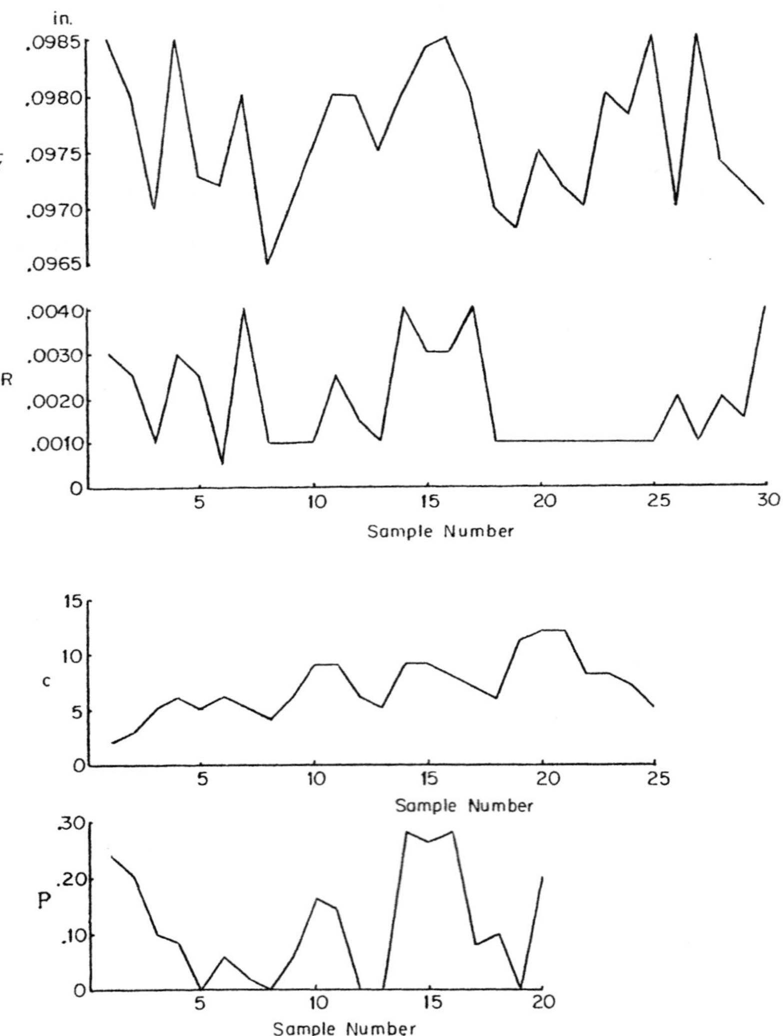

Figure 4.1 Four running record graphs of data in Table 4.1: x-bar's and R's for diameters of uncoated wire, counts c of "plug holes" nonconformities on planes and fractions nonconforming, p, for packing nuts.

type of variation is often said to come from *common causes*. The more closely we try to work, the more variation shows up. But on the other hand, there are some extreme variations which are not unavoidable, and their causes can be identified and appropriate action taken. These come from what is often called *special causes*.

4.3. UNUSUAL PERFORMANCE CALLS FOR ACTION

A significant rise in the sample fractions nonconforming, p, can signal the presence of some change in the process. We want to find out what the cause is so that we may eliminate it and obtain a lower process fraction defective p_0. Or in the case of a process being studied for nonconformities, c, we may note a significant drop in these. We want to know whether it is caused by lax inspection or by some change in production conditions, such as raw material or a better manufacturing method, which, if identified and incorporated, will continue to give better product.

When a quality characteristic is measurable, we have two charts to watch, x-bar and R, reflecting level and variability. Either or both may give evidence of process improvement or deterioration. Ranges may show worse or better within-sample variation. If better, we want to find out why so we can "do it again". If worse, we want to find the cause so we can eliminate it and obtain a more consistent, less variable product. Records of averages, x-bar, are very important, e.g., dimensions, strength, or chemical impurities. They may point the way to action for improvement.

The big word here is ACTION.

4.4. WHAT IS UNUSUAL?

This is the basic question in following a performance record. Just when does a point give a reliable indication that a change has come into the process? In Figure 4.1, is the low x-bar point of sample 8 a clear indication of some special cause in the process? What about the five x-bars of 0.0985 in.? Is the pattern of ranges, R, such as can be attributed to chance variation?

Are the high or low points on the running records for plug hole nonconformities, c, meaningful? Is there reliable evidence of a trend upward? Are the fractions defective of the packing nuts homogeneous, or do they provide reliable evidence of an erratic production process? These are the kinds of questions to which we need answers.

Can you tell by examining the running record charts? And would you get the same decision if we were to use a different scale, e.g., smaller? The authors have often shown in lectures a slide with two running records plotted one above the other. The listeners are asked to vote on which shows the better control, the most freedom from extraneous process changes. They all vote for the one drawn to a smaller scale (so it looks more *homogeneous*). The author then tells them that it is the same data merely drawn on two different scales. Even though both scales are shown, nearly everyone votes for the one drawn to the smaller scale. If you, in using graphs, make decisions that depend *to any degree* on the scale used in plotting the data, then your decisions cannot be very objective!

Now the truth of the matter is that one cannot be sure when looking at a graph whether a given extreme-looking point is really a reliable indication of a special cause or whether it should just be considered part of the natural unavoidable variability, a common cause, of the process. A good part of the trouble is that when you look at a running record graph (trend chart) you cannot readily take account of the sample size. Thus, in the fraction defective record for the packing nuts, the sample size was $n = 50$. Had it been 20 or 25, it is likely none of the points would clearly indicate a cause. Or if n were 200, a great many of the 20 points would have provided clear evidence of some special cause in the process. We therefore need a way to "play the odds'" well by taking into account the sample size and using appropriate distribution models.

4.5. TWO KINDS OF CAUSES

We shall soon be discussing how we can use the control chart approach to greatly sharpen our decision-making ability, and thereby to get the most out of our production processes and, for that matter, any process that we might be wanting to improve or control.

In the preceding section we have begun to consider two rather distinct types of causes of process variation. There are, on the one hand, the few factors that have a relatively large influence upon product performance. Again we will call this type of causes, special. They are very well worth seeking and doing something about. Moreover they are findable, that is, can be identified. On the other hand, there are a large number of causes (common causes), each of which has an almost negligible effect, but in aggregate they give rise to a *distribution* of outcomes. This is the random pattern due to "chance".

Let us summarize these distinctions by a little table:

	Number present	Effect of each	Worth seeking?
Assignable causes	Very few, perhaps only one or none	Marked	Well worth seeking out, and possible to find
Chance causes	Very large number	Slight	Not worth looking for

It is worthwhile for you, the reader, to think about a production process with which you are familiar, then to make a list of chance or common causes and potential assignable or special causes. By using your knowledge and imagination you should be able to list 20 or 30 chance causes which may slightly affect the result of, say, a dimension. For example, let us say we are making a measurement of a part with a micrometer. There can be variation in the material of which the measured part is made, vibration, tool speed, coolant, lubricant, positioning of a part in a fixture on the machine, dust present, or in the reading of a micrometer. Assignable causes might be a change in material lot or a change of a process setting or a broken cutting tool. Try listing assignable causes for your process.

One word more: factors affecting production quality are not really classifiable into two perfectly distinct categories.

To some extent, they may be usefully thought of as forming a distribution with their relative strength or influence on the outcome going from a *few* very influential ones which we call assignable or special causes, on down to the host of *small chance* or common causes. (Such a distribution is sometimes called a Pareto distribution.). Our job is best done by trying to find when an assignable cause is operating and what change in conditions brought it into action.

4.6. CONTROL CHARTS

Walter Shewhart, in wrestling with a problem made complicated by variation, came to realize that it should be possible to determine when a variation in product quality really is the result of some process change, and when the apparent variation should be merely attributed to chance. That is, to decide whether some assignable cause is operating. This required that one should be able to measure how much variation is only to be expected in the process. Then variations outside this distribution or pattern become clear indicators that an assignable cause has come into the process and we should seek to find it, whether it produces an undesirable or desirable effect.

Shewhart and his associates proceeded to experiment. They decided upon the use of limits set at three standard deviations above and below the average. The standard deviation is estimated for the distribution of product quality when chance causes only are operating. Then if chance causes only are at work, very rarely will we find a sample result outside the limits:

Average ± three standard deviations (i.e., three times per 1000 samples from Table A)

Hence if we do find such a point or result, it is a much better bet that it is due to the presence of some assignable cause, rather than being a result of chance causes only. Then we look into the process conditions for the assignable cause.

Let us now reconsider the data of Table 4.1 as plotted in run charts in Figure 4.1 We show in Figure 4.2 these same data now made up into control charts by the inclusion of a central or average line (solid) and control limits (dotted). Points outside the limits are to be regarded as clear indications of the presence of an assignable or special cause. On the other hand, points inside the control limits are attributed to chance or common causes and no action is taken. We are likely to be wasting our time looking for the cause of any fluctuations between the control limits. It is like looking for a needle in a haystack without using a magnet! In this way, we save our time from fruitless search so that we may use it in a more thorough study, when warranted, of assignable causes.

Looking at the control chart for x-bar's we see that there are five x-bar points on or above the upper contro1 1imit and two on or below the lower control limit. Each is to be regarded as a clear indication of some factor affecting the wire diameter. Process level is quite out of control. In the chart for ranges, R, there are three points above the upper control limit, indicating causes of excessive within-sample variability, that is, of too much jumping around. There is also quite a long run of R's of 0.0010 in. below the central line, which is a hopeful sign. Somebody must have been doing something right during that time and we could improve the process by capturing that knack. Maybe, it was a better lot of material or an adjustment of the fixture.

Looking at the chart for c's, plug hole defects, we see no points above the upper control limit. So under that criterion, the process is in control, and no action is indicated. However, there does seem some evidence of a trend toward higher counts of plug holes. The first nine counts are all below the central line. Then near the end there are three *consecutive* points all fairly close to the upper control limit. Further study is indicated.

Finally, we examine the p-chart. The three high fractions nonconforming lying above the upper control limit are a clear indication of an assignable cause. The other points are in control. The relatively large number of $p = 0.00$ points is an encouraging sign, however.

The details of figuring central lines and control limits will, of course, be given in subsequent chapters, specifically Chapters 5, 6, and 8.

4.7. INTERPRETATIONS OF POINTS AND LIMITS

Since these interpretations are so basic and important, let us set them down explicitly.

Interpretation of a point beyond a control limit. Any point beyond either control limit on a control chart is to be regarded as a clear indication of the presence of an assignable cause among the conditions under which the point arose. This includes production and measurement or test conditions.

Interpretation of a point between control limits. Any *individual* point between the control limits of a control chart is attributed to chance causes, since it provides no reliable indication of the presence of an assignable cause. Although an assignable cause may have lain behind the point, we are in general better off to attribute it to chance and wait for clear evidence.

There are several other indications of the presence of an assignable cause even though the individual points are each within the control limits. When a number of individual points exhibit a pattern which would lead believe that the pattern is *not* random we can be pretty sure that there is, indeed, an assignable cause at work. Examples of this might be:

1. A *run* of at least eight consecutive points above or below the center line of the chart.
2. Two out three consecutive points close to a single limit.
3. A *trend* of at least eight consecutive points steadily increasing or decreasing. This is often seen in tool-wear applications.
4. A series of *alternating* high and low points. This sometimes occurs when there are two machines or

two operators using the same machine, e.g., two shifts.

Most users of control charts whether they be engineers, clerks, or operators readily recognize these conditions. Why eight consecutive points for a run or trend? Opinion varies; however, the authors recommend eight because this gives clear evidence of the presence of an assignable cause (a change in the characteristic in question).

There are two other occasions when the user of a control chart should investigate further:

1. When the points all are very close to the center line. At first one would think that this is good! However, some thought would lead one to understand that the variability as measured by the range is very large compared to the variability of x-bar's. This is most often caused by multiple streams of product reaching the sampling station, e.g., parts made from a four-cavity injection molding machine where we are taking one piece from each of the four cavities in our sample of four. In this case there may be real differences in the parts, e.g., weight, from the four cavities. The average of the four parts does not change much unless the material, temperature, or pressure change quite a bit. Another example of this is found in the sampling of paper thickness. Samples taken from across the web (sheet) may vary quite a bit if there is a thickness profile from one edge to another, i.e., thicker on one side than the other or thicker in the middle than on either end. Here, again the average paper thickness coming from the machine varies little unless there are significant changes in the pulp mix, temperature, or roller tension. This type of pattern is often called *hugging*. It should be investigated to see if the sample is being affected by nonrandom variable.

2. Occasionally you might see that the control limits on a control chart do not fit the points, i.e., a large number of points are outside the limits. The authors have seen charts where fewer than half of the points are *inside* the limits. This is caused by having a range that is measuring a very small amount of variability compared to the overall process varia-

bility. This occurs often in chemical applications where taking multiple samples from a homogeneous batch give virtually the same values (variability mostly due to measurement error), while the batch-to-batch variability can be much larger. This could also occur in a continuous process where the very short time variability is small compared to the hour-to-hour flucuation of the process. In any case when the control limits do not fit the points, we must look for a cause and change the process, the manner of taking samples or the type of control chart used.

Additional information on the rules for identifying out-of-control conditions is found in Nelson (1984).

4.8. TWO PURPOSES OF CONTROL CHARTS

There are two purposes of control charts, which we consider briefly here.

Purpose 1 (Analysis of past data). The objective of analyzing past data is to test for control or uniformity. Could such results as have been observed *readily* have come by chance variations alone from a homogeneous process? Or is there reliable evidence of some nonchance-factor assignable cause entering into the process? If the latter, then action is suggested in order to approve control of the process. This is often done to set up a control chart on a new or existing process.

Purpose 2 (Control against standards). For this control chart purpose, some standard or standards are available. They may be from analysis of past data leading to a long run of good control or they may be from some goals or requirements. Then each sample result is compared with the standard via control limits to see whether it is compatible. If inside the limits and the random pattern is continuing, no action is indicated. But if outside the control limits or a nonrandom pattern is recognized, then appropriate action is taken.

Thus in Purpose 1, at least in the beginning, a preliminary run of, say, 25 samples is analyzed for homogeneity. Then each additional sample is plotted and interpreted at once, in relation to the previous data. In Purpose 2, we can set up control limits immediately, plot each sample as *it* arises and take appropriate action. After a while it may prove necessary or desirable to change the standards to reflect current process capabilities. Both purposes have their place in applications to process control.

4.9. PROCESS IN STATISTICAL CONTROL

A definition basic to our work is the following:

Definition 5.1. Process in statistical control. A process is said to be in statistical control if the sample results from the process behave like random drawings of chips from a bowl.

In this we usually think of the chips in a bowl as having numbers on each and forming a distribution for the bowl as a whole. Drawings are best done with replacement so that at each drawing the distribution of numbers in the bowl is always the same. Or the bowl could contain two different colors of chips or balls, one for good ones another for nonconforming units, so as to simulate a binomial population. By use of control charts, it has commonly been found possible (if sufficiently desirable) to get production processes into statistical control.

4.10. ADVANTAGES OF A PROCESS IN CONTROL

There are quite a few practical and lucrative advantages of a process in control (that is, in statistical control).

1. Ordinarily, when beginning to analyze a production process by a control chart and a preliminary run of data, we will find a lack of control. This is an indication that further improvement is possible through finding assignable causes and removing undesirable ones, or incorporating desirable ones such as, say, improved strength. Thus, the operation of

getting a process into good control commonly brings a marked improvement in quality.

2. A process in control is predictable. For example, we do not need to measure or test every piece or unit. We can accurately predict that the current production is like the past, as long as current sample results continue to show control relative to the limits. This is not the case if there is lack of good control; we are in a much weaker position relative to the prediction of quality.

3. With regard to advantage 2, when the test is destructive, i.e., blowing times of fuses or tensile strength of rivets, the only practical way to know what the untested product can do is to obtain test results from the destroyed pieces in a sample. If the process is in good statistical control, we can predict reliably how the remaining product would perform, if it were tested.

4. In line with advantage 2, we can often greatly reduce the amount of inspecting or testing needed, when we obtain good control of a process at a satisfactory quality level. If the process is not in statistical control, then we may well need to retain 100% inspection or at least heavy inspection, instead of small samples to compare with control limits.

5. If product from the supplier shows good statistical control at satisfactory quality performance, then receiving inspection may well be reduced to a minimum or even eliminated. The supplier's process is predictable.

6. In a similar manner, if our own final inspection or test results show good control at a satisfactory quality level, we can much more safely guarantee our output to the consumer.

7. A record of good statistical control of our processes is very helpful in dealing with customer complaints or, at the extreme, in liability cases.

8. A control chart on measuring instruments (using a *homogeneous* material as a reference standard) showing good control, provides confidence in the instrument and the technician. Also it enables an accurate determination of the measurement error. But if there is lack of control in measuring

the reference material, there can be little confidence in the measurements.

9. Good control of laboratory or research results give confidence in the adequate control of the experimental conditions. Poor statistical control of results indicates poor control of the experimental conditions.

4.11. SUMMARY

In this chapter we have been describing a powerful decision-making tool for process improvement and control. Various types of control charts are available for analyzing process data. In fact, virtually all varying data may be analyzed by some *type* of control chart.

The objective is to determine how much variation can be attributed to chance causes and thus to determine limits of normal variation. Then results outside such limits can be taken as reliable indicators of the presence of some assignable cause, which is to be sought among the process conditions operating while that sample was being produced and inspected or tested. Appropriate action can then be taken. In this way processes can be brought into statistical control, bringing advantages such as those in Section 4.10.

A short booklet *very* well worth reading to supplement this chapter is American Society for Quality Control Standard (1969).

4.12. PROBLEMS

4.1. For some production process with which you are familiar, make a list of chance causes and of potential assignable causes.

For Problems 4.2–4.4, plot the quality variables given; make any comments or speculations you think might justify further study. Preserve your running record charts for problems in Chapters 6 and 7.

4.2. Inspection of fiber containers for contamination, resulting from gluing, from samples of $n = 25$.

Sample number		1	2	3	4	5	6	7	8
Fraction nonconfiguring, p		0.04	0.20	0.04	0.04	0.00	0.08	0.08	0.08
Sample number		9	10	11	12	13	14	15	16
Fraction nonconfiguring, p		0.08	0.12	0.00	0.04	0.04	0.00	0.08	0.00
Sample number	17	18	19	20	21	22	23	24	25
Fraction nonconfiguring, p	0.00	0.08	0.00	0.00	0.00	0.12	0.04	0.00	0.00

4.3. For 21 production days the total nonconformities, c, found on the day's production of 3000 switches ran as follows for December:

30	56	47	86	44	23	16	64	80	54	73
65	76	69	53	58	30	91	90	36	57	

4.4. A transmission main shaft bearing retainer carried specifications of 2.8341–2.8351 in. In a production run, measurements were made in 0.0001 in. from the "nominal" 2.8346 in. Thus, the specifications were −5 to +5 in the coded units. The following data on averages and ranges were recorded. Plot the two running trend charts, x-bars above the corresponding R's.

x-bar	+3.4	+1.4	+5.6	−2.6	+1.2	−2.0	−8.2	−5.8	−7.8	4.0
R	6	3	24	1	6	3	1	6	2	6
x-bar	−5.0	−10.2	+2.2	+0.8	+1.0	+1.6	−7.8	−3.8	−2.0	0
R	5	9	7	8	2	4	3	7	2	0

REFERENCES

American Society for Quality Control Standard (B3) or American National Standards Institute Standard (Z1.3 - (1969)). *Control Chart Method of Controlling Quality During Production*. ASQ, P.O. Box 3005, Milwaukee, WI 53203.

JT Burr. (1994). *SPC for Everyone*. ASQ, Quality Press, Milwaukee, WI.

L Nelson. (1984). Technical Aids, *Journal of Quality Technology*, Vol. 16, No. 4.

WA Shewhart. (1931). *Economic Control of Quality of Manufactured Product*. Princeton, NJ: Van Nostrand-Reinold.

5

Control Charts for Attributes

This chapter covers the use of control charts for process control, where quality is measured by either a count of nonconforming pieces or a count of nonconformities in a sample of product. There are two distinct cases, as pointed out in Chapter 4: (1) analysis of past data, and (2) control with standards given. We first consider the analysis for counts of nonconforming pieces or units.

5.1. CHARTS FOR NONCONFORMING PIECES

A sample of n pieces or units is to be inspected for any units, which contain nonconformances. There may be just one kind of nonconformity being looked for; or we may be looking for any nonconformance within a class of nonconformities. For example, we may have a class of six different "major" nonconformances, a piece being a major nonconformity if it has one or more major of these six nonconformances. Or the class might be of "minor" nonconformances or of "incidental" nonconformances. If the nonconformance in question is a "critical" one, then 100%, 200%, or more inspection is used to prevent any critical nonconformances from getting out, and thus no sampling would be used.

Classification of nonconformities:

Critical—Some flaw or error in the product or service that would cause injury, death or major financial loss if the product or service is used.

Major—This is usually used to identify flaws or errors that would lead to substantial loss on the part of the manufacturer or provider in order to rectify the product and might cause the loss of a customer if not detected.

Minor—Those flaws or errors that the operator, provider, inspector or clerk can rectify on the spot with minimal cost to the company.

Incidental—Those flaws or errors that may be present but are very likely not to be noticed by the customer or user.

As you can see then, "nonconformance" spans a wide range of significance and importance. There are many classifications but the authors prefer this simple approach based on the economic results of the nonconformity. In any case, we may well wish to analyze the process for problems and take appropriate action.

5.1.1. The np or d Chart

As given before we use the notations:

$$n = \text{number of pieces in the sample} \tag{5.1}$$

$$d = np = \text{number of nonconforming pieces in the sample} \tag{5.2}$$

$$p = d/n = \text{fraction nonconforming for the sample} \tag{5.3}$$

$$p\text{-bar} = \text{average fraction nonconforming for a run of samples} \tag{5.4}$$

$$p_o = \text{true process fraction nonconforming or a standard} \tag{5.5}$$

Let us now take up the case of the standard given and then of analysis of past data, although in practice the latter analysis is likely to be used initially in the analysis of a process.

Standard given, p_o: We assume that somehow we have a *standard* process fraction nonconforming, p_o. This may come from a substantial run of homogeneous past production or it could be a goal or a requirement of some kind. (Much as we might like, p_o cannot in general be 0.) We assume also that we have a clear definition of each nonconformity in the class being studied and that the nonconformaties act independently. In general, we will take a random sample from production; but sometimes, we take all pieces arising from production, forming them into samples of n consecutive pieces. Further, accurate inspection is assumed and must be arranged for through adequate training and good measurement or inspection tools.

In either case, we are determining for each sample whether the count of nonconformances, d, is (1) compatible with the assumed standard p_o, or (2) the sample result gives reliable evidence that the *process fraction nonconforming*, p_o, was not at the assumed p_o, while this sample was being produced. This is done by setting appropriate control limits and noting whether $d = np$ lies (1) between the limits or (2) on or outside the limits. The control chart, therefore, provides a decision-making tool.

Now how are the limits set? Obviously, the assumed standard, p_o, and the sample size, n, should both be taken into account. Let us see how. The conditions given here justify our using the binomial distribution discussed (perhaps even cussed) in Chapter 3. There we saw in (3.4.1) that the expected or theoretical average number of nonconforming units in a sample is np_o. This we use for the central line.

$$np_o = \text{central line for np (standard } p_o \text{ given)} \tag{5.6}$$

Next, we have seen in Chapter 3 that the only-to-be-expected deviation from this average is $c_d = \sqrt{np_oq_o}$ from (3.8), where $q_o = 1 - p_o$. Now np or d counts will quite seldom lie as much as $3\sigma_d$ away from np_o. We thus take as limits:

$$\text{Limits}_{np} = np_o \pm 3\sqrt{np_o q_o} \quad \text{(standard } p_o \text{ given)} \tag{5.7}$$

In practice, be sure to remember the 3 and the radical. You do not want the assignable cause to be YOU. Since these are 3σ limits for d, as long as p_o remains constant, observed counts, d, will seldom lie outside the limits of (5.7). In practice, therefore, whenever a point does go outside the limits, we regard this as clear evidence that p_o is no longer at the assumed value. Thus we conclude that there is an assignable cause which has changed p_o.

Example 1. Let us analyze the following data obtained by experiment. They were taken randomly from a box, containing 1702 beads 1 cm in diameter of which 1600 were yellow (good pieces) and 102 red (nonconforming pieces). Therefore

$$p_o = 102/1702 = 0.060$$

A paddle with 50 holes was used for $n = 50$. (Technically, the "Hypergeometric distribution" of Section 3.6 applies rather than the binomial distribution; but since $N = 1702$ and $n = 102$, n is less than $0.05N$, the latter distribution provides a nearly perfect approximation to the former).

The sample results are given below for 30 samples:

Sample number	1	2	3	4	5	6	7	8	9
Nonconforming, $np = d$	1	3	2	3	3	3	2	3	3
Fraction nonconf., p	0.02	0.06	0.04	0.06	0.06	0.06	0.04	0.06	0.06
Sample number	10	11	12	13	14	15	16	17	18
Nonconforming, $np = d$	4	3	5	3	4	4	2	3	6
Fraction nonconf., p	0.08	0.06	0.10	0.06	0.08	0.08	0.04	0.06	0.12
Sample number	19	20	21	22	23	24	25	26	27
Nonconforming, $np = d$	3	7	2	3	3	3	3	3	4
Fraction nonconf., p	0.06	0.14	0.04	0.06	0.06	0.06	0.06	0.06	0.08
Sample number	28	29	30						
Nonconforming, $np = d$	2	4	4						
Fraction nonconf., p	0.04	0.08	0.08						

We could plot either of the two quality characteristics: number nonconforming, $np = d$ or the fraction nonconforming, p. In Section 5.1.1, we were concerned with the number nonconforming, np. Thus we have the first 30 points of Figure 5.1.

Let us now figure the control lines for the points, using the standard $p_o = 0.06$ and $n = 50$, from (5.6) and (5.7):

$$\text{Central Line} = np_o = 50(0.06) = 3$$

$$\text{Limits}_{np} = np_o \pm \sqrt{np_o q_o} = 5 \pm \sqrt{50(0.06)(0.94)}$$
$$= 3 \pm 5.04 = \_\_, 8.04$$

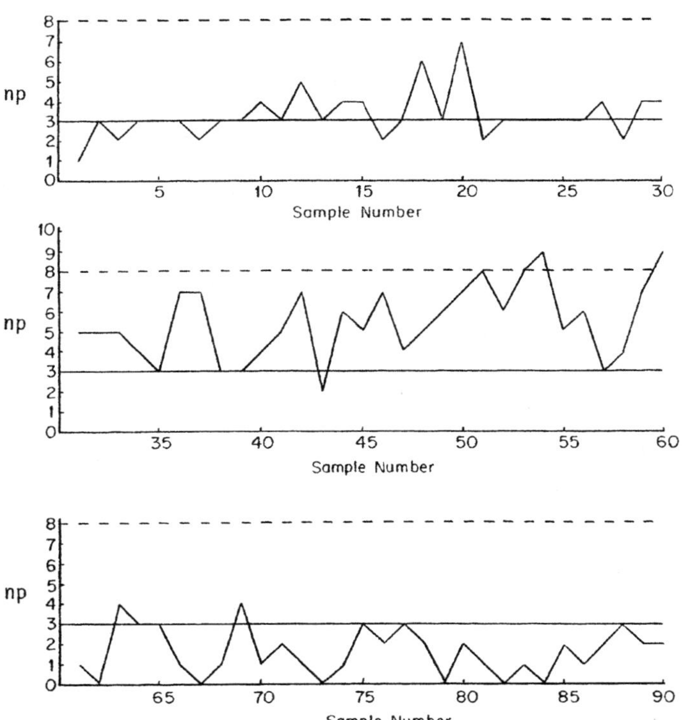

Figure 5.1 Chart for 90 samples, each of $n = 50$ pieces, with the number nonconforming, $np = d$, plotted. Central line and upper control limits given for $p_o = 0.06$. For samples 1–30, $p_o = 0.06$; 31–60, $p_o = 0.10$; 61–90, $p_o = 0.03$. Note increased average and variability in second part, and decreased average and variability in third part.

The average is thus three nonconforming per sample. How far off from this average can we be and still be in control? A count of eight nonconforming would be so very close to the upper control limit, 8.04 that in practical work we would investigate the conditions under which the product was produced and inspected or tested. Certainly $np = 9$ would be investigated. On the other hand, the arithmetic in carrying out (5.7) for the lower control limit would give -2.04. Since all counts np of defectives are 0 or more, the limit -2.04 is meaningless. All it means is that a count of no defectives would not be sufficiently rare to justify carrying on any investigation. But when a lower control limit is above 0, then points below such a lower control limit are worth investigating and can lead to real process improvement (unless it should prove to be due to lax inspection).

In Figure 5.1, we see that all points lie below the upper control limit. Therefore we say that the process is in control with respect to the standard $p_o = 0.06$. We have no reliable evidence that p_o is not 0.06. This is the conclusion we would expect since we know that p_o was 0.06 for these first 30 samples. We might mention that at the start we had quite a run of points on or below the central line, but two-thirds of these, namely six, were right on the central line.

For the next 30 samples, we have the following:

Sample number	31	32	33	34	35	36	37	38	39
Nonconforming, $np = d$	5	5	5	4	3	7	7	3	3
Fraction nonconf., p	0.10	0.10	0.10	0.08	0.06	0.14	0.14	0.06	0.06
Sample number	40	41	42	43	44	45	46	41	48
Nonconforming, $np = d$	4	5	7	2	6	5	7	4	5
Fraction nonconf., p	0.08	0.10	0.14	0.04	0.12	0.10	0.14	0.08	0.10
Sample number	49	50	51	52	53	54	55	56	57
Nonconforming, $np = d$	6	7	8	6	8	9	5	6	3
Fraction nonconf., p	0.12	0.14	0.16	0.12	0.16	0.18	0.10	0.12	0.06
Sample number	58	59	60						
Nonconforming, $np = d$	4	7	9						
Fraction nonconf., p	0.08	0.14	0.18						

Let us suppose that the quality control person still compares these results against $p_o = 0.06$, thus continuing the central line at 3 and the upper control limit at 8.04. He will find no single point indicative of a process change increasing p_o until he reaches samples 51, 53, and 54. These are not compatible with $p_o = 0.06$ and indicate a shift in p_o toward a higher value. Actually all of the first 12 points are on or above the central line of 3, and this "run" also signifies an increase in p_o. For samples 31–60, p_o was made equal to 0.10 by adding 16 red beads to the box, giving

$$p_o = 178/1778 = 0.100$$

Now suppose that p_o were to decrease. How will this show up on the chart? The points will tend to be lower. But in this example where there is no lower limit, we cannot get an indication by a point *below* the lower control limit. However, a long run of points below, or possibly on, the central line can be used as a clear indication of a lowering of p_o.

The data for samples 61–90 were obtained with

$$p_o = 49/1649 = 0.030$$

and the *np* counts are plotted as the last 30 points in Figure 5.1.

Sample number	61	62	63	64	65	66	67	68	69
Nonconforming, $np = d$	1	0	4	3	3	1	0	1	4
Fraction nonconf., p	0.02	0.00.	0.08	0.06	0.06	0.02	0.00	0.02	0.08
Sample number	70	71	72	73	74	75	76	77	78
Nonconforming, $np = d$	1	2	1	0	1	3	2	3	2
Fraction nonconf., p	0.02	0.04	0.02	0.00	0.02	0.06	0.04	0.06	0.04
Sample number	79	80	81	82	83	84	85	86	87
Nonconforming, $np = d$	0	2	1	0	1	0	2	1	2
Fraction nonconf., p	0.00	0.04	0.02	0.00	0.02	0.00	0.04	0.02	0.04
Sample number	88	89	90						
Nonconforming, $np = d$	3	2	2						
Fraction defective, p	0.06	0.04	0.04						

Starting with sample number 70, all of the remaining 20 points are either on or be-low the central line. We may regard the first 10 such points for samples 70–79 as providing a significant run.

This example illustrates a process in control relative to the actual true process average, p_o. It also shows how we may obtain an indication of (1) an increase in p_o and (2) a decrease in p_o. Further, it may be noted by the reader that the clear indication of a process change did not come very quickly. An *np* chart is not very sensitive to moderate changes in p, unless the sample size n is large. (How large is "large" depends somewhat upon the size of p_o.)

Analysis of past data by np charts: The objective is to test whether a run of samples from a production process is in control with respect to itself. Could the observed counts of nonconforming, $np = d$, have readily come from a process with a constant p_o, that is, could we attribute all of the variation to chance causes? Or is there evidence of one or more assignable causes coming in to create shifts in p?

For this we need a preliminary run of at least 20 samples, preferably all with a constant sample size n; although we can handle the data if n varies. (Sometimes it is desirable to regard all of the production obtained during an hour, shift or day as a sample resulting in a varying sample size.) If the sample size is constant we run an *np* chart on the preliminary series of counts nonconforming, *np*. We look for points out of the control limits for evidence of the presence of some assignable cause which should be hunted down from the process conditions at the time the point went out. If there are no such points then we say that the process is in control, i.e., p_o appears to be constant. This does not mean that p_o is necessarily satisfactory. It could in fact be poor. But if the process is in control, and p_o is too high for our purposes, then it will require a fundamental change in the process. This is because there is no evidence of any assignable cause that we can work on.

Now let us see the details of the analysis. We have just been talking about p_o for the process. But when we are performing an "analysis of past data", we do not have any known p_o. All we have is a series of $np = d$ counts over samples of n pieces. We are testing whether the process fraction defective p_o is constant. So until we disprove this, we will act preliminarily as though p_o were constant and proceed to estimate p_o from the data. We use as the estimate p-bar based on all the data as follows:

$$\bar{p} = \frac{\text{total number nonconforming}}{\text{total number inspected}} = \frac{\sum d}{\sum n} \tag{5.8}$$

Thus for Example 1, for samples 1–30, there were 98 nonconforming found in 30 samples, each of 50, or in 1500 pieces inspected.

Thus

$$p\text{-bar} = 98/1500 = 0.0653$$

This provisional estimate p-bar of the assumed constant process fraction nonconforming, p_o is now substituted for the p_o in (5.6) and (5.7). Then

$$\text{Central line for } np = n\bar{p} \text{ (analysis of past)} \tag{5.9}$$

$$\text{Limits}_{np} = n\bar{p} \pm 3\sqrt{n\text{p}(1-\bar{p})} \text{ (analysis of past data)} \tag{5.10}$$

As an illustration, let us calculate the control lines for samples 1–30 of Example 1. We have

$$\text{Central line for } np = 50(0.0653) = 3.27$$

$$\begin{aligned}\text{Limits}_{np} &= 50(0.0653) \pm 3\sqrt{50(0.0653)(0.9347)} \\ &= 3.27 \pm 5.24 = \text{—}, 8.5\end{aligned}$$

So an np of 9 or more would be out of control; an np of 8, in control.

Example 2. As an illustration, let us consider the following data on the final inspection of auto carburetors for all types of nonconformities, on a single production line for a single day January 15. Each carburetor containing any nonconformities was called nonconforming. A record sheet was being used upon which to record the various types of nonconformities, with 32 types listed, and further space was available to write in any nonconformity type not among the 32 types. Samples consisted of 100 consecutive carburetors, which, while not chosen at random, can be considered a sampling of production under the conditions in force while they were produced and inspected. There was one incomplete sample at the end of the day which we omitted here.

Sample number	1	2	3	4	5	6	7	8
Sample size, n	100	100	100	100	100	100	100	100
Nonconforming, d	10	4	6	12	6	8	10	12
Sample number	9	10	11	12	13	14	15	16
Sample size, n	100	100	100	100	100	100	100	100
Nonconforming, d	8	7	3	4	3	4	4	10
Sample number	17	18	19	20	21	22	23	
Sample size, n	100	100	100	100	100	100	100	
Nonconforming, d	7	5	3	6	8	10	4	

The numbers of nonconforming units, $d = np$, are plotted in Figure 5.2. For the control lines, we first find p-bar:

$$p\text{-bar} = 154/2300 = 0.0670$$

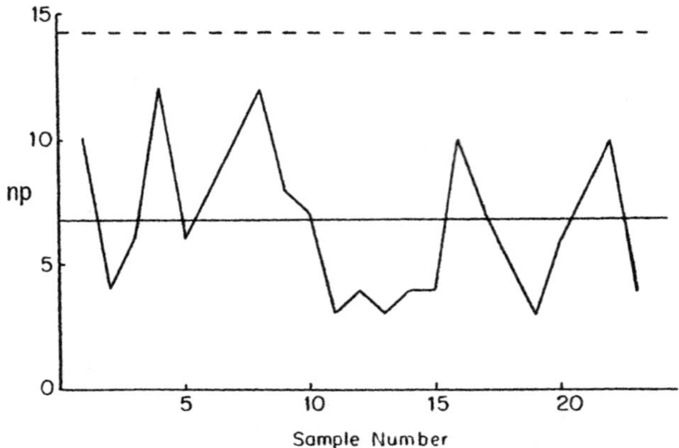

Figure 5.2 Numbers np of carburetors containing at least one nonconformity of any kind. Consecutive samples, each of $n = 100$, inspected at end of production line for all types of nonconformities represent a single day's production. Central line $np = 6.7$, upper limit 14.2.

from which is calculated:

$$\text{Central line for } np = 100(.0670) = 6.7$$

$$\begin{aligned}\text{Limits}_{np} &= 100(0.0670) \pm 3\sqrt{100(0.0670)(.9330)} \\ &= 6.7 \pm 7.5 = \_\_, 14.2\end{aligned}$$

These lines showed the process to be in control for nonconformity types taken as a whole. That is, there was no reliable evidence of any assignable cause affecting a large number of or all nonconformity types. But the p value was regarded as quite satisfactory.

Careful study of the tally sheet on types of nonconformities brought to light potential causes. Steps were taken to eliminate these causes. Also the nonconformities from these causes were eliminated from the data as being no longer typical of the process. This gave a revised p-bar $= 64/2300 = 0.0278$ as a provisional standard, p_o, for a goal for subsequent production. Later on in this book, we shall show some of the subsequent data.

It is a very desirable approach to periodically revise p_o, reducing the number of nonconformities that are considered to be eliminated by appropriate action and thus to set a provisional standard p_o for the next week or month. One plant making small motors, with which the author is well acquainted, achieved much success in this way.

If sample size varies: For an np chart, whether using p-bar or p_o, if the sample size varies, it is theoretically required that new limits be calculated for each different n. This is because n affects both the central line and the width of the control band, as may be seen in (5.6), (5.7), (5.9), and (5.10). However, if the n's are fairly uniform, we may use an average n in these formulas instead of using each individual n. Now what do we mean by "fairly uniform?" For a rough "rule of the thumb", we first find the average sample size, n-bar, and calculate $0.95n$-bar and $1.05n$-bar (5% variation from n-bar). Then, if all of our n's lie within these limits, we are safe in using the limits from n-bar for all the n's in the set. This can save us quite a bit of calculating and plotting time. If desired, we could find the exact limits for any point close to the n-bar limit, whether inside or out.

5.1.2. Fraction Nonconforming or p Chart

We now take up the fraction nonconforming chart, that is, where the quality variable being plotted is p rather than $d = np$. Basically, this is merely a change of vertical scale from the np chart. Thus consider the data of Example 2 of the preceding section. There we plotted the d counts 10, 4, 6, 12, and so on. Now we can find the fractions nonconforming, p, by dividing d by 100, which gives 0.10, 0.04, 0.06, 0.12, and so on. Each of these numbers is simply one-hundredth of the d numbers.

Case-Standard given p_o: Here we are in the fortunate position of having a standard value of the process fraction nonconforming, p_o. (From past data showing good control or some agreed-upon goal.) Then since we assume independence of nonconformance and constant p_o, we have the binomial distribution for p values, and can use the following formulas for the sample fraction nonconforming derived from dividing d by n and (3.8) by n;

$$E(p) = \frac{d}{n} = p_o \tag{5.11}$$

$$\sigma_p = \sqrt{\frac{p_o(1-p_o)}{n}} \tag{5.12}$$

$$\text{Central line for } p = p_o \tag{5.13}$$

$$\text{Limits}_p = p_o \pm 3\sqrt{\frac{p_o(1-p_o)}{n}} \tag{5.14}$$

Example 3. If we use the data of Example 2 with those nonconforming deleted which were a result of the causes presumed to have been identified and removed, the following data emerge:

Sample number	1	2	3	4	5	6	7	8
Sample size, n	100	100	100	100	100	100	100	100
Nonconforming, d	9	3	5	6	3	4	3	6
Fraction nonconf., p	0.09	0.03	0.05	0.06	0.03	0.04	0.03	0.06
Sample number	9	10	11	12	13	14	15	16
Sample size, n	100	100	100	100	100	100	100	100
Nonconforming, d	2	3	0	1	1	1	0	3
Fraction nonconf., p	0.02	0.03	0.00	0.01	0.01	0.01	0.00	0.03
Sample number	17	18	19	20	21	22	23	
Sample size, n	100	100	100	100	100	100	100	
Nonconforming, d	2	1	0	1	1	6	3	
Fraction nonconf., p	0.02	0.01	0.00	0.01	0.01	0.06	0.03	

These p values are plotted in Figure 5.3. Now we use the agreed-upon standard value, p_o, $= 0.0278$ from the purified data. Then (5.13) and (5.14) give:

Drawing in these control lines in Figure 5.3, we find that one point, for the first sample, is out of control. Six different kinds of nonconformities made up the total of nine nonconformaties, which does not exactly "pin-point" the trouble! Perhaps it was general laxness at start-up time in the shift.

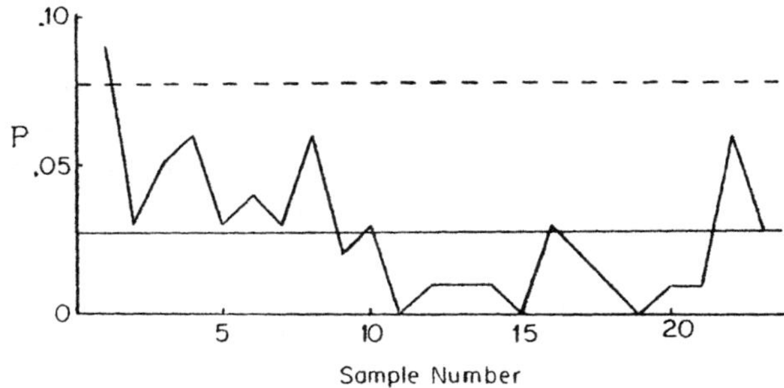

Figure 5.3 Data on carburetors containing one or more nonconformities from unknown causes. Nonconformities from known and removed assignable causes were eliminated from data of Example 2 to give data of Example 3. Fractions nonconforming, p, were plotted for the samples of $n = 100$. A standard value of $p_o = 0.0278$ was the central line, giving an upper control limit of 0.077.

As a further illustration of the p chart being essentially a change of scale from the np chart, consider the limits for the first 30 samples of the experimental data of Example 1. The control lines set from $p_o = 0.060$ were

$$\text{Central line for } np = 3$$

$$\text{Limits}_{np} = \text{—}, 8.04$$

Let us now use (5.13) and (5.14) for control lines for p values:

$$\text{Central line for } p = p_o = 0.06$$

$$\begin{aligned}\text{Limits}_p &= p_o \pm 3\sqrt{\frac{p_o(1-p_o)}{n}} = 0.06 \pm \sqrt{\frac{0.06(0.94)}{50}} \\ &= 0.06 \pm 0.1008 = \text{—}0.1608\end{aligned}$$

Now since the p values are each $d/50$, we ought to be able to divide the control lines (central line and limits), for np by 50 and find those for p. Thus

$$3/50 = 0.06$$

$$8.04/50 = 0.1608$$

Case-analysis of past data: Again we consider the common case of analysis of a preliminary run of sample fractions nonconforming, p. Since we do not know p_o, we use the provisional estimate p-bar found by use of (6.8). Then substitute p-bar for p_o in (5.11) and (5.12), obtaining

$$\text{Central line for } p = \bar{p} \quad \text{(analysis of past data)} \qquad (5.15)$$

$$\text{Limits}_p = \bar{p} \pm 3\sqrt{\frac{\bar{p}(1-\bar{p})}{n}} \text{ (analysis of past data)} \qquad (5.16)$$

Example 4. Table 5.1 gives the fractions nonconforming for an electrical equipment given a "B test" and failing. The p values shown are plotted in Figure 5.4.

Table 5.1 *B*-Test Failures of Electrical Equipment for September. Eighteen Fractions Defective for the Month. Daily Production about 1650

Day	Fraction defective	Day	Fraction defective
1	0.0256	11	0.0082
2	0.0245	13	0.0084
3	0.0225	14	0.0139
4	0.0104	15	0.0090
6	0.0191	16	0.0356
7	0.0042	17	0.0186
8	0.0086	18	0.0221
9	0.0195	20	0.0180
10	0.0183	21	0.0110

Now let us calculate the control lines. Overall, the original data showed 485 failures or nonconformities out of 29,690. By (6.8), this gives

$$\bar{p} = \frac{485}{29,690} = 0.0163$$

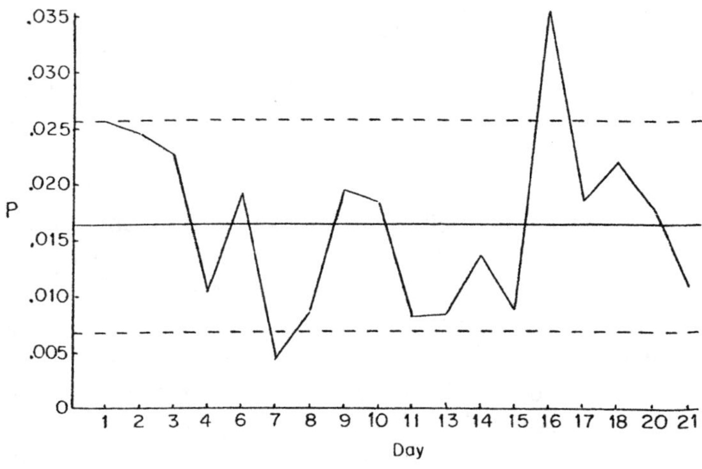

Figure 5.4 Fraction nonconforming control chart on electrical equipment from *B*-test failures in September. Daily production is about 1650. Control lines set from data in Table 5.1.

Using the average production of 1650 as n, we find by (5.16)

$$\text{Limits}_p = 0.0163 \pm 3\sqrt{\frac{0.0163(0.9837)}{1650}}$$
$$= 0.0163 \pm 0.0094 = 0.0069, 0.0257$$

Here it can be seen that we have two real control limits, the lower one being above zero. Thus we might have points out of control on either the bad side or the good side; that is, there is a possibility of finding evidence of a "good" assignable cause on the seventh sample.

If sample size varies: Sometimes sample sizes vary, for example, the p values may represent the entire production over equal blocks of time. Then we may have to take this into account. However, variations in n are less upsetting to the p chart than to the np chart. This is because for the p chart the central line is either p-bar or p_o (central line for p chart) which does not involve n at all, whereas for the np chart, the central line varies directly with n, thus

np - bar or np_o(central line for np chart)

Therefore, no matter how much n varies, the p values will always have the same central line about which they vary. On the other hand, the central line for the np or d counts of nonconforming units will have differing central lines, if n varies much at all.

Now it is true that the *width* of the control band will vary as n varies, for both kinds of charts. For the p chart, the width varies *directly* with the square root of n as we see in (5.14) or (5.16). For the np chart, the width of the control band varies directed within as (5.7) and (5.10) show. Thus the same percentage variation in n will bring similar percentage changes in the widths of the control bands of the two kinds of charts, but in opposite directions. So the big difference is that the central line is unaffected in a p chart by changes of n but is affected in the np chart.

Let us reconsider Example 4. Actually, the n's ran from a minimum of 1390 to a maximum of 2000. Let us figure the limits from n of 1650 and then for $n = 1390$ and 2000, for both kinds of charts and see whether we could reasonably use n-bar instead of the actual n's of each sample, despite the considerable variation in the n's. We have

$$0.0163 \pm 3\sqrt{\frac{0.016 \pm 3(9837)}{n}} = 0.0163 \pm \frac{0.380}{\sqrt{n}}$$

into which we substitute $n = 1390$, 1650, 2000

$$0.0163 \pm 0.0102 = 0.0061, 0.0265$$
$$0.0163 \pm 0.0094 = 0.0069, 0.0257$$
$$0.0163 \pm 0.0085 = 0.0078, 0.0248$$

In spite of the sizable variation in the sample sizes, the limits are not radically different, and they all have the same central line, p-bar $= 0.0163$. On the other hand, for an np chart, we use (5.10) for limits as follows:

$$22.66 \pm 14.16 = 8.5, 36.8$$
$$26.90 \pm 15.43 = 115, 42.3$$
$$32.60 \pm 16.99 = 15.6, 49.6$$

Certainly these differ too much for us to use just one set of limits based on n-bar $= 1650$. The trouble is largely from variation in the central line.

As a rough *rule of thumb* for a p chart when the n's vary, we can find n, and then if all n's lie within $0.7n$-bar and $1.3n$-bar, use n-bar to set one pair of limits for all samples. Sometimes we may need to make up several groupings of n's, within each grouping the n's being fairly uniform. This saves on arithmetic and plotting time.

Percent nonconforming chart: This variation of a p chart is sometimes desired, especially by executives who like to deal with percentages and with operators who do not like to deal with decimal points. The authors suggest that the quality control person make all the calculations in decimals, as we have been doing for a fraction nonconforming chart, but then merely revise the vertical scale of the chart to read in percents rather than decimals.

Chart for nonconforming units, *np*, or for fraction nonconforming, *p*?: Which of these two kinds of charts to use in your plant is really a matter of personal preference. Use the type of chart, which will be most easily understood by your associates and those who will be taking action on the processes.

Advantages of np chart for nonconforming units

1. There is no need to divide each d by n to find p for the sample.
2. It is easier and quicker to plot whole numbers, np, than decimals p.
3. The arithmetic for limits is a trifle easier.

Advantages of p chart for fraction nonconforming

1. It is probably more easily understood, either in decimal form or in percentages.
2. When sample sizes vary, the central line remains the same, and only the width of the band varies, whereas in an np chart even the central line varies.
3. We can permit much more variation of n's around n-bar and still use limits based only on n-bar.

5.1.3. Instructions for use of *p* and *np* charts

1. The point in the process to take *samples* of pieces or possibly to inspect runs of the *entire production* must be decided. This should usually be as near as feasible to where trouble may well be arising. This enables you to be closer physically and timewise to changes in the process which may prove to be assignable causes.
2. Workable definitions of each nonconformity, which will cause a piece to be called nonconforming, must be delineated. This is a vital and often difficult step. But insofar as possible, the definitions should be so clear-cut that everyone involved would call each given imperfection a nonconformity or defect or else everyone would call it a good piece. (For example, the

author came across a case where the same 60 food-jar caps were examined by many people for "functional" nonconformities. The number declared functional nonconforming by the various people ran from 3 to 60!) Good gauges or other equipment may be needed. Also a set of "limit samples" for visual inspection may be developed, to show for each type of imperfection a limit, so that a larger imperfection is a nonconformity whereas a smaller one is permitted. Or limits may separate major nonconformities, minor nonconformities and no nonconformities.

3. Individuals should be properly trained for inspecting or testing the samples of product, so that results are reliable.
4. Sample size must be determined. It should not be so large that many changes in process conditions may occur while a given sample of product is being produced. But the sample size should be large enough so that d-bar is at least one.
5. If a sample is to be drawn from, say, an hour's production, steps should be taken to ensure that the sample is drawn randomly and without bias, insofar as possible.
6. It is helpful to make up a list of potential assignable causes which might be a cause of nonconformities. Enlist the help of everyone connected with the process.
7. It is useful to have a log of the process conditions, so that for each sample there is a record of the time, production line operator, inspector or quality person, test set or gauge used, set-up person, source of raw materials, and so on.
8. Previous data may be available and useful, but it is advisable to be skeptical unless you know the conditions under which they were collected. Usually it is desirable to get new sample data, perhaps collecting more rapidly than you expect to in the future.
9. Plot the p or np points for the preliminary run of samples, and then calculate and draw the control lines.

10. Show the chart and explain it with great care to the process operator, supervisor, and others involved. Look for any indicated assignable causes. Here is where teamwork really counts.
11. If an assignable cause for bad product has been found and eliminated, delete all the sample results occurring while this cause was at work in the process. This may include points within the control band as well as points above the upper control limit.
12. If an assignable cause for increasing the production of good product, e.g., the source of raw material, is found, consider incorporating it in the process, if feasible.
13. From p-bar possibly revised as in 11 above, set control lines to be extended for new sample data. Examine each sample point as soon as it arises for possible action on the process.
14. Periodically revise p-bar, not retaining obsolete data.
15. When reasonable control at a satisfactory quality performance is obtained, we can call p-bar a standard fraction nonconforming, p_0. It may then be possible to take samples less frequently, or to take smaller samples. But some, continued follow-through or control is desirable.
16. It is helpful to develop and print forms with spaces for information and data, and perhaps even a place to plot points.

5.2. CHARTS FOR NONCONFORMITIES

We shall now take up the second general type of attribute control chart, namely, charts for nonconformities, rather than for nonconforming units. We might have a single kind of nonconformity under study, or we may have a class of several different kinds of nonconformities, e.g., nonconformities that might be classified as "major nonconformities". If a single piece, unit, article, or subassembly may well contain more than one nonconformity, we may wish to count the number of nonconformities per sample of n pieces. Here n could even be one, as in final inspection of trucks or lawn mowers. Or n could be 24 for a case of bottles. By counting the number of nonconformities, we may obtain a better idea of how nonconforming the product is. This is especially true when a large or complicated product or assemblies are being inspected and tested, and nearly every one has one or more nonconformities. It does little good to plot a series of fractions nonconforming, p, running 1.00, 1.00, 0.98, 1.00, 0.96, 0.98, 1.00, and so on. Instead we learn more by counts of nonconformities in samples, especially using a tally sheet or Pareto chart listing the nonconformities in the class under study.

Once more we emphasize that the nonconformities of interest may vary greatly as to importance and seriousness, from one study to another.

5.2.1. The *c* Chart for Nonconformities

Let us repeat the notations as given before in Chapter 3:

$$c = \text{number of nonconformities in a sample of product} \tag{5.17}$$

$$c_0 = \text{standard or true process average of nonconformities per sample} \tag{5.18}$$

$$c\text{-bar} = \text{average of a series of counts-}c \tag{5.19}$$

The objective of a c chart is to plot and analyze a series of counts of nonconformities, c, taken over uniform sample sizes of product.

Standard given, c_0: For the following analysis of a c chart, we make these assumptions:

1. The samples provide equal "areas of opportunity" for nonconformities to occur, that is, not both small and large subassemblies or varying numbers n of pieces for a sample.
2. Nonconformities act independently, not tending to come in bunches. Thus the occurrence of one

nonconformity does not affect the probability for another nonconformity of that type or some other type to occur.
3. All types of defects under study are unequivocally defined.
4. Inspection is thorough insofar as possible.
5. Provisionally, c_o is constant for the process.
6. The *possible* number of nonconformities is far above c_o.

Granted these assumptions, then the Poisson model for the distribution of nonconformities is applicable.

We may now immediately set up the control lines from the assumptions from Section 3.5 and (3.12). Thus

$$\text{central line for } c \text{ values} = c_o \text{ (standard given)} \tag{5.20}$$

$$\text{limits}_c = c_o \pm 3\sqrt{c_o} \text{ (standard given)} \tag{5.21}$$

For an example, let us suppose that considerable past data have shown rather good control for the number of "seeds" in cases of 24 glass bottles, with c-bar $=4$. ("Seeds" are bubbles of at least a certain size.) We might then call $c_o=4$ against which to check future production. We show 100 c-value counts in Table 5.2. The counts are plotted in Fig. 5.5.

Let us use (5.20) and (5.21) to set the control lines with $c_o=4$:

$$\text{central line} = 4$$

$$\text{Limits}_c = 4 \pm 3\sqrt{4} = \text{—}, 10$$

Among the first 50 counts of seeds in cases of 24 bottles, there is one count of 10. This is on the control limit and would be considered an indication of an assignable cause. The production conditions under which this case was produced should be studied.

Now the c values of Table 5.2 actually were not counts of seeds; this interpretation was just a possibility for this sampling experiment. Instead the first 50 counts, c, were from

Table 5.2 Counts, c, of "Seeds" in Cases of 24 Glass Bottles

$c_o = 4$		$c_o = 4$		$c_o = 8$		$c_o = 2$	
Case	c	Case	c	Case	c	Case	c
1	6	26	5	51	4	76	2
2	5	27	10	52	8	77	2
3	8	28	8	53	9	78	2
4	4	29	3	54	10	79	0
5	4	30	1	55	5	80	3
6	3	31	4	56	10	81	0
7	5	32	6	57	10	82	1
8	5	33	4	58	7	83	0
9	2	34	3	59	7	84	7
10	4	35	2	60	9	85	2
11	2	36	2	61	9	86	2
12	5	37	4	62	11	87	0
13	2	38	3	63	7	88	0
14	3	39	5	64	8	89	3
15	9	40	4	65	7	90	6
16	1	41	3	66	7	91	1
17	2	42	2	67	5	92	2
18	4	43	2	68	9	93	4
19	3	44	3	69	4	94	4
20	2	45	5	70	5	95	3
21	4	46	3	71	7	96	1
22	5	47	5	72	9	97	4
23	2	48	2	73	8	98	1
24	3	49	3	74	5	99	2
25	3	50	5	75	10	100	3

a bowl of 500 chips numbered according to a Poisson distribution with $c_o = 4$. Thus the $c = 10$ point being on the control limit was mere chance, an event that will occur by chance only quite rarely. Let us just see how rare it was. For this, we consult Table B, entering with $c_o = 4$.

$$P(10 \text{ or more}) = 1 - P(9 \text{ or less}) = 1 - 0.992 = 0.008$$

Therefore we could expect to draw a c value of 10 or more, eight times in 1000, or once in 125. We had one among the first 50, which is not too surprising. But in practice any

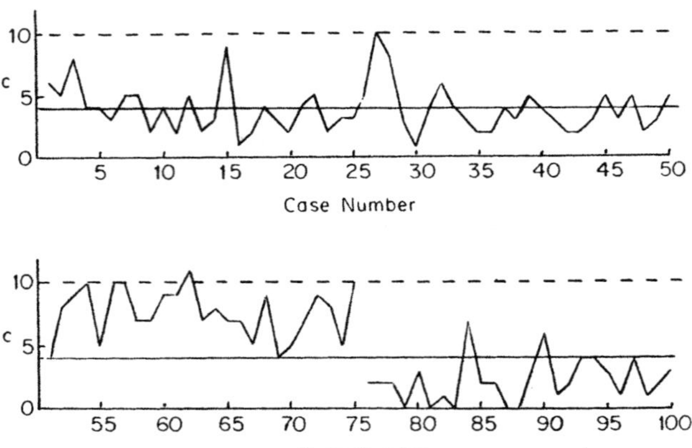

Figure 5.5 Experimental data from Poisson populations, interpreted as counts of "seeds" in cases of 24 glass bottles. Data are from Table 5.2. The first 50 c values are from $c_0 = 4$; cases 51–75, $c_0 = 8$; and cases 76–100, $c_0 = 2$. Control lines shown for all 100 cases are from $c_0 = 4$.

such point would be regarded as an indication of an assignable cause.

Now consider cases 51–75. They were actually drawn from a Poisson population with $c_0 = 8$, that is, twice as high as for the original 50. How soon do we find evidence of an assignable cause? The first definite indication is on case 54, where a count 10 occurs. Thereafter among these 25 c counts, there are three more 10's and a 11. Each one *by itself* is sufficient to be regarded as evidence of an assignable cause. Note also that all but two counts, c, lie above the central line. Of course, a jump of c_0 from 4 to 8 is quite a change. If c_0 went from 4 to only 5 or 6, it might take quite a few samples before a definite indication shows up.

The last 25 counts were drawn from a Poisson population having $c_0 = 2$. Now how can we get an indication of an improvement, that is, of a *lowered* c_0 value? It cannot come through a point on or below the lower control limit, because with $c_0 = 4$, there is no lower limit. (Not until $c_0 = 9$, do we have a meaningful lower control limit.) Thus the only way we can obtain an indication of improvement is by a long run of c's on or below the central line, or perhaps by a generally lower set of points. In the last 25 c values, only two counts are above the central line and the first eight are all below the central line. Then at the end there are 10 in succession all on or below the central line; surely a significant run.

We usually use the analysis with standard given, only after a reasonably long run of in-control conditions at satisfactory quality. Then we can set a meaningful c_o from the process average c-bar, and be checking for control against a standard c_o.

Analysis of past data: In this case, the purpose is to find out whether or not the process seems to be in control or whether there is evidence of assignable causes. If the latter, then we want to take appropriate action. This will in general yield considerable process improvement. The objective is always to achieve a process in control at a satisfactory quality level.

In analyzing a run of, say, 25 c-values as preliminary data, we do not have a standard c_o from which to work. Thus we cannot use (5.20) and (5.21). However, we are assuming preliminarily that the process is in control, that is, has just one average c_o, even though we do not know what it is. But for our estimate of what this c_o is, we take the average of our c values, namely c-bar. Thus instead of (5.20) and (5.21), we must substitute c-bar for c_o and obtain:

$$c\text{-bar} = \text{central line for } c \text{ values (analysis is of past data)} \quad (5.22)$$

$$\text{Limits}_c = \bar{c} \pm 3\sqrt{\bar{c}} \quad \text{(analysis of past data)} \quad (5.23)$$

Example 5. In inspecting consecutive large aircraft, the following counts of missing rivets were made:

Plane	1	2	3	4	5	6	7	8	9	10	11	12	13
c	11	5	5	5	21	13	17	8	3	2	9	6	8
Plane	14	15	16	17	18	19	20	21	22	23	24	25	26
c	5	10	16	8	6	5	6	7	8	4	1	4	3

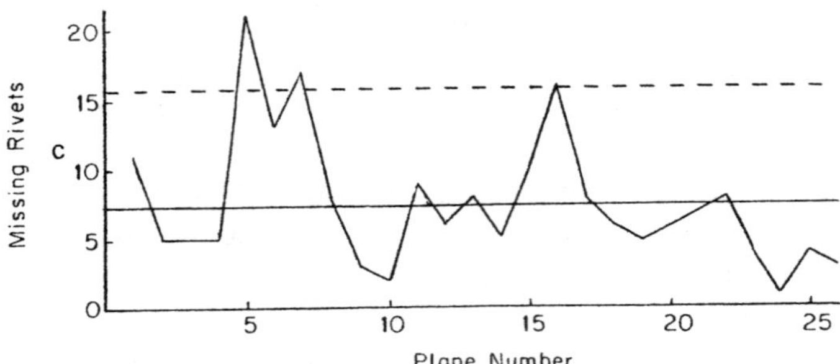

Figure 5.6 Missing rivets from inspection of 26 large aircraft, analyzed as preliminary data. Central line is 7.54 and the upper control limit is 15.8. Three high points are out of control.

These counts of missing rivets are shown in Fig. 5.6. For control lines, we have

$$c\text{-bar} = 196/26 = 7.54$$

$$\begin{aligned}\text{Limits}_c &= c - \text{bar} \pm 3\sqrt{c} - \text{bar} = 7.54 \pm 3\sqrt{7.54} \\ &= 7.54 \pm 8.24 = \text{—}, 15.8\end{aligned}$$

Three points lie above the upper control limit, indicating the presence of some assignable cause(s). Some process improvement seems to have been made toward the end.

The psychological effect of control charts, such as this one, should not be underestimated. It enables everyone to see "How are we doing?" and encourages them to do their best and then improve on that best! Great gains were made in studies of which these data were a part. See also Problem 5.15.

5.2.2. Charts for Nonconformities per Unit

This type of control chart is also concerned with nonconformities. It is especially useful to condense data on nonconformities, in order to give a broad picture, and/or to eliminate too much detail. A second purpose is to take account of varying opportunities for defects, e.g., varying areas of cloth, painted surface or paper, or varying lengths of insulated wire for breakdowns of insulation.

Example 6: As an example, in the final inspection of completed trucks just driven off the line, the daily production was 125. It would be possible to make a c chart for the individual trucks, where c would be the number of nonconformities or errors per truck. This would give 125 individual c-points to plot per day. Such a chart, while feasible, tends to supply too much detail and fails to give the overall picture. Another possible chart would be to count up the total number of nonconformities or errors on the daily production of 125 trucks, to call this number c, and thus to have just one summary c-point per day reflecting the total errors. Such a chart gives the overall picture. But it does not show very clearly how much *average* trouble there has been *per truck* during the day. Thus we find it convenient to plot the average errors per truck, found by dividing the total number of errors on all trucks during the day by 125. Now we see Table 5.3. In the third and seventh columns are the total errors c noted for each of the 20 days. These total errors c were then divided by 125 to give the number of errors per truck for

Table 5.3 Errors per Truck at A, Assembly Line, and B, Station. Production 125 Trucks per Day. Total Errors and Average Errors per Truck

Day	Total errors, c	Average errors, u	Day	Total errors, c	Average errors, u
Jan 13 M	218	1.74	Jan 27 M	226	1.81
Jan 14 T	200	1.60	Jan 28 T	175	1.40
Jan 15 W	165	1.32	Jan 29 W	181	1.45
Jan 16 Th	195	1.56	Jan 30 Th	191	1.53
Jan 17 F	209	1.67	Jan 31 F	159	1.27
Jan 20 M	250	2.00	Feb 3 M	162	1.30
Jan 21 T	188	1.50	Feb 4 T	171	1.37
Jan 22 W	150	1.20	Feb 5 W	183	1.46
Jan 23 Th	219	1.75	Feb 6 Th	170	1.36
Jan 24 F	252	2.02	Feb 7 F	199	1.59
				3863	30.90

the day. For example, for January 13

$$u = c/n = 218/125 = 1.74$$

We shall use the following notations:

$$n = \text{number of units of product per sample} \quad (5.24)$$

$$c = \text{total number of nonconformities in a sample of } n \text{ units} \quad (5.25)$$

$$u = c/n = \text{average number of nonconformities } \textit{per unit} \text{ over } n \text{ units} \quad (5.26)$$

$$\bar{u} = \frac{\sum u}{k} = \text{average values of } u \text{ for } k \text{ equal-sized samples} \quad (5.27)$$

$$u\text{-bar} = \left(\sum c\right) \Big/ \sum n = \text{average of } u \text{ values if } n \text{ values vary} \quad (5.28)$$

$$u_o = \text{population or true process average nonconformities per unit} \quad (5.29)$$

The 20 errors per truck from Table 5.3. are plotted in Fig. 5.7. We now need control lines for the errors per truck data. These are provided by the following:

$$u_o = \text{central line} \qquad \text{(standard given)} \quad (5.30)$$

$$\text{Limits}_u = u_o \pm 3\sqrt{u_o/n} \qquad \text{(standard given)} \quad (5.31)$$

$$u\text{-bar} = \text{central line} \qquad \text{(analysis of past data)} \quad (5.32)$$

$$\text{Limits}_u = \bar{u} \pm 3\sqrt{\bar{u}/n} \qquad \text{(analysis of past data)} \quad (5.33)$$

Since in our example we do not have u_0 as a given standard, we must use (5.32) and (5.33). For (5.32) since the sample sizes are equal we may use (5.27), finding

central line $= u$-bar $= 30.90/20 = 1.55$

We could also have used (5.28):

central line $= u$-bar $= 3863/[20(125)] = 1.55$

Then by (5.33),

$$\text{Limits}_u = 1.55 \pm 3\sqrt{1.55/125} = 1.55 \pm 0.33 = 1.22, 1.88$$

These three control lines are shown in Fig. 5.7. Points above the upper control limit indicate the presence of one or more assignable causes of trouble for the day in question. Meanwhile, points below the lower control limit are indicative of a "good" assignable cause, which may help improve quality if incorporated. There are two of the former and one of the latter in the 20 days.

In this plant, nonconformities per unit charts with the accompanying tally table (Pareto chart) of the kinds of errors led to large improvements in the production lines, very big savings on repairing and fixing up trucks and fostered quality mindedness and pride of workmanship.

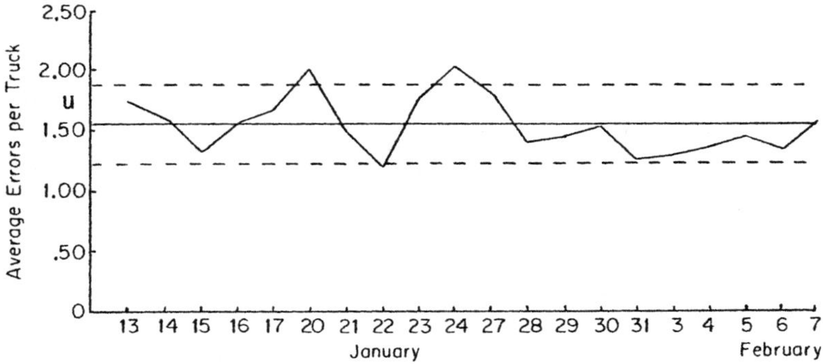

Figure 5.7 Control chart for average errors per truck from 20 days' production, each of 125 trucks. Data are shown in Table 5.3 and was analyzed as preliminary data for control.

A final point worth considering in this example: Would we arrive at the same conclusions if we had plotted the total nonconformities per day in an ordinary c chart instead of the nonconformities per unit, u? Yes. We would use (5.22) and (5.23):

$$\text{central line} = c\text{-bar} = 3863/20 = 193.2$$

$$\text{Limits}_c = \bar{c} \pm \overline{3\sqrt{\bar{c}}} = 193.2 \pm 3\sqrt{193.2} = 193.2 \pm 41.7$$
$$= 151.5, 234.9$$

How do these compare with the control lines for u values, namely 1.55, 1.22 and 1.88? If we divide the line values for c's by the sample size, $n = 125$, we obtain exactly the line values for the u's (apart from round-off errors). Thus since the c's divided by 125 are the u values, the two charts give identical decisions.

Example 7. Let us consider briefly the following hypothetical example for breakdowns of insulation under an excessive test voltage for insulated wire:

Length (m)	Breakdowns c	Units (500 m) n	Breakdowns per unit $u = c/n$	$\text{Limits}_u = \bar{u} \pm 3\sqrt{\bar{u}/n}$
500	2	1	2.00	—, 6.37
500	0	1	0.00	—, 6.37
1500	5	3	1.67	—, 4.55
500	3	1	3.00	—, 6.37
1000	3	2	1.50	—, 5.10
500	4	1	4.00	—, 6.37
1500	8	3	2.67	—, 4.55
1000	5	2	2.50	—, 5.10
500	1	1	1.00	—, 6.37
1000	4	2	2.00	—, 5.10
	35	17		

For u-bar we use (5.28) since the areas of opportunity, n, for defects vary. Here we use a 500-m length as one unit for n. (We could have used 1000 or 1500 m.) We thus assume that in 1000 m, there is twice the opportunity for breakdowns as in 500 m, which seems reasonable. The central line for all points is the same:

$$u\text{-bar} = 35/17 = 2.06$$

But the *width* of the control bands will depend upon n by (5.33).

$$\text{Limits}_u = 2.06 \pm 3\sqrt{\frac{2.06}{n}}$$

$$n = 1 \quad 2.06 \pm 4.31 = \text{—}, 6.37$$

$$n = 2 \quad 2.06 \pm 3.04 = \text{—}, 5.10$$

$$n = 3 \quad 2.06 \pm 2.49 = \text{—}, 4.55$$

All 10 points lie within the respective limits. Actually they should be expected to because for 500-m lengths we drew from a Poisson population of chips with $c_o = 2$; for 1000 m, $c_o = 4$; and for 1500 m, $c_o = 6$.

In summary, we can effectively use a nonconformities-per-unit control chart if (1) we want an overall chart to summarize a number of units of production to avoid too much detail or (2) the size of the samples or opportunity for nonconformities varies. Such a chart is really just a change of scale from a c-chart for the total nonconformities per sample. The u chart is sometimes called a c-bar chart because u is in reality an average of n c-values.

5.2.3. Instructions for Use of *c* and *u* Charts

The instructions given in Sec. 5.1.3 for p and np charts apply with the obvious modification of np to c and of p to u. In the outline there given, suggestion 2 is important, since an inspector may have to be looking for 15 or 20 different visual nonconformities, for example. Unless the definitions of the various nonconformities are clear and steps taken to be sure that accurate inspection is made, results are likely to be meaningless. In suggestion 4, unless c-bar is at least one, the chart is not very powerful at indicating assignable causes. In suggestion 13, revised c-bar or u-bar is used to extend control lines. In suggestion 15, we would be setting standards c_0 or u_0. In suggestion 16, the development of a list of possible nonconformities, including some blank spaces for tallying nonconformities, is highly desirable.

5.3. SUMMARY

We have described and illustrated two distinct types of control charts each of which finds many applications in industry. Both are concerned with counted or attribute data. Likewise both involve nonconformities which must be clearly defined and soundly inspected for. On the one hand, we have charts for nonconforming units, in which each piece or unit is declared to be nonconforming if it contains at least one nonconformity in the class in question. The appropriate charts are the *np* and *p* charts. Such charts are mostly used where the process fraction nonconforming is small, say 0.10 or less, although the mathematics place no such restriction. On the other hand, we have the *c* and *u* charts for nonconformities. They involve counts of nonconformities (in some class) found on a sample of units or material. Such counts of nonconformities are especially useful when the chance of a single unit having more than one nonconformity is not negligible. Then counts of nonconformities tell *how nonconforming* the sample or unit is.

We recall that *np* and *p* charts are basically the same chart, but merely with a different vertical scale. Similarly, the *c* and *u* charts are also basically the same, but simply using a different vertical scale. When sample sizes vary, however, we find it more convenient to use, respectively, the *p* chart and the *u* chart, for then at least the central line will be unaffected by the sample size and with only the width of the control band depending upon the square root of n.

The *np* and *p* charts are based upon the binomial distribution, whereas the *c* and *u* charts are based upon the Poisson distribution. Both of these distributions assume independence of nonconformities, i.e., that the occurrence of nonconformity or a nonconforming unit does not affect the chances on subsequent pieces drawn for the sample. Also if the charts are to be based upon samples, rather than 100% of the output, then the samples are to be randomly drawn. Further, at the outset when analyzing past data, it is provisionally assumed that p_o or c_o for the process is constant. In the case of c charts, it is assumed also that the maximum *possible* value of c is far greater than the *average* value, c_o.

Some further modifications and comments on attribute charts are given in Chapter 8.

Again we mention that nonconformities can run anywhere from highly critical to quite incidental.

5.4. PROBLEMS

5.1. For the experimental data in Example 1, see Sec. 5.1.1, for samples 31–60, analyze by an np control chart under the case analysis of past data. How is control? What would the control lines be for the case control against standard $p_o = 0.10$. Are the results compatible with $p_o = 0.10$?

5.2. For the experimental data in Example 1, see Sec. 5.1.1, for samples 61–90, analyze by a p control chart under the case analysis of past data. How is control? What would the control lines be for the case control against standard $p_o = 0.03$. Are the results compatible with $p_o = 0.03$?

5.3. For the packing nut data of Table 4.1, find the central line and control limits for a p chart, analyzing as past data. Comment on control. In March samples of 50 packing nuts gave the following counts of $np = d$: 0, 0, 4, 0, 1, 4, 5, 0, 0, 7, 0, 0, 0, 0, 0, 0, 2, 3. Make an appropriate chart and comment. Such charts were of great value to the company involved in this case.

5.4. For the data plotted in Problem 4.2, complete the p chart by finding and drawing in the control lines. Comment. (The next 25 samples showed $d = 1$, 0, 0, 0, 0, 0, 1, 1, 0, 0, 1, 0, 0, 2, 1, 0, 0, 1, 0, 0, 0, 0, 0, 0, 1.)

5.5. The present problem is on auto carburetors inspected at the end of the assembly line for all types of nonconformities. These data are for January 27, and follow the data of Example 2, Sec. 5.1.1. For a long time, a goal of $p_o = 0.02$ had been set. Plot the given data, either p or np, and set central line and control limits from $p_o = 0.02$. Has the goal been reached? Thirty-five d values with $n = 100$ follow: 4, 5, 1, 0, 3, 2, 1, 6, 0, 6, 2, 0, 2, 3, 4, 1, 3, 2, 4, 2, 1, 2, 0, 2, 3, 4, 1, 0, 0, 0, 0, 1, 2, 3, 3.

5.6. The data here shown are for 100% inspection of lots of malleable castings for cracks by magnaflux. The lot size varies considerably, so a p chart is used instead of an np chart. By some care, one can use only four separate n's. Make the control chart and comment.

Lot number	Lot size n	Nonconformities d	Fraction nonconforming p
1	1138	143	0.126
2	600	30	0.050
3	700	298	0.426
4	500	70	0.140
5	750	141	0.188
6	600	26	0.043
7	750	130	0.173
8	775	141	0.182
9	750	17	0.023
10	850	66	0.078
11	775	126	0.163
12	750	20	0.027
13	675	100	0.148
14	1395	32	0.023
15	116	16	0.138
16	750	116	0.155
17	780	132	0.169
18	1400	40	0.029
19	424	50	0.118
20	384	16	0.042
21	286	22	0.077
22	1000	52	0.052
23	360	51	0.142
24	1100	38	0.035
	17608	1873	0.1064

5.7. For the data of Problem 5.6, do you think that the occurrences of cracks in castings would be independent, that is, if one casting has a crack, this defect is unrelated to whether the next one has a crack? Might this account for the lack of control noted? Would you still investigate points outside of limits?

5.8. In Problem 5.6, p-bar $= 0.1064$ was found by use of (5.8), that is, 1873/17,608. Do you think you would obtain the same p by adding the 24 p values and dividing by 24? This would be an unweighted average of p's. If we substitute np for d in (6.8) we have $(\sum np)/\sum n$. Thus p is a weighted average of p values with weights n.

5.9. Samples of 39 articles each, from an optical company, were examined for breakage. Forty-three such samples over two months time yielded $d = 2$, 1, 1, 0, 2, 3, 2, 1, 1, 2, 1, 1, 1, 2, 1, 1, 0, 1, 0, 1, 0, 2, 0, 1, 1, 0, 1, 0, 7, 10, 6, 1, 2, 2, 2, 1, 1, 0, 2, 2, 2, 5, 2. Analyze by an appropriate control chart. Would you prefer a p or an np chart? Why? If these samples are in fact lots, can you think of any practical reason for the size 39?

5.10. For the data on total nonconformities found on daily production of 3000 switches (Problem 4.3), find the central line and control limits, analyzing as past data. Comment on control. Make chart if not previously done.

5.11. The following data were observed just 1 year before the data of Problem 4.3 at the start of the control chart program. Analyze as past data, the 25 production counts of total nonconformities in 3000 switches per day: 450, 454, 564, 369, 294, 358, 343, 227, 263, 248, 692, 314, 247, 521, 435, 1054, 727, 282, 647, 400, 372, 203, 160, 275, 244 (total 10,143). Comment.

5.12. The following data are for 29 samples each of 100-yd lengths of woolen goods. The woolen goods pass slowly across a table and nonconformities noted are marked by passing a bit of yarn of contrasting color through the goods. Nonconformities: 2, 0, 0, 1, 1, 2, 2, 1, 0, 1, 1, 0, 2, 1, 1, 5, 0, 3, 1, 1, 1, 2, 2, 1, 0, 0, 0, 1, 4. Analyze the data by a c chart. Why could we not treat as d values with $n = 100$ and make an np chart?

5.13. For the experimental data on "seeds" in Table 5.2, analyze samples 51–75 as past data by a c chart. Does the process show control, that is, homogeneity? Also analyze for control against the standard $c_o = 8$.

5.14. Follow the same procedure as for Problem 5.13 but use samples 76–100 and $c_o = 2$.

5.15. For large aircraft assemblies, there were 18 categories of nonconformities. (The data were taken subsequently to those in Table 4.1.) Analyze by a c chart for whatever nonconformities are assigned and comment.

Plane	Alignment	Tighten	Cello seals	Foreign matter	Replace	Plug holes
1	5	91	3	26	2	7
2	9	151	7	23	5	4
3	3	149	8	29	4	4
4	4	205	12	20	18	2
5	6	185	5	26	4	3
6	7	106	2	13	9	5
7	14	171	1	13	2	4
8	18	113	8	16	2	2
9	11	171	5	21	8	6
10	11	162	16	16	5	7
11	11	148	8	18	13	6
12	8	117	14	13	6	5
13	10	134	12	11	9	3
14	8	116	7	27	3	7
15	7	112	3	23	5	9
16	16	161	3	34	14	10
17	13	125	14	14	10	7
18	12	145	9	7	17	3
19	9	129	6	25	11	8
20	11	151	5	18	6	3
21	11	140	19	15	15	6
22	8	118	6	18	8	1
23	8	126	3	13	9	5
24	9	101	4	1	1	1
25	4	110	2	8	4	1
26	13	142	7	1	4	0

5.16. The following were the number of breakdowns in insulation in 5000-ft lengths of rubber covered wire: 0, 1, 1, 0, 2, 1, 3, 4, 5, 3, 0, 1, l, 1, 2, 4, 0, 1, 1, 0. What were the central line and control limits for the number of breakdowns? What would be the central line and limits for the average number of breakdowns per length for samples of five lengths?

5.17. The following data show daily averages of errors per truck over the daily production of about 125 trucks at final inspection. These data precede those of Table 5.3.

Day	Total errors, c	Average errors, u	Day	Total errors, c	Average errors, u
Nov 21 Th	156	1.25	Dec 5 Th	175	1.40
Nov 22 F	198	1.58	Dec 6 F	189	1.51
Nov 25 M	281	2.25	Dec 9 M	135	1.08
Nov 26 T	312	2.50	Dec 10 T	159	1.27
Nov 21 W	256	2.05	Dec 11 W	148	1.18
Nov 29 F	182	1.46	Dec 12 Th	174	1.39
Dec 2 M	192	1.54	Dec 13 F	178	1.42
Dec 3 T	178	1.42	Dec 16 M	260	2.08
Dec 4 W	196	1.57	Dec 17 T	231	1.85
				3600	28.80

a. Plot the average errors per truck for the days.
b. Draw in control lines from the previous u-bar $= 1.97$ and comment.
c. Analyze these data as a preliminary run of data, checked for control.
d. Suggest a u-bar value for subsequent data.

6

Control Charts for Measurements: Process Control

We now take up the study of a most powerful set of tools for process control, when the quality characteristic is measurable. Any of the huge variety of product measurements may be analyzed by such control charts. Typical industrial measurements are length, thickness, diameter, width, weight of an item, density, chemical composition, percent impurity, hardness, tensile strength, package content weight, resistance, voltage, color characteristics, angle of bend, length of life, ultimate strength, corrosion resistance, thickness of coating, starting torque, horsepower, and acidity. Usually, quality measurements carry some unit. Also the production process manufacturing the product has requirements or goals set in the form of a minimum or a maximum limit, or else two specifications, between which the measurements are supposed to lie with high probability. The objective is to obtain "satisfactory, adequate, dependable, and economic quality" from the process, in the words of the quality control pioneers at the Bell Telephone Laboratories.

6.1. TWO CHARACTERISTICS WE DESIRE TO CONTROL

We have seen in Chapter 2 that when we have a sample of data either measurement or attribute, we can begin an analysis of the data by tabulating them into a frequency table and then drawing a histogram or a frequency polygon, such as Figure 2.1 shows.

Now as we discussed before, there are two characteristics of data in general. For example, with the weights of charge given in Table 2.1, the specifications were 454 ± 27 g. The average content weight for the process is to be at least 454 g, and the customer is anxious to have none below 427 g, while the manufacturer is desirous of having none above 481 g, because of excessive overfill. We must therefore be concerned with both the average charge weight of the process and the variability. It would be undesirable to have the average charge weight right at 454 g if we have so much

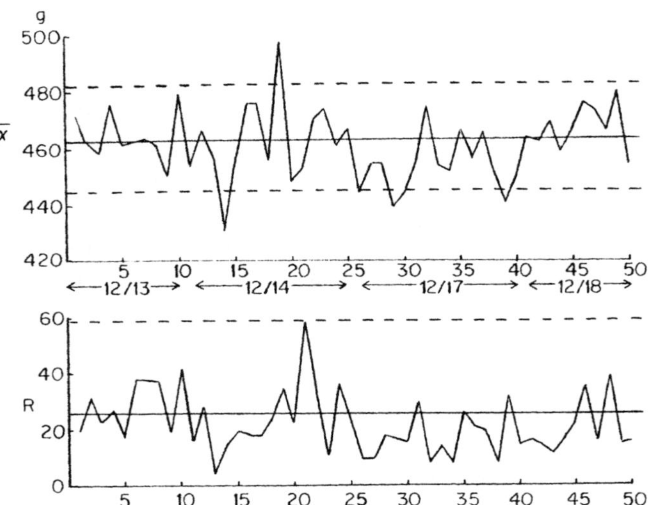

Figure 6.1 Charge weights in grams of insecticide dispensers. Averages, $\bar{x}$, and ranges, R, plotted for samples of $n = 4$ from Table 2.1. First 25 samples were the preliminary data. Specifications for x: 454 ± 27 g.

Table 6.1 Charge Weights of Insecticide Dispenser in Grams Taken in Samples of Four. Average, $\bar{x}$, and Range, R, Listed for Each Sample. First 25 Samples, Preliminary Data. Specifications for x: 454 ± 27 g. Stand Number 4

Sample number	Date	Observed charge weights x_1	x_2	x_3	x_4	Average $\bar{x}$	Range R	Remarks
1	12/13	476	478	473	459	472	19	
2		485	454	456	454	462	31	
3		451	452	458	473	458	22	
4		465	492	482	467	476	27	
5		469	461	452	465	462	17	
6		459	485	447	460	463	38	
7		450	463	488	455	464	38	
8		Lost	478	464	441	461	37	Sample of 3
9		456	458	439	448	450	19	
10		459	462	495	500	479	41	
11	12/14	443	453	457	458	453	15	
12		470	450	478	471	467	28	
13		457	456	460	457	458	4	
14		434	424	428	438	431	14	
15		460	444	450	463	454	19	
16		467	476	485	474	476	18	
17		471	469	487	476	476	18	
18		473	452	449	449	456	24	
19		477	511	495	508	498	34	
20		458	437	452	447	448	21	
21		427	443	457	485	453	58	
22		491	463	466	459	470	32	
23		471	472	472	481	474	10	
24		443	460	462	479	461	36	
25		461	476	478	454	467	24	End of
26	12/17	450	441	444	443	444	9	Preliminary
27		454	451	455	460	455	9	run
28		456	463	Lost	445	455	18	Sample of 3
29		447	446	431	433	439	16	
30		447	443	438	453	445	15	
31		440	454	459	470	456	30	
32		480	472	475	472	475	8	
33		449	451	463	453	454	14	
34		454	455	452	447	452	8	
35		474	467	477	451	467	26	
36		459	457	465	444	456	21	

(*Continued*)

Table 6.1 (*Continued*)

Sample number	Date	Observed charge weights x_1	x_2	x_3	x_4	Average $\bar{x}$	Range R	Remarks
37		465	475	456	468	466	19	
38		458	450	451	451	452	8	
39		447	417	449	445	440	32	
40		453	442	456	453	451	14	
41	12/18	471	467	461	455	464	16	
42		462	454	462	468	462	14	
43		474	471	471	463	470	11	
44		461	454	468	452	459	16	
45		473	453	465	475	466	22	
46		474	455	486	490	476	35	
47		466	471	482	474	473	16	
48		447	454	476	486	466	39	
49		473	488	482	475	480	15	
50		460	450	461	445	454	16	

variability that perhaps 10% of the charge weights are below the 427 g minimum. Such a situation would be: satisfactory average, excessive variability. Also it would be undesirable to have the average charge weight be 447 g, even though, say, no charge weights are outside the limits 427, 481 g. This situation would be: low average, satisfactory variability. Thus, we will need to pay close attention to both the average and the variability whenever the quality characteristic is measurable.

In Chapter 2, we were mostly concerned with sample characteristics. Thus, for a sample of measurements, x, the ordinary average was called x-bar (the sum of the numbers, Σx, divided by n, the number of them). The typical amount of deviation of the sample x's from their average, $\bar{x}$, was called the "standard deviation", s, given in (2.4). We might also use another measure of sample variability, the range, $R = x_{\max} - x_{\min}$. The latter is only used for quite small samples up to $n = 10$. Thus, we may describe our sample of x's by x-bar and R, or by x bar and s.

But we also must consider the true process characteristics. For the present we shall assume that the process is uniform or steady, that is, in control, even though there is variability in the product. How can this be? What we mean is that the process average and process variability are both constant. By now it must be apparent that we need symbols for these.

We shall call the true process average, if constant, by the small Greek letter mu, μ:

$$\mu = E(x) = \text{expectation or average of } x\text{'s for process} \tag{6.1}$$

The sample averages, x-bar, are estimates of μ when the process is in control, that is, stable. We have already been using the small Greek letter sigma, σ, for process or population standard deviation: (2.4), (2.5), and (2.8). Thus, we naturally use for the population standard deviation of the x's

$$\sigma = \sigma_x = \text{standard deviation of } x \text{ for process or population} \tag{6.2}$$

Now if we know μ and σ for a process, we know quite a bit. For example, if the distribution is reasonably well behaved, that is, with the greatest frequency toward the middle of the range of x values and with symmetrically decreasing frequencies on each side, we can make the following statements:

$\mu - \sigma$ to $\mu + \sigma$ include about $65 - 70\%$ of the x's

$\mu - 3\sigma$ to $\mu + 3\sigma$ include at least 99% of the x's

These statements come from the so-called normal distribution, a widely applicable model (see Section 2.9).

Our objective of control charts for measurements is to determine whether the process is in control, that is, has constant μ and σ. If it is in control, we want to estimate μ and σ and compare them with requirements. If the process is not in control, we want to take steps to get it into control, at a satisfactory combination μ and σ. And then we want to maintain control.

6.2. AN EXAMPLE, *x*-BAR, *R* CHARTS FOR PAST DATA

The data on charge weights of an insecticide dispenser, given in Table 2.1, are here reproduced again along with 25 additional samples in Table 6.1. Also for each sample of four charge weights x, there is given the average x and the range R. These charge weights were obtained by the quality control man upon his return from a short course in statistical quality control. The problem was that inventories showed that the monthly overfill (above the specified average of 454 g) was running at $14,000. That is, the manufacturer was giving away this much over and above the required average. And yet despite this average overfill, some charge weights were running below the minimum of 427 g. Clearly, improvement was desirable.

The quality control man used the first 25 samples, that is, December 13 and 14, as the preliminary run. These are plotted as the first 25 x-bar points and R points of Figure 6.1. After plotting the preliminary points the next step is to calculate the control lines (central line and limits) and draw them on the chart. The case of control here is *analysis of past data*, i.e., there were no available standard values, μ and σ, from which to work. Instead the average of averages, x-double bar, and the average range, R, are used for establishing the control lines

$$\bar{\bar{x}} = \frac{\Sigma\bar{x}}{k} = \text{overall average of } k \text{ sample averages}, x\text{-bar} \tag{6.3}$$

$$\overline{R} = \frac{\Sigma R}{k} = \text{average of } k \text{ sample ranges, } R \tag{6.4}$$

The process level is reflected by x-double bar, whereas R-bar is a measure of process variability, These are taken as central lines, for the respective charts.

$$x\text{-double bar} = \text{central line for } x'\text{s (analysis of past data).} \tag{6.5}$$

$$R\text{-bar} = \text{central line for } R'\text{s (analysis of past data)} \tag{6.6}$$

For control limits, we use these two provisional averages x-double bar and R-bar as follows:

$$\text{Limits}_{\bar{x}} = \bar{\bar{x}} \pm A_2\overline{R} \text{ (analysis of past data)} \tag{6.7}$$

$$\text{Limits}_R = D_3\overline{R}, D_4\overline{R} \text{ (analysis of past data)} \tag{6.8}$$

The quantities A_2, D_3, and D_4 are "control chart constants," calculated mathematically from the normal distribution model, which is discussed in Section 2.9. The quantities depend upon the sample size, n, for the samples (see Table C in the back of the book).

Now for the first 25 samples of Table 6.1, we find

$$x\text{-double bar} = 11,589/25 = 463.6\,\text{g}$$

$$R\text{-bar} = 644/25 = 25.8\,\text{g}$$

Using Table C, we also have from (6.7), (6.8) and Table C for the row $n = 4$:

$$\text{Limits}_{x\text{-bar}} = 463.6 \pm .729(25.8) = 463.6 \pm 18.8$$
$$= 444.8, 482.4\,\text{g}$$

$$\text{Upper control limit}_R = D_4\ R\text{-bar} = 2.282(25.8) = 58.9\,\text{g}$$

$$\text{Lower control limit}_R = D_3\ R\text{-bar} = 0(25.8) = 0\,\text{g}$$

(The lower control limit for ranges, while calculated at zero, is really meaningless, just as we had seen for the lower limit for some p, np, and c charts. Only if n exceeds 6, do we have a real lower control limit for R' s.)

Upon drawing in these control lines, we find two x-bar points were out of the control band—one high, one low. Also there is one very high range, almost on the control limit, indicating excessive variability within sample 21. Investigations were begun at once on the conditions under which these three samples were produced. Also the control lines were extended into the next week' s data and watch kept after each sample x-bar and R were plotted. Three more x-bar points below the lower control limit were found. Such points give real concern as to whether the lower specification, 427 g, was being met. In fact, even among the few charge weights of these samples there were two below 427 g. Some improvement in variability seems to have occurred because in the second 25 R's only five are above the central line.

After the first 50 samples, the control lines were revised for subsequent data analysis as follows:

$$\bar{\bar{x}} = \frac{23,066}{50} = 461.3$$

$$R = \frac{1,091}{50} = 21.8$$

$$\text{Limits}_{\bar{x}} = 461.3 \pm .729(21.8) = 445.4, 477.2$$

$$\text{Upper control limit}_R = 2.282(21.8) = 49.7$$

These control lines would be extended.

Some of the assignable causes which were actually found were obstructions in the tubes, nonconforming cutoff mechanisms and poor control in timing and pressure. Within a month, the rate of loss from overfill per month was down to about $12,000 and inside of three months it was down to about $2000 per month, along with much improved capability of meeting of the minimum specification.

Some readers may be concerned about the two samples of $n = 3$ which we treated as though $n = 4$. Some variation of treatment is theoretically called for but there were only two such samples and the control chart constants are not radically different; so the easy approximate approach was used.

6.3. AN EXPERIMENTAL EXAMPLE, *x*-BAR AND *R* CHARTS FOR PAST DATA

Let us now conduct an experiment to illustrate x-bar and R charts by drawing numbered beads from a bowl. We shall try to estimate the contents of the bowl from our sampling data of x-bar and R. Later on the exact contents will be provided.

For the preliminary run of 25 samples, each of $n = 5$, in Table 6.2, we first plot the x's and R's as shown in Figure 6.2. We note one seemingly extremely low x, for sample 9, at -2.6. In practice, is this a reliable indication of an assignable cause? The R's appear to be rather homogeneous. Let us find the control lines by (6.5)–(6.8)

$$\text{Central line}_{\bar{x}} = \frac{\Sigma\bar{x}}{25} = \frac{+0.6}{25} = +0.02$$

$$\text{Central line}_R = \frac{\Sigma R}{25} = \frac{94}{25} = 3.76$$

Table 6.2 Record Sheet for Measurements
Material or Product: piston Rings
Characteristic Measured: Edge Width
Unit of Measurement: in 0.0001 from 0.2050 in.
Specifications: 0.2050 ± 0.0005 in.
Plant: No. 1
Data Recorded by: I. W. Burr

Series number	Date (and hour?) produced	Measurements of each of five itmes in series *a*	*b*	*c*	*d*	*e*	Average for series	Range for series	Standard deviation
1		+1	0	−2	−1	0	−0.4	3	1.14
2		+1	+1	+3	0	−1	+0.8	4	1.48
3		+1	+2	−1	+4	+2	+1.6	5	1.82
4		−1	−2	+1	−2	−2	−1.2	3	1.30
5		0	0	+3	0	0	+0.6	3	1.34
6		+2	−1	+1	−1	−1	0	3	1.41
7		−1	0	−1	+1	−2	−0.6	3	1.14
8		+1	−1	−1	−1	−2	−0.8	3	1.10
9		−4	−3	0	−3	−3	−2.6	4	1.52
10		+1	−1	−1	−2	+2	−0.2	4	1.64
11		−2	0	+1	−1	+3	+0.2	5	1.92
12		0	+2	+2	+1	−2	+0.6	4	1.67
13		−2	−2	0	−3	0	−1.4	3	1.34
14		−2	−2	0	+3	+3	+0.4	5	2.51
15		+1	+1	0	+2	0	+0.8	2	0.84
16		0	0	+2	−2	+3	+0.6	5	1.95
17		0	0	−3	0	+1	−0.4	4	1.52
18		0	+1	+2	+2	+1	+1.2	2	0.84
19		−1	+4	+1	0	+1	+1.0	5	1.87
20		0	0	+2	0	0	+0.4	2	0.89
21		−1	−3	−1	−1	−4	−2.0	3	1.41
22		0	+3	0	−3	+1	+0.2	6	2.17
23		+5	+1	−2	0	0	+0.8	7	2.59
24		0	+1	0	+1	+3	+1.0	3	1.22
25		0	+2	−1	0	−1	0	3	1.22
26		−1	+1	+3	−1	+2	+0.8	4	1.79
27		0	0	0	0	+1	+0.2	1	0.45
28		+2	0	+3	+1	+1	+1.4	3	1.14
29		+1	−1	+1	−1	−3	−0.6	4	1.67
30		+1	+1	−3	+2	+2	+0.6	5	2.07
31		+2	+2	−2	−1	+1	+0.4	4	1.82
32		−1	+2	−1	+1	+2	+0.6	3	1.52
33		+1	+2	−1	+3	0	+1.0	4	1.58

(Continued)

Table 6.2 *(Continued)*

Series number	Date (and hour?) produced	Measurements of each of five itmes in series *a*	*b*	*c*	*d*	*e*	Average for series	Range for series	Standard deviation
34		−1	−1	0	−1	−2	−1.0	2	0.71
35		0	0	−3	+2	+2	+0.2	5	2.05
36		+3	+4	−1	+2	0	+1.6	5	2.17
37		+2	+1	0	0	0	+0.6	2	0.89
38		+2	+3	−2	−1	0	+0.4	5	2.07
39		−1	+1	−1	−2	0	−0.6	3	1.14
40		+1	+4	−4	+2	−4	−0.2	8	3.63
41		+1	+1	−2	0	−2	−0.4	3	1.52
42		0	−5	0	+5	−1	−0.2	10	3.56
43		−2	0	+3	+1	0	+0.4	5	1.82
44		−1	−1	0	0	0	−0.4	1	0.55
45		−2	−1	+1	0	+1	−0.2	3	1.30
46		+2	+2	−3	−1	+1	+0.2	5	2.17
47		−1	+1	0	−1	−1	−0.4	2	0.89
48		0	0	−2	0	0	−0.4	2	0.89
49		0	+3	0	0	−2	+0.2	5	1.17
50		+1	0	0	−1	−1	−0.2	2	0.84

$$\text{Limits}_{\bar{x}} = \bar{\bar{x}} \pm A_2\overline{R} = +0.02 \pm 0.577(3.76)$$

$$= +0.02 \pm 2.17 = -2.15 + 2.19$$

$$\text{Upper limit}_R = D_4\overline{R} = 2.115(3.76) = 7.95$$

$$\text{Lower Limit}_R = D_3\overline{R} = 0(3.76) = \underline{}$$

Drawing in the control lines, we find for sample 9 that x-bar is below the lower control limit. Therefore, in practical application, we would examine the conditions surrounding sample 9 for some assignable cause of off-level performance (although such a point can possibly occur through an increase in the process variability). With the grinding of piston rings by disc grinders, such a low x-bar could be caused by softer stock, thinner piston rings from a previous grind, stowage of

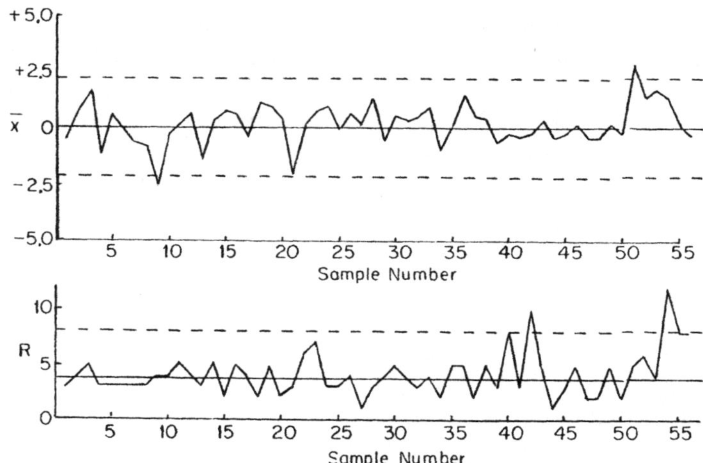

Figure 6.2 Averages, x-bar, and ranges, R, from Table 6.2. Experimental data interpreted as though edge-widths of piston rings were subject to specifications of 0.2050 in. Samples 1–50 were from $\mu = 0$, $\sigma = 1.715$; 51–53 from $\mu = +2$, $\sigma = 1.715$; and 54–56 from $\mu = 0$, $\sigma = 3.47$. Analyzed as past data, preliminary run, 1–25. Control lines revised after sample 50 and extended.

feed rate, or by some temporary excessive pressure upon the grinding discs. Since in our experiment all of the numbered beads were being chosen randomly one by one with replacement after each, this excessively low x-bar is simply a rare accident of sampling. (The authors make no apology for such a point occurring. We plunge into such a sampling experiment never knowing what will occur and have only probability to guide us, much like the great 19th century preacher, Henry Ward Beecher, who "plunged into every sentence trusting God almighty to get me out of it.").

Not having found the assignable cause (which we cannot find in this random sampling experiment), we extend the limits already found, and plot the new sample results x-bar and R as they come. The extended limits are shown in Figure 6.2. On sample 40, we find an R on the upper control limit, a cause for action to look for something which might give greater variability. (A change in average *level* would not increase the process *variability*. Loosening of disc grinders might increase variability.) Then on sample 42 there is a very high R point. Again the cause is sought. But we know that it is here simply a very rare accident in this sampling experiment.

After the first 50 samples, we again calculate control lines. If, as here, we have not found the cause for out-of-control points, we use all of the 50 samples

$$\text{Central line}_{\bar{x}} = \frac{+4.6}{50} = +0.09$$

$$\text{Central line}_{R} = \frac{190}{50} = 3.80$$

$$\text{Limits}_{\bar{x}} = +0.09 \pm 0.577(3.80) = +0.09 \pm 2.19$$
$$= -2.10 + 2.28$$

$$\text{Upper limit}_{R} = D_4\overline{R} = 2.115(3.80) = 8.04$$

$$\text{Lower limit}_{R} = D_3\overline{R} = 0(3.80) = \_\_$$

Can we at this point estimate what the population in the bowl was like? Yes, it can be estimated rather tentatively, because control is not perfect. However, it was not so very bad, even though in practice every x-bar and R point outside the limits would be studied.

In American Society for Quality Control Standard (1969, p. 16), a guide is given as to when in practice we may regard a high degree of control as having been attained: (1) none out of 25 successive points, (2) not more than 1 out of 35, or (3) not more than 2 out of 100, fall outside 3σ limits. Here, despite one x-bar point and two R points out of control limits, control is quite good. Nevertheless, even just one point outside the control band is worth investigating.

Let us now use x-double bar and R-bar to make estimates of the population or process distribution; that is, we wish to estimate the population average μ and standard deviation, σ.

x-double bar estimates μ if reasonable control shown (6.9)

R-bar divided by d_2 estimates σ if reasonable control shown (6.10) where d_2 is another control chart constant (see Table C). Using these we have the estimates

$$\mu = x\text{-double bar} = +0.09$$

$$\sigma = R\text{-bar}/d_2 = 3.80/2.326 = 1.63$$

Thus, we can expect at least 99% of the measurements, x, to lie within the limits

$$\mu \pm 3\sigma = +0.09 \pm 3(1.63) = -4.80 + 4.98$$

Therefore, we would seem to be just capable of meeting specifications of -5 to $+5$.

In practice, we would extend the limits found from the 50 samples and plot each x-bar and R as they arise, watching for evidence of assignable causes.

Now let us change the distribution of numbers in the bowl and plot x-bar and R for each sample and see how soon we obtain a warning of a change having occurred. Sample 51 is

$$+1, +1, +6, +2, +5 \quad x\text{-bar} = +3.0, \quad R = 5$$

Plotting these sample results on Figure 6.2, we find the x-bar point to lie well above the upper control limit, whereas the R point is in control. Thus, the sample shows evidence of an assignable cause, that is, something has changed the distribution of individual x's for the process. Note also the $+6$, which is out of specifications. Since x-bar went out, we expect it to be a result of a shift upward in the process mean μ (although it could be from a sizable increase in the process standard deviation σ, even with no change in μ). It would seem that the process level should be reset. But by how much? A sample of 5 is rather too few to determine how far off μ is. If one makes a practice of resetting or adjusting a process on too few measurements they will only tend to increase the variability of which the process is capable. Thus, we take two more samples

52	+1	+2	+2	+4	−2	x-bar = +1.4, $R = 6$
53	+1	+4	0	0	+4	x-bar = +1.8, $R = 4$

These x-bar's are also high though not beyond the control limit. We average the three x-bars's:

$(+3.0 + 1.4 + 1.8)/3 = +2.07$

Thus, we adjust the process level down by 2. Note that all three ranges are well inside the limit, and so we are confident that the assignable cause affected μ and not σ.

Next the bowl distribution was again changed. The first sample follows:

54 $-1+2-4+2+8$ x-bar $= +1.4$, $R = 12$

The x-bar point is in control, but the range is way above the upper control limit, signaling that something has caused an increase in σ. So steps are taken to find the assignable cause. Causes that increase σ are of major importance, in general, and their correction deserves high priority. Two more samples were drawn to further illustrate the change:

55	-4	$+1$	$+4$	$+1$	-1	x-bar $= +0.2$, $R = 8$
56	$+5$	$+1$	-3	-3	-1	x-bar $= -0.2$, $R = 8$

Both of these ranges are at the upper control limit. Note also the $-+8$ in sample 54, which was out of specifications.

It is desirable to mention that in our experiment, in each case when we changed the bowl, the very first sample gave a signal by a point outside of the control band. This does not always occur. For example, we might just as well have found samples 52 and 53 before finding one like 51. But the stronger the assignable cause is, the more quickly we are likely to obtain a warning signal.

6.4. SOME POPULATION DISTRIBUTIONS FOR SAMPLING EXPERIMENTS

Sampling experiments, with samples drawn from a bowl containing a population, can be a great aid in getting the "feel" of sampling results. This enables the student or reader to experiment as much as he wishes, until he becomes familiar with the workings of chance and gains confidence.

We list in Table 6.3 some convenient distributions for sampling experiments. These are the same approximately normal distributions as were extensively used in the famous

Table 6.3 Approximately "Normal" Population for Sampling Experiments or we can code as in (2.8) but with the modification as in the last of (6.11)

Number	Populations						
x	A	B	C	D	E	F	G
+11				1			
+10			1	1			
+9			1	1	1		
+8			1	3	3		
+7		1	3	5	10		
+6		3	5	8	23		
+5	1	10	8	12	39		
+4	3	23	12	16	48		
+3	10	39	16	20	39	1	
+2	23	48	20	22	23	3	
+1	39	39	22	23	10	10	1
0	48	23	23	22	3	23	3
−1	39	10	22	20	1	39	10
−2	23	3	20	16		48	23
−3	10	1	16	12		39	39
−4	3		12	8		23	48
−5	1		8	5		10	39
−6			5	3		3	23
−7			3	1		1	10
−8			1	1			3
−9			1	1			1
−10			1				
N	200	200	201	201	200	200	200
μ	0	+2	0	+1	+4	−2	−4
σ	1.715	1.715	3.47	3.47	1.715	1.715	1.715

War Production Board courses in quality control by statistical methods (Working and Olds, 1944). Such distributions can be marked on fiber discs or chips. Such pieces can sometimes be obtained free from stamping holes in fiber sheets.

The smaller the diameter and the thicker the fiber, the more easily they may be mixed. Different colored inks can be used to distinguish the different distributions, as can circling and underlining. Or the chips may be dyed different colors. The author uses different colored 1-cm-diameter beads, punched with metal punches and marked with India ink. Plus and minus signs can be filed out of a large nai1, which is then heated and quenched to harden the metal. Alternatively, a less elegant method is to use round, metal rimmed key tags, marking first one side with one distribution with a marker, e.g., blue. Randomly lay out the tags face down and mark the blank side with a second distribution on this side in another color, e.g., red. One author has used this method for Distributions A and B (Table 6.3) in seminars and demonstrations for many years and finds that it is easier to maintain the distribution' s integrity. For further description of this see Burr (1993).

Another available, quite normal distribution is obtainable from the total showing on a roll of three dice. It is the following running from 3 to 18

x	3	4	5	6	7	8	9	10	11	12	13	14	15	16	17	18	Total
f	1	3	6	10	15	21	25	27	27	25	21	15	10	6	3	1	216

For this, one may find $\mu = 10.5$ by symmetry, and $\sigma = 2.96$. The average μ may be adjusted by adding on some constant to each x.

Take note that when we have a finite population of numbers and wish to find the *population standard deviation*, we do not use $N - 1$ in the denominator as one might expect in looking at (2.4). Instead we use either of

$$\sigma = \sqrt{\frac{\Sigma(x-\mu)^2 f}{N}} = \sqrt{\frac{N\Sigma(x^2 f) - (\Sigma xf)^2}{N}} \tag{6.11}$$

In Section 6.3, the first 50 samples were drawn from Population A of Table 6.3, having $\mu = 0$ and $\sigma = 1.715$. Note that 100% of this population lies within -5 to $+5$, two of the $N = 200$, being those extremes. Then for samples 51–53, we used Population B having the same $\sigma = 1.715$, but with μ shifted up to $+2$. Finally, for samples 54–56, we used Population C having μ again centered at 0, but with about double the variability, namely $\sigma = 3.47$. When the author obtained the range R of 10, he got interested in how rare an event this was, since in many similar experiments he never saw such a range. Is it not rare to have the *one* +5 and the *one* −5 in the same little sample of $n = 5$ x's? So he calculated it, finding the probability of such a sample to be about 0.0005, or about 1 in 2000!

6.5. CONTROL CHARTS FOR *x*-BAR AND *R*, STANDARDS GIVEN

When a process has shown good control with satisfactory quality performance, then you would usually wish to set standards values of average, μ, and standard deviation, σ. This would be done by use of (6.9) and (6.10). Whenever we have standard values μ and σ, we can use them to set control lines for x-bar and R charts as follows:

$$\text{Central line}_{x\text{-bar}} = \mu \tag{6.12}$$

$$\text{Limits}_{x\text{-bar}} = x\text{-bar} \pm A\sigma \tag{6.13}$$

$$\text{Central line}_R = d_2\sigma \tag{6.14}$$

$$\text{Limits}_R = D_1\sigma, D_2\sigma \tag{6.15}$$

As an example, let us set control lines for x-bar and R for the experimental data of Section 6.3, Table 6.2, which were drawn from Population A of Table 6.3. For this $\mu = 0$, $\sigma = 1.715$, $n = 5$. Thus, using constants in Table A:

$$\text{Central line}_{x\text{-bar}} = 0$$
$$\text{Limits}_{x\text{-bar}} = 0 \pm 1.342(1.715) = \pm 2.30$$
$$\text{Central line}_R = d_2\sigma = 2.326(1.715) = 3.99$$
$$\text{Limits}_R = 0(1.715), 4.918(1.715) = \_\_, 8.43$$

The one x-bar point, sample 9, is still out, but now only one range, $R = 10$, for sample 42, is out of control.

One advantage in control against standards is that we are ready to interpret each sample as soon as it arises and need not wait for a preliminary run of samples.

The reader may have noticed that appropriate formulas for measurement control charts such as (6.12)–(6.15) are given at the bottom of Table C.

6.6. CONTROL CHARTS FOR STANDARD DEVIATIONS, *s*

As we have seen, the variability between the measurements x within a sample may be described by either the range R or the standard deviation s. Up to now we have been using range charts. We now briefly cover control charts for values of s.

Why might we wish to use s charts for studying the variability within samples? There are several reasons:

1. The standard deviation, s, even for small samples, is a slightly more reliable measure than is the range. This is in part because it makes full use of all of the measurements, x, instead of just the two extreme x's. Thus, s makes more complete use of the information in a sample.
2. As the sample size increases above, say, 10, the range loses out rather rapidly in reliability in comparison with the standard deviation. Hence, above 10 we will always use s rather than R.
3. If a computer is being used we might as well use the "best" measure available. (Actually, also it is more difficult to program a computer to find R than to find s).

Let us therefore give the formulas appropriate (which may also be found at the bottom of Table C).

Case (Analysis of past data).

$$\text{Central lines}_s = s\text{-bar} \tag{6.16}$$

$$\text{Limits}_s = B_3 s\text{-bar}, B_4\, s\text{-bar} \tag{6.17}$$

Case (Standard σ given).

$$\text{Central line}_s = c_4\sigma \tag{6.18}$$

$$\text{Limits}_s = B_5\sigma, B_6\sigma \tag{6.19}$$

For illustration take the first 50 s values of the experiment (Table 762, last column). They are plotted in Figure 6.3. Also plotted above for comparison are the corresponding 50 ranges. Note how they fall and rise in the same way as the

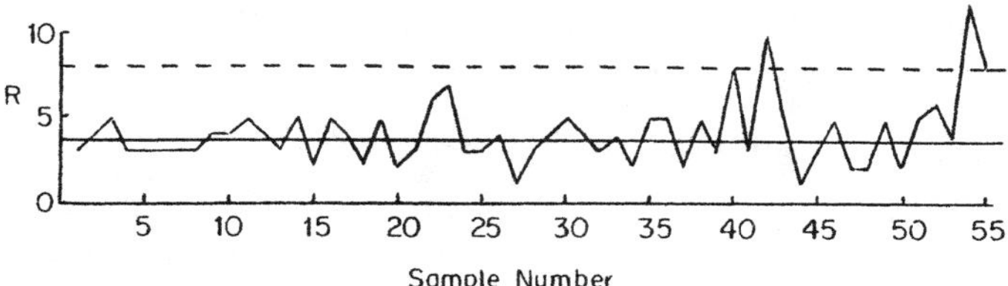

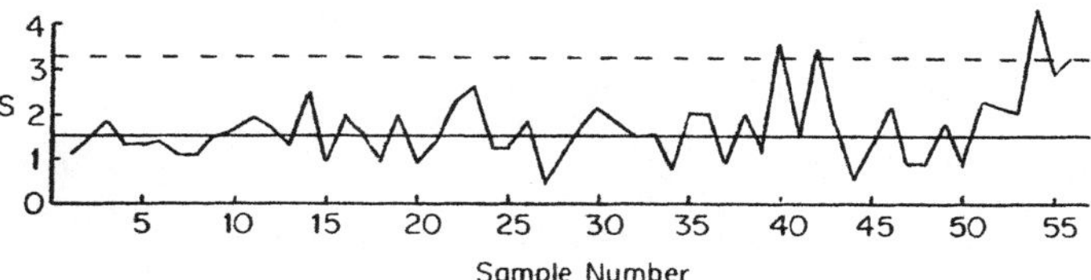

Figure 6.3 Comparison of range, R, and standard deviation, s, charts. Experimental data are from Table 6.2, plus the six other samples for other populations. Note the similarity between the two charts.

standard deviations. We also draw in the control lines from the past data using s-bar and R-bar for the 50 samples. Thus,

$$\begin{aligned} \text{Central line}_s &= s\text{-bar} = 77.78/50 = 1.556 \\ \text{Limits}_s &= B_3\, s\text{-bar}, B_4\, s\text{-bar} \\ &= 0(1.556), 2.089(1.566) = \text{—}, 3.25 \end{aligned}$$

The s values for samples 40 and 42 are out of control, just as the R values were.

Let us also find control limits from the standard value $\sigma = 1.715$, for Population A (Table 6.3) from which the samples were drawn. Using (6.18) and (6.19):

$$\begin{aligned} \text{Central line}_s &= c_4\ \sigma = 0.940(1.715) = 1.61 \\ \text{Limits}_s &= B_5\sigma, B_6\ \sigma \\ &= 0(1.715), 1.964(1.715) = \text{—}, 3.37 \end{aligned}$$

Again the s values for samples 40 and 42 are out of control. So we see that the R and s charts tell the same story as regards within-sample variability.

In analysis of past data, we use R in setting limits for x-bar, via (6.7). Analogous to (6.7) is

$$\text{Limits}_{x\text{-bar}} = x - \text{double bar} \pm A_3\, s\text{-bar}. \qquad (6.20)$$

Using it with s = 1.556, x-double bar = +0.09 we obtain

$$\begin{aligned}\text{Limits}_{x\text{-bar}} &= +0.09 \pm 1.427(1.556)\\ &= +0.09 \pm 2.22 = -2.13, +2.31\end{aligned}$$

while using $R = 3.80$ in (6.7) gives

$$\text{Limits}_{x\text{-bar}} = -2.10, +2.28$$

Surely from a practical standpoint these two sets of limits for x-bar are virtually identical.

A third use of R *or* s-bar is to estimate σ after reasonable control of variability has been demonstrated. Thus, we could use (6.10) or the analogous

s-bar/c_4 estimates σ if reasonable control shown (6.21)

Using (6.21), we obtain

$$\sigma = 1.556/.940 = 1.66$$

whereas by (6.10) the estimate for σ was 1.63. Again R and s-bar did a quite similar job in estimating $u = 1.715$.

Comparison of R' s and s' s from small samples – three uses.

1. The R chart and s chart give practically identical pictures of control of sample variability. When one is in control, the other will be also, and points out of control on one chart will correspond to those out on the other chart.
2. Under analysis of past data, use of R-bar and s-bar lead to practically the same limits on x-bar charts.
3. Estimates of σ, when reasonable control of variability is shown, will be very nearly the same using R and s.

Thus, when working with small samples, say up to $n = 10$, we are at liberty to use either R or s. Take your choice. But, of course, we will not use both R and s. Unless a calculator programmed for s is used, almost everyone uses range charts.

6.7. COMPARISON OF A CONTROL CHART WITH SPECIFICATIONS

It is important to point out immediately in the discussion of control charts for measurements that one must be careful to distinguish between control and capability. Control is the ability of the process to be consistent. Capability is the ability of the process to provide a product that meets the requirements of the customer. When using a measurements control chart with samples of two or more, one must be careful not to fall into the trap of comparing this chart to specification limits. To the uninitiated, it is not obvious that the control chart is a distribution of averages while the specifications are set on the distribution of individuals. Our customers are not purchasing products which are averages of a distribution; rather they are purchasing individual items.

A much expanded discussion of this is in Chapter 7.

6.8. CONTINUING THE CHARTS

In the early stages of a control chart application, we are primarily concerned with letting the process do the talking, telling us how it is doing. That is, we collect the sample data, plot them, calculate the central lines and control limits, draw them in, and interpret the results to all involved. As a consequence of indicated lack of control, we seek out the assignable causes responsible and may possibly find one or more from the preliminary run of data. In any case, we face the problem of continuing the charts. Usually, we merely extend the control lines on both the x-bar and R (or s) charts. There arises the question, however, as to whether to include the data corresponding to points that lay outside the control band. Data produced while assignable causes were operating should be eliminated, and x-double bar and R-bar revised only if *both* of the following are met: (1) The assignable cause for such performance was found, and (2) it was eliminated. Now if the assignable cause behind the unusual performance was not found, or having been found, nothing was done to remove it, then such data are still as typical of the process as any other and should be retained. In fact, whenever any significant change is made in the process, all data produced before the change are no longer typical of the revised process and should be discarded. In line with this too, we should, in general, be periodically revising the control lines from only fairly recent data.

6.9. WHEN AND HOW TO SET STANDARD VALUES

Standard values of μ and σ are commonly only set after good control has been achieved and the process is producing a satisfactorily high percentage of product within the specification limits for x's. In this desirable situation we set standard values by

$$x\text{-double bar becomes standard } \mu \tag{6.23}$$

$$R\text{-bar}/d_2 \text{ becomes } \sigma \tag{6.24}$$

or

$$s\text{-bar}/c_4 \text{ becomes standard } \sigma \tag{6.25}$$

Such standard μ and σ must be such that the natural x limits

$$\mu \pm 3\sigma \text{ lie on or inside specification limits} \tag{6.26}$$

Then at least 99.7% of x values will be meeting specifications as long as control is maintained.

There is an intermediate situation which is practical. It is useful in processes where the process mean is relatively easily adjusted. Then if there are two specifications L and U for x, we need the σ set from (6.24) or (6.25) to be small enough so that

$$6\sigma \leq U - L$$

Thus, σ is set from data from the process. But we might set σ, not from x-double bar, but instead at some desirable level.

Thus if 6σ is close to the tolerance $U-L$, we might well take μ at the nominal (middle of the specification range) $(U+L)/2$. Or if σ is small enough so that the natural process spread 6σ is considerably smaller than the tolerance $U-L$, then we might wish to place μ at a point 3σ above L, or at 3σ below U. These latter alternatives may be used also when there is only one limit, either L or U, but not both. This may enable saving on material, fill weight, or machining or grinding time. Or it may be useful as a starting process setting in machining outside diameters, say. Here the aim would be to start a run at a process average level $L+3\sigma$, and then as the cutting tool wears and the process average increases, end the run when it reaches $U-L$. See Figure 6.6. for illustration of cases.

We now use the standard values of μ and σ to determine control lines against which to compare future sample data x-bar and R or s, having confidence that as long as control is maintained, we are meeting specifications.

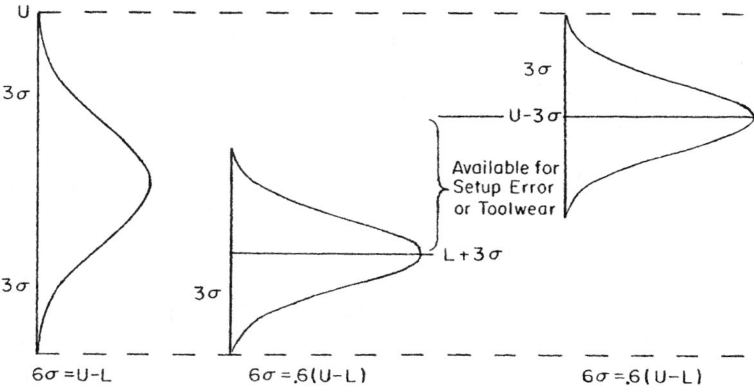

Figure 6.6 Natural process spread $6\sigma_x$, in relation to tolerance $U-L$ and desired process averages. In the first section, $6\sigma_x$ fills the tolerance, and the process average must be close to the nominal $(U+L)/2$ in order to be safely away from the specification limits. For the second section $6\sigma_x = 0.6\ (U-L)$. The lower distribution has its average at $L+3\sigma_x$ and thus almost no pieces will lie below L. The upper distribution has its average at $U-3\sigma_x$, and thus almost no pieces will lie above U. Thus, when $6\sigma_x = 0.6\,(U-L)$, there is available considerable room for set-up error and/or tool wear, in fact. $4(U-L)$.

6.10. EXAMPLES

Example 1. This example is on the *yield* of steel from an open hearth furnace, that is, tons of steel divided by tons of metallics charged. Yield improvement studies frequently disclose evidence of careless practices. In such cases, it is desirable to have operators and supervision have a graphical picture of their progress. Once such careless practices are eliminated by intensive study and action, the tendency is for vigilance to relax and the gains thus made to be lost. A constant check must be maintained to ensure maximum results and to show when undesirable factors are creeping in.

In the present example, a typical one, period A is a 20-day history period, giving yields for each of the three daily heats of steel, some 200 tons each. Period B gives typical data from a period 3 weeks after control charts became a subject for daily discussion at operating meetings. Period C follows period B by 60 days.

For Period A, the initial history period, we find (see Table 6.4)

$$x\text{ - double bar} = 93.8 \text{ and } R\text{-bar} = 3.22$$

and we use $n = 3$ in Table C to find $A_2 = 1.023$ and $D_3 = 0$, $D_4 = 2.575$ and thus:

$$\text{Limits}_{x\text{-bar}} = 93.8 \pm 1.023\ (3.22) = 93.8 \pm 3.3 = 90.5, 97.1$$

$$\text{Limits}_R = 0(3.22), 2.575(3.22) = \_\_, 8.29$$

Plotting the x-bar and R points and drawing in the control lines we have the first part of Figure 6.7. Day 6 has an x-bar right on the lower control limit, from two yields in the 80s, while the 14th day has an x-bar well below the limit, as well as an R out of control. These are a result of one very low yield of 83.0. It would seem that this yield may have acted as a stimulus, because the next 6 days show some improvement.

The 20 days of Period B give

x-double bar $= 94.8$ Limits$_{x\text{-bar}} = 92.3,\ 97.3$
R-bar $= 2.48$ UCL$_R = 6.39$

Table 6.4 Three Periods of a Yield Study on an Open Hearth Furnace, Making Steel, Three Heats per Day. Period A Initial History; period B, 3 Weeks Later; Period C, 60 Days Later

	Period A					Period B		Period C	
	% yield, 3 Heats								
Day	x_1	x_2	x_3	Average, $\bar{x}$	Range, R	Average, $\bar{x}$	Range, R	Average, $\bar{x}$	Range, R
1	95.6	94.8	94.8	95.1	.8	95.0	3.8	94.7	1.0
2	94.5	91.0	96.8	94.1	5.8	95.9	.3	95.3	1.0
3	96.1	97.3	93.4	95.6	3.9	92.4	7.4	95.2	1.9
4	95.0	94.0	94.7	94.6	1.0	95.3	5.2	95.2	.3
5	94.0	94.8	93.7	94.2	1.1	95.9	2.1	95.6	.6
6	88.5	94.0	89.2	90.6	5.5	95.2	2.7	96.2	.6
7	91.3	94.2	91.6	92.4	2.9	96.0	2.4	96.4	.8
8	96.7	94.8	93.4	95.0	3.3	95.9	1.2	94.4	2.6
9	90.5	91.1	94.7	92.1	4.2	94.8	.5	96.0	.7
10	90.3	94.0	94.6	93.0	4.3	95.8	1.4	95.2	2.1
11	91.7	91.6	95.9	93.1	4.3	93.7	3.4	94.2	6.4
12	93.2	95.7	93.3	94.1	2.5	95.0	1.9	96.3	1.3
13	90.7	95.0	93.7	93.1	4.3	95.2	.4	94.3	2.1
14	92.4	92.6	83.0	89.3	9.6	95.6	1.0	97.1	.8
15	96.0	95.9	95.7	95.9	.3	92.0	4.1	95.5	2.0
16	95.9	95.7	95.5	95.7	.4	94.5	2.9	96.9	.2
17	96.2	93.6	93.6	94.5	2.6	94.8	3.2	96.1	1.2
18	95.6	94.6	93.0	94.4	2.6	95.6	.4	95.4	2.2
19	94.3	94.0	94.6	94.3	.6	95.0	.5	95.3	1.5
20	96.8	92.3	95.5	94.9	4.5	92.1	4.7	96.6	.5
				1876.0	64.5	1895.7	49.5	1911.9	29.8
				$\bar{\bar{x}} = 93.8$	$\bar{R} = 3.22$	$\bar{\bar{x}} = 94.8$	$\bar{R} = 2.48$	$\bar{\bar{x}} = 95.6$	$\bar{R} = 1.49$
			UCL	97.1	8.29	97.3	6.39	94.1	3.84
			LCL	90.5	–	92.3	–	97.1	–

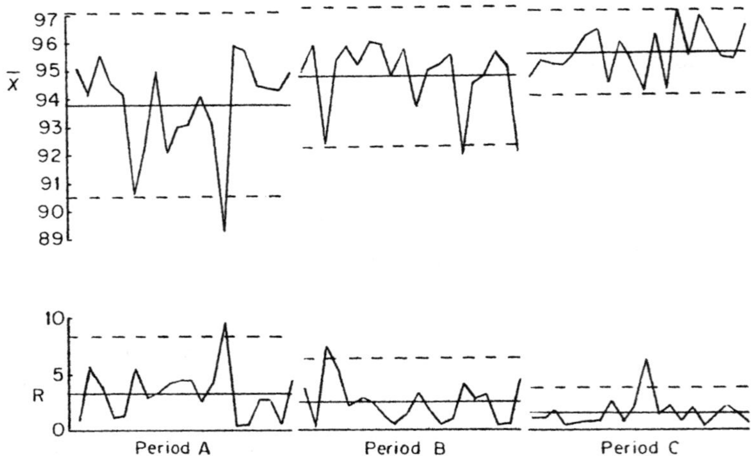

Figure 6.7 Three periods of percent yield for an open hearth furnace, 3 heats/day. Period A represents the initial history; period B, 3 weeks later; and period C, 60 days later. Note increasing average and decreasing variability over the 3 periods. One percent improvement in yield means a gain of about 2 tons of usable steel per heat. Data in Table 6.4.

Thus, there has been an increased yield and more consistency (lower ranges) than in Period A. Day 3 has its x-bar practically on the lower limit and day 15 has x-bar definitely below. The former contained an 88.7 yield, and the latter had three yields all rather low, including a 90.1. Also the range for the third day was high.

For Period C, the 20 days gave

x-double bar = 95.6 Limits$_{x\text{-bar}}$ = 94.1, 97.1
R-bar = 1.49 UCL$_R$ = 3.84

Therefore, there has been a further increase in yield, and an improvement again in consistency. There are two x-bar's practically on the lower control limit and a high range on day 11 mostly because of a low yield of 91.1.

The three periods give a most encouraging picture of improvement. The gain from A to C is 1.8%, or 3.6 tons more steel for each heat, on the average, from the same amount of metallics used. Continuing occasional evidence of lack of control gives hope of still further improvement. Continuation of the charts can help this and also act as a follow-through monitor.

Example 2. In Table 6.5, there are given measurements x, x-bars, and R' s, for two periods of a machining process. Let us consider the first period, with points as plotted in Figure 6.8.

x-double bar = 0.19219; R-bar = 0.000455; $\text{limits}_{x\text{-bar}} =$ 0.19193, 0.19245; $\text{UCL}_R = 0.00096$

For the 20 samples, we have

$$x\text{-double bar} = 0.19238$$

$$R\text{-bar} = 0.000825$$

$$\text{Limits}_{x\text{-bar}} = 0.19238 \pm 0.577(0.000825) = 0.19190, 0.19286$$

$$\text{UCL}_R = 2.115(0.000825) = 0.00174 \text{ (all in inches)}$$

Drawing in the control lines, we find only one range out of control, in fact way outside. But the x-bar chart is in very poor control, although shift one is in much better control than shift two. Obviously assignable causes are at work, affecting the level. This might have been a result of poor set-up, or set-up changes made on insufficient evidence. But there were also causes affecting the process level μ.

The question arises as to whether specifications were being met. A frequency tabulation of the 100 diameters shows none outside of the specifications limits of 0.1900 to 0.1940 in. However, measurements tended to concentrate on whole and half thousandths, such as, 0.1930, 0.1935, and 0.1940, and thus the measuring was apparently not done fully to the nearest 0.0001 in. Moreover, there were eight at 0.1940 in., some of which might well have been above 0.1940 in. if more accurately measured. Nevertheless, we might regard this as a case of situation 2: specifications are met fairly well but the process is out of control. So effort was made to improve the process, for greater security and assurance.

Table 6.5 Production Data on Outside Diameter of a Brass Low- Speed Plug, Blue Print Specifications. 1900, 1940 in

	x_1	x_2	x_3	x_4	x_5	x-bar	R	Comments
October 7								
11:30 AM	0.1920	0.1930	0.1921	0.1923	0.1923	0.19234	0.0010	
12:45 PM	0.1916	0.1925	0.1926	0.1928	0.1925	0.19240	0.0012	
1:30 PM	0.1927	0.1927	0.1923	0.1913	0.1920	0.19220	0.0014	(a)
1:45 PM	0.1924	0.1922	0.1922	0.1927	0.1924	0.19238	0.0005	
2:15 PM	0.1919	0.1917	0.1918	0.1924	0.1916	0.19188	0.0008	(b)
2:55 PM	0.1918	0.1917	0.1924	0.1919	0.1917	0.19190	0.0007	
3:15 PM	0.1936	0.1937	0.1932	0.1930	0.1938	0.19346	0.0008	(c)
4:23 PM	0.1925	0.1930	0.1930	0.1925	0.1925	0.19270	0.0005	
4:35 PM	0.1930	0.1927	0.1915	0.1902	0.1910	0.19168	0.0028	
5:08 PM	0.1920	0.1922	0.1920	0.1917	0.1910	0.19178	0.0012	
5:25 PM	0.1925	0.1925	0.1920	0.1925	0.1915	0.19220	0.0010	
5:45 PM	0.1930	0.1930	0.1935	0.1930	0.1935	0.19320	0.0005	
6:05 PM	0.1940	0.1935	0.1940	0.1940	0.1940	0.19390	0.0005	(b)
7:40 PM	0.1915	0.1905	0.1915	0.1915	0.1915	0.19130	0.0010	(c)
8:13 PM	0.1940	0.1940	0.1940	0.1935	0.1940	0.19390	0.0005	(b)
8:35 FM	0.1930	0.1930	0.1930	0.1931	0.1929	0.19300	0.0002	
9:02 PM	0.1915	0.1915	0.1920	0.1915	0.1920	0.19170	0.0005	
9:25 PM	0.1915	0.1915	0.1915	0.1915	0.1919	0.19158	0.0004	
9:50 PM	0.1920	0.1915	0.1920	0.1920	0.1920	0.19190	0.0005	
10:16 PM	0.1925	0.1921	0.1920	0.1925	0.1920	0.19222	0.0005	
						3.84762	0.0165	

x-double bar = 0.19238; R-bar = 0.000825; $\text{Limits}_{x\text{-bar}}$ = 0.19190, 0.19286; UCL_R = 0.00174

	x_1	x_2	x_3	x_4	x_5	x-bar	R	Comments
December 10								
11:30 AM	0.1920	0.1921	0.1917	0.1920	0.1920	0.19196	0.0004	
12:20 PM	0.1921	0.1916	0.1921	0.1918	0.1920	0.19192	0.0005	
12:45 PM	0.1921	0.1922	0.1920	0.1924	0.1922	0.19218	0.0004	
1.05 PM	0.1920	0.920	0.1919	0.1921	0.1918	0.19196	0.0003	
1:20 PM	0.1920	0.1916	0.1920	0.1917	0.1916	0.19178	0.0004	
1:43 PM	0.1917	0.1923	0.1921	0.1923	0.1922	0.19212	0.0006	
2:10 PM	0.1917	0.1919	0.1918	0.1918	0.1917	0.19178	0.0002	
2:30 PM	0.1918	0.1922	0.1920	0.1917	0.1918	0.19190	0.0005	
2:48 PM	0.1918	0.1922	0.1918	0.1918	0.1920	0.19192	0.0004	
3:05 PM	0.1922	0.1921	0.1922	0.1923	0.1921	0.19218	0.0002	
2:28 AM	0.1923	0.1926	0.1926	0.1929	0.1929	0.19266	0.0006	
2:50 AM	0.1922	0.1926	0.1924	0.1928	0.1928	0.19256	0.0006	
3:05 AM	0.1924	0.1927	0.1926	0.1931	0.1928	0.19272	0.0007	

(*Continued*)

Table 6.5 (*Continued*)

	x_1	x_2	x_3	x_4	x_5	x-bar	R	Comments
3:25 AM	0.1928	0.1927	0.1928	0.1924	0.1930	0.19274	0.0006	
3:45 AM	0.1923	0.1928	0.1925	0.1926	0.1923	0.19250	0.0005	
4:10 AM	0.1924	0.1924	0.1928	0.1927	0.1928	0.19262	0.0004	
4:30 AM	0.1920	0.1918	0.1920	0.1919	0.1920	0.19194	0.0002	
4:48 AM	0.1923	0.1920	0.1921	0.1918	0.1919	0.19202	0.0005	
5:05 AM	0.1923	0.1918	0.1919	0.1920	0.1921	0.19202	0.0005	
5:25 AM	0.1921	0.1919	0.1924	0.1924	0.1925	0.19226	0.0006	
						3.84374	0.0091	

Comments: (a) out of stock; (b) adjustment made; (c) sharpened tool.

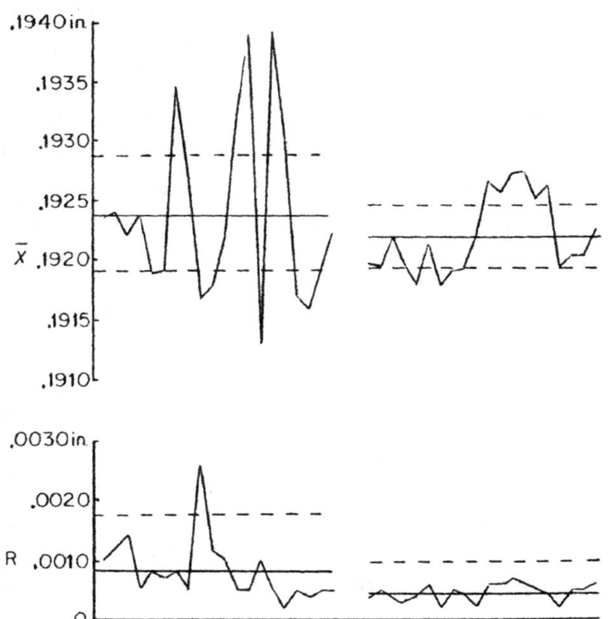

Figure 6.8 Control charts for average $\bar{x}$ and range R for the outside diameter of a low-speed plug. Data are from Table 6.5. Specifications are 0.1900 to 0.1940 in. for periods before and after action taken. In second period, $\bar{x}$ values are still out of control, but variability is much improved, so that now specifications are being safely met.

Actions taken were (1) developing a new cam and tool layout, (2) changing the sequence of operations, and (3) adding a steady rest and skiving tool and holder.

The second set of data was taken 3 months later. First, note the range chart. R-bar is about one-half of that for the earlier period showing much more uniformity around the current level, whatever it is. Moreover, the R chart shows perfect control. The x-bar points for December 10 would have been in control relative to the x-bar control limits for October 7. But because R-bar is so much smaller, the x-bar limits from the present data show much lack of control. The first shift is well centered on the nominal 0.1920 in., but some x-bar's are below the lower limit because x-double bar includes the high x-bar's of the second shift. The operator on the second shift must have changed the process setting on insufficient evidence, in distrust of the first shift operation. One often sees such intershift rivalry.

Now for maximum security against the specifications, the plant could have used the nominal of 0.1920 in. as the average and limits, around this standard of $\pm A_2R$. That is

$$\begin{aligned}\mu \pm A_2R\text{-bar} &= 0.1920 \pm 0.577(0.000455)\\ &= 0.19200 \pm 0.00026 = 0.19174, 0.19226\end{aligned}$$

Here we have used a desired *standard* μ and used *past data* to set the width of the control band, a mixture of the two control chart purposes. All of the x-bar's for shift one lie inside these limits showing good adherence to the nominal. The first 6 x-bar's of the second shift are all outside this control band; the last 4 x-bar's are inside. This is a problem of set-ups rather than an assignable cause affecting the mechanics of the process.

Now is this so very bad? How about specifications? Is the process meeting specifications, even when at the high level of the first 6 points of the second shift? Let us see. Whenever the process average is at some level μ, within what limits will the individual x's lie? This requires us to know or else estimate σ for the x's, as we have seen in Section 6.7. Here, using

$d_2 = 2.326$ from Table C:

$$\sigma_x = R\text{-bar}/d_2 = 0.000455/2.326 = 0.000196$$

So the natural limits of individual x's around any process average μ are:

$$\mu \pm 3\sigma_x = \mu \pm 3(0.000196) = \mu \pm 0.00059$$

Now does μ have to be right at the nominal 0.1920 in. in order to avoid trouble with the specification limits 0.1900, 0.1940 in.? No (see Figure 6.9). How low and how high could μ be and still have us meeting the limits? If μ is as high as

$$\text{Max}\,\mu = \text{U} - 3\sigma_x = 0.19400 - 0.00059 = 0.19341\,\text{in.}$$

or as low as

$$\text{Min}\,\mu = L + 3\sigma_x = 0.19000 + 0.00059 = 0.19059\,\text{in.}$$

We are still meeting specification limits. Even for the first 6 points of the second shift, the process average, μ, seems to be only about 0.19265 in., and so we are safely meeting the

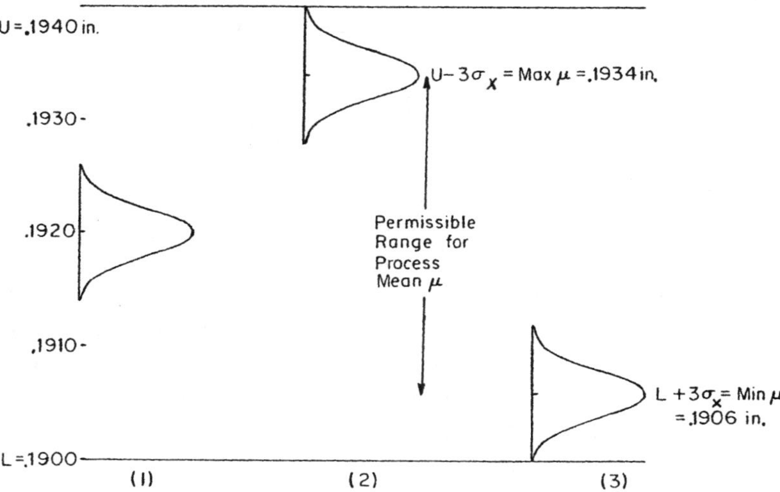

Figure 6.9 Distributions for individual x's for diameter of low-speed plug. (1) Distribution centered, the ideal. (2) The highest safe distribution. (3) The lowest safe distribution. Shows the permissible variation for the process mean μ. σ_x was estimated from $\bar{R}$ by $\bar{R}/d_2$.

specification limits for x's, since the R chart shows perfect control. (If it were not in control, we could not be so sure.) Thus if desirable, we could simply aim to try to maintain the true process average somewhere between 0.19059 and 0.19341. Such an approach does give up, however, on the program of looking for assignable causes and merely concerns itself with checking the process setting. It is up to management as to whether to permit μ to vary between such limits, or to maintain somewhat tighter control.

6.11. SOME BACKGROUND OF CONTROL CHARTS

In all of the control charts described in Chapter 5 and the present chapter, we have been making use of theory of sampling from distributional models. These models are the binomial distribution for nonconforming units, d, and fraction nonconforming, p, the Poisson distribution for nonconformities, c, and nonconformities per unit, u, and the normal distribution for measurements, x. To find the behavior of sample characteristics such as d, c, x-bar, R, or s, draw a large number of random samples from the assumed population distribution, which might have been placed on chips or beads in a bowl. We have seen a *small* run of such sample results from each of the three distribution models. By taking a *large* number of samples and tabulating the results, we would have a good picture of the way the sample characteristic in question behaves, i.e., the sampling distribution.

However, there are obvious limitations to such an experimental approach, even if we use a computer. Fortunately, available theory enables us to determine the exact sampling distributions. In fact for d and c, they are, respectively, the binomial and Poisson distributions for the samples. For drawings from a normal distribution, theory provides the exact sampling distributions of x, R, and s.

Now in all these cases, the exact sampling distributions are fully determined

average $\pm$ 3 (standard deviation)

for the characteristic will contain virtually all of them. Hence if we find a sample value outside such 3σ limits we conclude that the population has somehow changed (rather than that a rare accident of sampling has occurred). Thus, we conclude that there has been a change in p_o, c_o, μ, or σ from the standard value, and we seek the assignable cause for the change in the process.

Now if no standard was given, we need a preliminary run of samples upon which to base our analysis. Provisionally assuming that results behaving like drawings from one bowl), we proceed to estimate the standard by

p-bar for p_o

c-bar for c_o

x-bar for μ

R-bar$/d_2$ or s-bar$/c_4$ for σ

Then these estimates are used as though they were the standard values.

The control chart constants and formulas given in Table C all come from theory on the normal distribution. Moreover, even if the actual distribution of x departs quite a little from the symmetrical normal distribution, the control chart constants still apply well. Thus, we use

Standard given: μ, σ		Analysis of past data
Central line$_{x\text{-bar}}$:	μ	x-double bar
Limits$_{x\text{-bar}}$:	$\mu \pm A\sigma$	x-double bar $\pm A_2R$-bar (or $\pm$ A_3 s-bar)
Central line$_R$:	$d_2\sigma$	R-bar
Limits$_R$:	$D_1\sigma$, $D_2\sigma$	D_3R-bar, D_4R-bar
Central line$_s$:	$c_4\sigma$	s-bar
Limits$_x$:	$B_5\sigma$, $B_6\sigma$	B_3s-bar, B_4s-bar
Central line$_x$:	μ	x-bar or x-double bar
Limits$_x$:	$\mu \pm 3\sigma$	x-bar $\pm$ $3R$-bar$/d_2$ or x-bar $\pm$ $3s$-bar$/c_4$

The center line and limits for individual x's are used in determining the "natural process limits", but only when the x-bar and R (or s) charts show good control.

Again let us emphasize that limits for x-bar's and x's must be carefully distinguished. Limits for x's are always narrower than those for x's, by a factor of $\sqrt{n}$. Thus,

$$\text{Limits}_x\text{: } \mu \pm 3\sigma$$

$$\text{Limits}_{x\text{-bar}}\text{: } \mu \pm 3\sigma_{x\text{-bar}} = \mu \pm 3\sigma_x/\sqrt{n}.$$

The x limits are the ones that we must use to determine the capability of the process to meet specification limits. Remember that we buy individual items from the store, not average items. On an average a box of corn flakes may contain 10 ounces but an individual box that we buy may have only 9.5 ounces (or maybe 10.5 ounces).

6.12. SUMMARY

In this chapter, we have studied control charts for measurements x, for the two characteristics of sample data, namely, level by x-bar and variability by R or s. We thus use two charts for measurement data. It seems unnecessary to list again the appropriate formulas which have been scattered throughout the chapter and collected in the preceding section, and also in Table C.

These methods may be used to analyze practically any series of varying results or data. It is, however, desirable to collect the data in *small* samples so that conditions have little chance to vary while the n number of x values for the sample are taken. In this way all of the variation between these x's can be attributed to chance causes. Between this sample and the next, there will be a time interval and conditions might vary. Thus, we aim at within-sample variation being by chance causes alone, whereas sample-to-sample variation may also contain assignable causes as well as chance causes.

Typical types of analyzable data are all kinds of dimensions such as diameters and lengths, electrical characteristics, timing such as relays and blowing time of fuses, production and unit cost figures, weights of contents or pieces, placement of holes, thicknesses of paper or rubber, chemical compositions of all kinds, percent impurities, lengths of life, bending, tensile or breaking strengths, thickness of coating, mechanical properties such as horsepower or torque, bounce of balls, color and taste properties, lengths of yard goods measured vs. ordered, temperature controls, and resistance to corrosion.

More examples of applications and some of the theory are in Burr (1976). Many good case histories are given in Grant and Leavenworth (1972). Another useful reference is American Society for Testing Materials Manual with many numerical examples.

6.13. PROBLEMS

6.1 Percentage moisture content of an oat cereal was subject to a maximum specification limit of 4.0. Final checks were made to the nearest 0.1%. Samples 1, 2, and 20 follow: 4.0, 4.2, 3.6, 3.8, 3.6; 3.3, 3.7, 3.4, 3.6, 3.7; 4.2, 4.9, 3.6, 3.5, 4.0. The 20 x-bars and R's were:

Sample	1	2	3	4	5	
Average, x-bar	3.84	3.54	3.96	3.58	3.50	
Range, R	0.6	0.4	1.1	0.3	0.9	
Sample	6	7	8	9	10	
Average, x-bar	3.34	3.76	3.74	3.56	3.58	
Range, R	0.6	0.5	0.9	0.5	1.0	
Sample	11	12	13	14	15	
Average, x-bar	3.70	3.64	3.54	3.72	3.58	
Range, R	0.9	0.1	0.6	0.6	1.1	
Sample	16	17	18	19	20	Total
Average, x-bar	4.04	3.42	3.86	3.94	4.04	73.88
Range, R	0.6	0.9	1.3	0.5	1.4	15.4

(a) Make x-bar and R charts and comment. (b) Is the process meeting the specification limit? (c) If justified set the natural limits for the x's. (d) What action would you suggest? A frequency tabulation gives the following:

x	3.0	3.1	3.2	3.3	3.4	3.5	3.6	3.7	3.8	3.9
f	5	3	4	5	6	10	13	11	9	10
x	4.0	4.1	4.2	4.3	4.4	4.5	4.6	4.7	4.8	4.9
f	11	2	5	4	1	0	0	0	0	1

(e) Comment on this distribution.

6.2 This problem considers wire, cold drawn to size, then galvanized and fabricated into field fence. The present data are for size 12-1/2, carrying specifications on diameter of 0.096, 0.102 in. Wire is produced by the ton, coating is applied by ounces, and the final fence is sold by the rod. It is obvious that holding to the low side of the diameter specification will produce more miles of finished wire per ton of raw stock, and hence more rods of final fence. The following x and R values are for diameter, with $n = 7$.

(a) Draw x-bar and R charts, check control, comment. (b) Can you tell from the data what the precision of measurement was? (c) What can you say about meeting specifications for the diameter? Decimal points omitted for the diameters in inches:

Sample	1	2	3	4	5	6	7	8	9	10
x-bar	0985	0980	0970	0985	0973	0972	0980	0965	0970	0975
R	0030	0025	0010	0030	0025	0005	0040	0010	0010	0010
Sample	11	12	13	14	15	16	17	18	19	20
x-bar	0980	0980	0975	0980	0984	0985	0980	0970	0968	0975
R	0025	0015	0010	0040	0030	0030	0040	0010	0010	0010
Sample	21	22	23	24	25	26	27	28	29	30
x-bar	0972	0970	0980	0978	0985	0970	0985	0974	0972	0970
R	0010	0010	0010	0010	0010	0020	0010	0020	0015	0040
	Total									
	2.9288									
	0.0570									

6.3 This problem concerns the same wire discussed in Problem 6.2. The data are on the weight of coating in ounces per square foot of area. There is a *minimum* specification of 0.250 oz/ft. It is obvious that holding as closely as possible to the minimum coating weight, and as uniformly as possible, will save considerable coating materials. Values of x-bar and R are given below on coating weight with $n = 2$. (a) Make x-bar and R control charts, check control and comment. (b) What can you say about meeting the minimum specification on weight of coating? (c) What action would you suggest? (d) In the first sample of 2, x-bar $= 0.255$, R-bar $= 0.010$, can you then determine the two x's? Decimal points omitted for the coating weights in ounces per square foot area:

Sample	1	2	3	4	5	6	7	8	9	10
x-bar	255	330	280	215	220	223	255	235	222	315
R	010	020	100	010	040	000	020	030	020	030
Sample	11	12	13	14	15	16	17	18	19	20
x-bar	380	280	260	270	410	310	250	275	275	280
R	010	040	020	050	150	100	040	030	030	040
Sample	21	22	23	24	25	26	27	28	29	30
x-bar	275	300	360	270	290	300	260	280	285	295
R	030	040	000	010	020	010	010	020	010	010

Σx-bar $= 8.455$ $\Sigma R = 0.950$

6.4 The following data give results for $n = 10$ measurements of corrosion resistance for each of 12 minor additions of an element to alpha aluminum bronze in mg/dm^2/day (Hoefs, 1953). Twelve elements were chosen randomly from the 22 in the study. x's and s's are given below.

(a) Make x-bar and s charts. Is there control? (b) Would you expect control? Would you want it?

Element x-bars		
Antimony	29.50	2.13
Arsenic	29.89	1.51
Cadmium	28.94	2.05
Beryllium	25.87	3.89
Chromium	32.13	1.20
Zinc	29.14	.65
Zirconium	26.38	1.13
Phosphorous	28.64	1.42
Silicon	30.10	2.24
Calcium	28.63	1.41
Lead	31.67	.97
Tin	29.48 1.29	
	350.31	20.49

6.5 The table below gives information on a machining process, three phases of actual data. Treating each set of 20 samples as a run of past data, draw the x-bar and R charts, plotting the 60 points (x-bar and R) in succession and drawing the respective control lines. Comment on the three phases with respect to control, meeting of specifications, and need for action.

UNIT OF MEASUREMENT: 0.0001 in. from 1.1760 in.
SPECIFICATIONS: 1.1760 – 1.1765

Comments: Samples 1–20 first stage—experimental data
Samples 21–40 second stage—after some action was taken
Samples 41–60 third stage—after more assignable causes were eliminated

Sample	x_1	x_2	x_3	x_4	x_5	x-bar	R	Notes
1	+3	0	−3	+4	+4	+1.6	7	
2	0	+3	+3	+2	+2	+2.0	3	
3	+5	+3	+5	+6	+5	+4.8	3	
4	+3	+2	+3	+2	+1	+2.2	2	
5	+3	+2	+2	+1	+1	+1.8	2	
6	+1	+1	0	−3	+3	+0.4	6	
7	+2	−4	+1	−2	+2	−0.2	6	
8	+2	+4	+3	+3	+2	+2.8	2	
9	+3	+2	+4	+4	+3	+3.2	2	
10	+3	+4	+3	+1	+3	+2.8	3	
11	+3	+2	+4	+4	+5	+3.6	3	
12	+2	+4	+2	+1	+5	+2.8	4	
13	+1	+3	+3	+2	+3	+2.4	2	
14	−5	−6	0	+4	+4	−0.6	10	
15	+1	0	+4	+4	+5	+2.8	5	
16	+4	+4	+3	+5	+1	+3.4	4	
17	+6	+6	+6	+8	+4	+6.0	4	
18	+3	+8	−2	+4	+3	+3.2	10	
19	−5	+3	−3	+1	+2	−0.4	8	
20	+1	+7	+8	−2	0	+2.8	10	
						+47.4	96	
21	+4	+3	+3	+2	+1	+2.6	3	
22	+1	−1	+5	+3	+4	+2.4	6	
23	+1	+2	−1	0	+2	+0.8	3	
24	+1	+4	+4	+2	+2	+2.6	3	
25	+4	+4	+2	+4	0	+2.8	4	
26	0	+7	+4	+6	+3	+4.0	7	

(*Continued*)

Sample	x_1	x_2	x_3	x_4	x_5	x-bar	R	Notes
27	+5	+4	+6	+5	+4	+4.8	2	
28	+4	+1	+3	0	+1	+1.8	4	
29	0	+1	0	+4	0	+1.0	4	
30	+1	+3	+4	0	0	+1.6	4	
31	+2	+5	+2	+2	+4	+3.0	3	
32	+7	+3	+3	+5	+4	+4.4	4	
33	+4	+3	+4	+4	+1	+3.2	3	
34	−2	+4	+3	−4	+5	+1.2	9	
35	+4	+5	−2	+7	+4	+3.6	9	
36	0	+4	+4	0	+1	+1.8	4	
37	+3	+1	+1	+3	+1	+1.8	2	
38	+2	+3	+2	+6	+5	+3.6	4	
39	+3	+3	+4	+5	+2	+3.4	3	
40	+4	+4	+1	+5	+4	+3.6	4	
						+53.0	85	
41	+3	+4	+3	0	+3	+2.6	4	
42	+4	+1	+1	+3	+3	+2.4	3	
43	+3	+2	+3	+3	+3	+2.8	1	Attention focused on
44	+3	+2	+2	+2	+2	+2.2	1	process. Dirt found
45	+3	+3	+3	+3	+1	+2.6	2	on holding device
46	+2	+2	+3	+2	+3	+2.4	1	
47	+4	+3	+3	+4	+3	+3.4	1	
48	+3	+2	+3	+3	+2	+2.6	1	
49	+3	+3	+2	+3	+3	+2.8	1	
50	+3	+2	+3	+3	+2	+2.6	1	
51	+3	+3	+2	+3	+2	+2.6	1	Operator dressed wheel
52	+3	+3	+2	+3	+2	+2.6	1	
53	+2	+3	+3	+2	+3	+2.6	1	
54	+2	+2	+2	+3	+3	+2.4	1	
55	+2	+3	+3	+2	+3	+2.6	1	
56	+3	+3	+2	+2	+3	+2.6	1	
57	+3	+3	+3	+2	+2	+2.6	1	
58	+3	+3	+3	+3	+3	+3.0	0	
59	+4	+3	+3	+3	+3	+3.2	1	
60	+3	+3	+3	+3	+3	+3.0	0	
						+53.6	24	

6.6 The minimum distance between rubber gasket and the top of a metal cap for glass jars to contain food is an important characteristic. If too small, the cap may come off; if too large, the cap may leak and lose the vacuum. Given here are 24 values of x-bar and R for this distance, which has specification limits of 0.117 to 0.133 in. Make x-bar and R charts, and comment on control and the meeting of specifications. What action would you suggest? X-bar and R are in 0.001 in. units:

Sample	1	2	3	4	5	6	7	8
x-bar	120	122	120	118	120	120	124	113
R	20	10	20	20	20	30	15	15
Sample	9	10	11	12	13	14	15	16
x-bar	120	108	117	118	122	118	127	124
R	0	20	10	20	20	20	10	15
Sample	17	18	19	20	21	22	23	24
x-bar	127	122	123	127	118	122	118	122
R	10	15	10	10	10	10	20	20

The chuck was adjusted after sample 10.

6.7 The following data are shown on a data sheet form of the kind which is useful in collecting relevant information on a study. Complete information including notes is of great importance. A form can easily be made up for your own company. Samples 1–20 were taken as preliminary data. After considerable work seeking assignable causes and taking appropriate action, samples 21–40 were taken. Treating each as a separate run of past data, construct the two sets of x-bar and R charts in succession on the same scales. Comment on control and meeting of specifications (± 5 in coded units).

DATA SHEET

PART NAME: Transmission Main Shaft Bearing Retainer
PART NUMBER: 6577- D
OPERAT I ON ‘NUMBER: 27 MACHINE NUMBER: TC 3677
DEPARTMENT ‘NUMBER: 64
CHARACTERISTIC MEASURED: Inside Diameter
METHOD OF MEASUREMENT: Dial Indicator
UNIT OF MEASUREMENT: .0001 in. from 2.8346 in.
SPECIFICATIONS: 2.8341–2.8351 in.

DATA RECORDED BY: Davis DATE:

Sample number	Measurements of each of five items in sample a	b	c	d	e	Average for sample	Range for sample	Sample number	Average for sample	Range for sample
1	+4	+4	+6	+3	0	+3.4	6	11	−5.0	5
2	+2	+3	+2	0	0	+1.4	3	12	−10.2	9
3	+20	−4	−1	−3	+16	+5.6	24	13	+2.2	7
4	−3	−5	−5	+2	−2	−2.6	7	14	+0.8	8
5	0	−2	+1	+4	+3	+1.2	6	15	+1.0	2
6	−3	−3	−2	−2	0	−2.0	3	16	+1.6	4
7	−4	−7	−8	−11	−11	−8.2	7	17	−7.8	3
8	−2	−4	−7	−8	−8	−5.8	6	18	−3.8	7
9	−9	−7	−8	−7	−8	−7.8	2	19	−2.0	2
10	−4	−7	−1	−6	−2	−4.0	6	20	0	0

Sample number	Average for sample	Range for sample	Sample number	Average for sample	Range for sample
21	+1.2	2	31	−4.0	2
22	+0.8	3	32	−4.4	8
23	+0.6	3	33	+3.0	4
24	−0.4	1	34	−0.8	3
25	−0.2	1	35	−0.4	4
26	+0.6	1	36	+0.4	1
27	+0.2	2	37	+1.2	2
28	+0.2	1	38	+0.8	2
29	0	2	39	−1.0	3
30	−0.2	2	40	−1.0	2

6.8 The data given below are on thickness of wax on the inside of cartons (in 0.001 in.). The desired thickness is 5, with specification limits of 3 and 1. Write a brief report on your findings. Sample size was $n = 3$.

Date	Time	x-bar	R	Date	Time	x-bar	R
2/1	7:10 AM	5.23	0.9	2/3	7:40 AM	5.90	1.0
	1:40 PM	5.03	0.5		9:40 AM	5.53	.1
	5:30 PM	4.20	0.9		11:40 AM	4.90	.5

(*Continued*)

Date	Time	x-bar	R	Date	Time	x-bar	R
	9:45 PM	5.20	0.6		1:30 PM	4.80	.4
	1:15 AM	5.40	1.0		7:30 PM	5.20	.5
	2:10 AM	4.63	0.8		10:50 PM	5.00	.9
	4:50 AM	5.10	0.5		2:10 AM	5.70	1.0
	6:45 AM	4.70	1.0		4:10 AM	5.70	1.0
2/2	8:05 AM	3.97	0.9	2/4	9:05 AM	5.03	.7
	10:00 AM	4.13	0.9		11:40 AM	4.90	.2
	1:30 PM	4.50	0.1		1:30 PM	3.93	.4
	5:00 PM	5.33	0.2		5:30 PM	3.43	.9
	7:30 PM	5.00	0.8		7:40 PM	2.93	.4
	12:05 AM	4.73	0.1		11:30 PM	4.87	.3
	2:10 AM	5.23	0.8		2:00 AM	5.30	.5
	5:20 AM	5.73	0.9		4:20 AM	4.90	.3

6.9 Three heats of steel are made each day. Chemical analysis and tapping and pouring temperatures must be carefully controlled. Values of x-bar and R for samples of the $n = 3$ heats per day, for manganese in 1045 steel are given here. The data are in hundredths of a percent. Specifications are 70–90 in 0.01%.

Day	x-bar	R	Day	x-bar	R	Day	x-bar	R
1	76.3	5	10	77.7	6	19	79.0	8
2	73.0	15	11	77.3	6	20	74.3	4
3	77.0	5	12	79.3	2	21	85.3	23
4	80.0	16	13	75.7	11	22	81.3	15
5	72.0	18	14	78.3	4	23	80.0	20
6	72.7	11	15	79.3	9	24	79.0	22
7	78.7	5	16	78.0	8	25	84.0	8
8	80.3	6	17	83.3	9	26	75.0	10
9	77.7	17	18	80.7	18	Sum	2035.2	281

(a) Check for control by x-bar and R charts and comment. (b) What can you say about meeting specifications? (c) Do you have any recommendations? (d) Can you set natural limits for x for the process? If so, set them?

6.10 A drill depth dimension on a heater flange carried specification limits of 1.217, 1.222 in. Trouble with shallow holes had been encountered for years on this part. Control charts were run on x-bar and R, the latter being in control. Control limits for x-bar were 1.215, 1.221 in. for samples of 5. (a) What should this information immediately tell you? But now how about control ? Of the first 25 samples, all but three x-bar values were on or below the central line. But of samples 28–34, all but one were outside of the upper control limit. So control was poor giving even more trouble meeting specifications. In samples 1–25, holes more shallow than 1.217 were very common requiring 100% inspection and reworking. Some of the deep-hole pieces had to be scrapped.

As a result of this control chart study, the quality control manager made a number of recommendations which were adopted.

A subsequent run of 38 samples had x-double bar = 1.2194 in. and $R = 0.0026$. The R chart was in good control. (b) Therefore, estimate σ_x for the process. The nominal (middle of specification range) is 1.2195 in. Thus, the process is now well centered. (c) If the x-bar's are in control, what are the natural limits for x? (d) How do these compare with the specifications? But actually the x-bar's were not in control, out of 38 x-bar values there being 5 above the upper control limit and 2 at or below the lower control limit. The trouble would seem to have been poor adjustment of the process particularly in the form of overcompensating on insufficient evidence. If the x-bar's are brought into full control, with the same R-bar, then use of σ_x from (b) and $\mu = 1.2194$ in. along with the normal curve, Table A shows that the percent outside specification limits is about 2.6%:

$$z = \frac{1.2220 - 1.2194}{.00112} = +2.32 \qquad 0.0102 \text{ above}$$

$$z = \frac{1.2170 - 1.2194}{.00112} = -2.14 \qquad \underline{0.0162 \text{ below}}$$

$$0.0264 \text{ outside}$$

This was a very profitable and successful project.

6.11 What would you say to a shop man or union steward who contends that your x-bar limits, which must be well inside the specification limits U and L for x, mean that you are cutting the tolerance and making the job harder and that therefore the pay should be greater?

6.12 Which is likely to be more difficult to correct: out-of-control conditions on x-bar or on the R chart?

REFERENCES

American Society for Quality Control Standard (B3) or American National Standards Institute Standard (Z1.3—1969). Control Chart Methods of Controlling Quality During Production, ASQ, P.O. Box 3005, Milwaukee, WI, 53203.

American Society for Testing Materials Manual, Quality Control of Materials, ASTM, Philadelphia, PA.

IW Burr. Statistical Quality Control Methods. New York: Marcel Dekker, 1976.

JT Burr. SPC Tools for Everyone. ASQ, Milwaukee, WI: Quality Press, 1993.

EL Grant, RS Leavenworth. Statistical Quality Control. New York: McGraw-Hill, 1972.

RH Hoefs. The effect of minor alloy additions on the corrosion resistance of alpha aluminum bronze. PhD thesis, Purdue University, Lafayette, Ind. 1953.

H Working, EG Olds. Manual for an Introduction to Statistical Methods of Quality Control in Industry, Outline of a Course of Lectures and Exercises. Office of Production Research and Development, War Production Board. Washington, DC: 1944.

7

Process Capability

Is the Process by which we are making or providing a product for our customer capable of always meeting the customer's requirements? We will set some requirements for the use of the measures of process capability, discuss the applicability of the various measures and then suggest what needs to be done if the process is *not* capable. We also need to distinguish between *machine* capability and *process* capability. Furthermore, we need to think about how *measurement error* can affect our decisions against specification limits.

Let us, however, first take a look at specifications (or tolerances) and see how they are set up before we see how the process fits into these limits.

7.1. SPECIFICATION OR TOLERANCE LIMITS

Any discussion of the capability of the process to meet some specification limit(s) is based on the assumption that the specification limits *mean* something. All too often specification limits or tolerances are set arbitrarily. For example, "we always use ± 0.005 in." or "we are not sure what we need for tolerances, so let's be safe and set them at ± 0.001 in." or worse yet "we know our process can make a part ± 0.0025 so let's use that as the tolerance.".

Specification limits or tolerance limits must be set on the needs of the customer. These limits are used to distinguish between *good* and *bad* product, i.e., conforming (to limits) and nonconforming (to those same limits). They must, therefore, be realistic limits. Let us look at several different cases.

Case (1). When our customer can tell us what the real needs are for a product, we are in the strongest position to meet those needs. Let us say that we have one of two mating parts, say, a pin that will go into a sleeve. It is obvious that if the pin diameter is too *large* it will not fit into the sleeve. The author has seen an assembly person hammer a pin into a sleeve! Needless to say the assembly did not work well. Contrary to this, if the pin diameter is too *small*, it will wobble and the usable life of the pin will be reduced through excessive wear or breakage.

In this case, we can construct a set of tolerances for the pin which are unambiguous and are realistic, i.e., if the tolerance limits in either direction are exceeded, the part is not useable and is nonconforming. Note that the tolerance limits for the pin will be affected by the variability and tolerance limits of the sleeve. These two tolerances are dependent on one another. The statistics of this situation will be covered in Chapter 12.

Case (2). In the situation, where a chemical is being produced, we usually set specification limits on per cent purity or on the per cent of an impurity. Often this limit is single sided where there is either a *minimum* requirement as in the case of per cent purity or a *maximum* requirement in the case of the per cent impurity. For an illustration, let us say that we are making a chemical dye and we have felt that we need to have it at 98% purity. Therefore, we set the minimum specification limit at 98.0%.

Now comes the dilemma. What if the purity of a batch of this chemical dye is tested to be 97.8%. It is nonconforming to the specification but is it *unuseable*? The chances are that we could use it particularly if we could blend it with another batch that was 98.2% or higher. (Blending chemical batches covers a large multitude of bad manufacturing practices but the costs in inventory, handling, and the blending are very high.) If we do use this batch of 97.8% purity what are we communicating to both the operator and to the unit or company supplying the product? We are telling both that the specification limit is not really valid. One result which can occur is that the operator may become a little less careful in the manufacture and produce a chemical dye at even lower per cent purity. At some point we have to draw the line again, let us say at 97.0%. But what about a batch at 96.8%? Here we go again!

Case (3). In some cases, we have no idea what the tolerance or specification limits should be. This often happens in the development or design of a new product. In these cases, we must:

1. fall back on the experience (or gut feelings) of the developer or designer,
2. adapt limits that have been effective for similar products and characteristics, or
3. set the limits based on what we expect we will be able to meet.

In each of these (1), (2), or (3) above, we do not have realistic specifications. Only through use of the product or designing an experiment (all to often the last resort), can we begin to define the real needs that the customer has for the product limits. It is suggested by the authors that in design, development, and early production phases, interim tolerance or specification limits be used and evaluated periodically, say, every 12 months.

In talking about the cases (1) of a pin and (2) of a chemical dye, we have assumed that the measurement of the pin diameter and the per cent purity are made without error. This we know is not true. Later in this chapter, we will discuss the effect of measurement error on the decisions we have to make against tolerance or specification limits.

7.2. MANUFACTURING TOLERANCES VS. SPECIFICATION LIMITS

In some companies, it is a common practice to establish *manufacturing* tolerances or limits. This is done to assure that the *specification* limits or tolerances are always met. These manufacturing limits are always set tighter (or much tighter) than the specification limits, e.g., in the case of the pin mentioned above we might have a realistic specification limit for the diameter of ± 0.002 in.; so the company authorities set up a manufacturing limit of ± 0.001 in. which if held, will never produce a nonconforming pin.

There are two major faults with this way of thinking.

1. If the operators are not told of the situation and have it explained well, they will perceive eventually that parts outside the "limits" are acceptable product and will not strive to continue to meet them. *Note:* to an operator the limits that you give them are "specification or tolerance limits" no matter what you call them. They are the limits that the operators have to meet or they get marked up for a "nonconforming" product. When the operators do find out what the real specification or tolerance limits are, they may very easily become angry and frustrated with a system that is "putting the screws to them" by forcing them to make product to tighter limits than are required by the customer.

Let us step back for a moment and consider the difference between this philosophy of tight limits and manufacturing to nominal. In the latter case, we have realistic specifications; but through educating the operator in process control and by reducing sources of variability in the process, we begin to manufacture much closer to nominal. This will greatly improve our ability to meet the specifications, improve quality, and reduce manufacturing costs. In the global economy of the 21st century, this will be the most effective (or perhaps only) way of successfully competing for market share.

2. In some cases, the tighter manufacturing limits can lead to excessive production costs. Let us say that we have a set of specification limits that would only require production equipment capable of meeting ± 0.0025 in., e.g., five spindle screw machine. But in their wisdom, the company decides to set manufacturing limits at ± 0.001 in. This is at the limit of capability of the screw machine; so we may have to go to an NC lathe which is computer controlled, usually slower, longer setups and much more expensive to purchase. We therefore decide to use the five spindle screw machine. The operator will now have to more carefully control the process which may not even be capable of meeting the manufacturing limits under the best conditions. This can lead to the operator making adjustments when they are not needed (*over control*) or sorting and removing parts that are "nonconforming" to the manufacturing limits (although useable by the customer). Either way we end up with a frustrated operator.

Rules for establishing and using specification or tolerance limits:

1. Specification limits must be realistic, i.e., distinguish between good and bad product.
2. Use interim limits until you have established the realistic limits.
3. Make sure everyone is aware of the specification limits, i.e., do not use manufacturing limits.
4. Rigidly adhere to the limits set, i.e., reject, scrap, or rework all product outside the limits no matter how close it is to the limits.
5. If a waiver is obtained from the customer, tell the person making the product what happened, why it was waived and explain the need to continue to stay within the specification limits.
6. Know the variability of your measurement process.

If the specification limits cannot be met then you have a number of options:

1. Get the customer to widen the specification limits. This may not be possible, particularly if the current specification limits are realistic.
2. Get a deviation. Again this should not happen on a regular basis for it erodes the effectiveness of the specification limits, affects the motivation of the operator to try to meet specification limits and increases the cost of the goods manufactured by increasing the paper work.
3. Redesign the process. This can entail purchase of alternative manufacturing equipment, overhauling existing equipment, retraining the operators, performing more quality checks, implementing statistical control, and identification of sources of variability.
4. Redesign the product. This obviously can be very expensive to both the producer as well as the user. In the latter case, they may have to perform a complete re-qualification of the product.
5. Initiate 100% inspection. Very expensive and, at best, only about 85% effective.
6. Outsource the product, i.e., get some other company to make the product for you. This has a negative impact on your profit.
7. Discontinue the product from your sales line. If it were a minor portion of the company's sales and not worth the effort, this is a viable option; however, it does affect the confidence that your customers have in your capability.

7.3. NATURAL MACHINE AND PROCESS TOLERANCES

All machines and processes have natural tolerances. These are the limits of normal variation due to a myriad of causes, e.g., set up, vibration, temperature, wear, voltage fluctuations, humidity, time of day (operators may be weary or bored by the end of the shift), operator differences, reading of methods, flow rates, raw materials, sub components, etc.

The natural *machine* tolerance is defined as the inherent best that the machine can perform. In order to check this, we would obtain the best set up person that we have to do the set up, the best starting materials available (most uniform), the best operator (thoroughly trained if this is a new piece of equipment), run the machine until it has settled down and then take a sample of 40 consecutive pieces to minimize the effect of time. If under these circumstances, the machine cannot meet the specification limits of the product that it is to produce, then it never will meet those requirements no matter how much we do to get the operator to work harder and smarter. Again we will have a very frustrated operator. The second author consulted with a company that had three colored stickers for the status of their metal working machines, green—capable of meeting specifications, yellow—just barely capable of meeting specifications but must be kept on nominal and parts sorted, and red—not capable of meeting specifications. Any machine with a red sticker could not be used until it had been repaired and re-qualified as green.

The natural *process* tolerance includes all the variability that naturally occurs in the workplace, i.e., environment, operators, raw material, measurement, etc., as we indicated above. Since the natural process tolerances include all these sources of variability, they may be much wider than those of the natural machine tolerances.

7.4. MEASUREMENT ERROR AND ITS EFFECT ON DECISIONS AGAINST TOLERANCES

To many industrial, academic, and administrative people, a measured number is an absolute value, i.e., it is a true value without variability. As persons now steeped in statistics, you all now know that there is variability in everything we do or measure. Whenever we make a measurement, there is some level of variability; for if we make that measurement again and again, we will likely come out with different results. Hopefully these results will be slight and not affect our decisions on the quality or disposition of the product.

The second author became aware of a company that was rejecting almost 6% of its product. This product had to be reworked and usually passed after the first or second rework operation. A new quality manager decided to investigate this problem. First, he found that the tolerances were all about ± 0.0025 in. Second, he found that micrometers, height gages, depth gages, etc. were being used by some seven technicians. Third, he designed an experiment (Gage Repeatability and Reproducibility study) to determine the variability of the measurement process. Each technician was required to make several measurements on each of several dimensions on a variety of products. Fourth, the conclusion of this experiment was that the amount of variability ($\pm 3\sigma$) for each of the technicians was about ± 0.003! This means that any single technician measuring the same dimension of the same item several times might reject the part about 6% of the time or more. The inference is obvious. The manufacturing process was doing great but the measurement process was rejecting 6% of the product. Rework was doing nothing but wasting time and may even have increased the likelihood that bad product was being sent to the customer. The quality manager contracted with a local technical college to train his technicians in how to make measurements and use the test equipment. This resulted in the reduction of measurement error to about 10% of total tolerance, i.e., ± 0.00025.

In this example, we saw the effect of measurement variability on the decisions against tolerances; but measurement variability can also affect decisions on control. If the measurement variability is large, the R-bar will be larger than it would have been with a more precise measurement. The larger R-bar will, of course, make the x-bar control limits wider and result in a reduced ability to spot a change that has taken place in the process mean.

In comparing measurement variability to tolerances, we obtain a good estimate of the variability as was done above, e.g., Gage R & R study. Then we compare the six sigma limits to the total tolerance. Many companies require that the ratio be 10%; however, 20% is not too bad and some discrimination of quality can be obtained if the ratio is even 50%.

An example of the effect of the measurement variability on decisions against tolerances is illustrated in Fig. 7.1. The reader should remember that the distribution of the

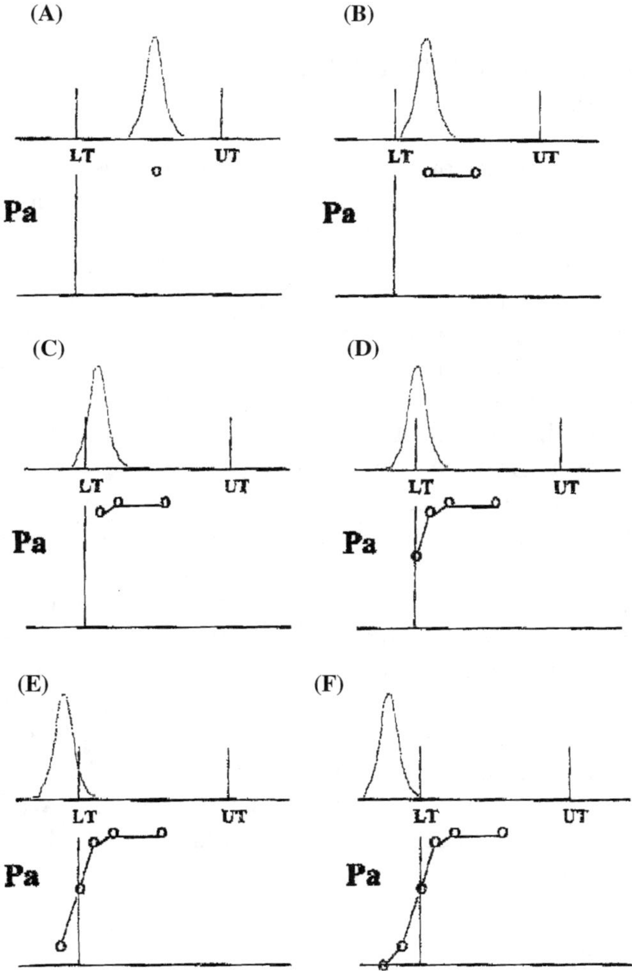

Figure 7.1 Operating characteristic curve for a measurement made with an instrument having the plotted repeatability ($\pm 3\sigma$). A, B and C show little or no ambiguity in our decision with respect to tolerances. D says that we have a 50/50 chance of a good or bad unit. E and F clearly show that the part is likely to be out of tolerance.

measurement variability is centered around the *value obtained* in the measurement, not around the true value for the item measured. The true value is somewhere inside the $\pm 3\sigma$ limits of the distribution (99.73% of the time).

7.5. COMPARISON OF A PROCESS WITH SPECIFICATIONS

In most industrial production, the basic question is whether the pieces or product satisfy requirements. Since it is impossible to produce pieces all exactly alike (except to some very gross measurement scale), we are forced to deal in *distribution of product characteristics*. Is the *distribution* satisfactory? This is the reason why specifications are commonly in terms of maximum and minimum limits for *individual measurements*, x. Or, there may be a *minimum* limit, say on strength or the length of life of a part. Or again there may be a *maximum* limit on, say, blowing time of a fuse or root-mean-square finish. But in any case, we seek an acceptable distribution. Of course, if the distribution of x's is not yet satisfactory, we may well have to sort the parts 100% to remove those outside of limits. But this is expensive and time consuming, possibly throwing the production schedule badly out of balance. The aim should be to *make it right in the first place*. This means to so produce that *all* parts or product lie inside specifications, or else all but some acceptable very small percentage outside. These concepts are basic in all production.

Let us now discuss briefly four situations we may well have in a production line. Failure to distinguish among these cases can be devastatingly expensive:

1. Process in control and meeting specifications acceptably.
2. Process not in control but meeting specifications.
3. Process in control but not meeting specifications.
4. Process not in control and not meeting specifications.

Situation 1 is, of course, the desirable one. Control charts are useful in a continuing check or follow-through on maintenance of this desirable condition. Also we may be able to decrease gradually the frequency of taking samples for charts and/or decrease any audit inspection. Or we may be able to run closer to a minimum fill-weight, thereby safely saving on material.

Situation 2 is also desirable, because of meeting specifications. But the lack of control is a danger signal that there are assignable causes at work, which may suddenly get much worse. At the very least, a careful watch should be maintained. It may well be desirable to find out the assignable causes and take appropriate action, perhaps bringing in some of the advantages mentioned under situation 1.

Situation 3 may merely require an adjustment of level x-bar which in dimensional measurements, for example, may sometimes be easily accomplished; also this can occur in some chemical processes. But in such things as tensile strength, hardness, or surface finish, adjustment of the mean level may not be so easily accomplished. A fundamental change in the process may be required; just tampering will not help.

Situation 4 calls for an aggressive campaign to seek out the offending assignable cause(s). Often when it or they are found, *and action taken*, it will be found that the process improvement brought about makes the process fully capable of meeting specifications. So seek control in this case.

Note the difference between situations 3 and 4. In the former, we need a fundamental change in the process, or a relaxation of specifications, and meanwhile must sort by 100% inspection, which is expensive, time consuming, and wasteful. But in situation 4, the watchword is to seek control.

Conditions when the process is in control: If the process is not yet in control (x-bar chart and/or R chart out of control), we cannot properly talk about the distribution of product measurements. This is because there is not just one distribution. Whenever an assignable cause becomes operative it changes the distribution, possibly drastically, from what it was while the assignable cause was not present. This is true in situations 2 and 4.

What we wish to consider now are various cases when the process is in control, especially with the R (or s) chart in control. Then we can obtain a good estimate of σ by (6.10) or (6.21). Consider some of the various conditions as shown in Fig. 7.2. In (1), the distribution just fits within the specification limits L and U. This was drawn with $\mu \pm 3\sigma$ right at the two limits. Very few pieces will be outside, but to prevent pieces outside, the process level μ must be carefully controlled. In (2), however, the natural process limits $\mu \pm 3\sigma$ are comfortably inside L to U. There is some latitude for μ to be permitted to vary, perhaps as in tool wear.

Now look at (3). The variability is satisfactory, but the level μ needs to be lowered for better centering to cut the percentage above U. This may or may not be easily done.

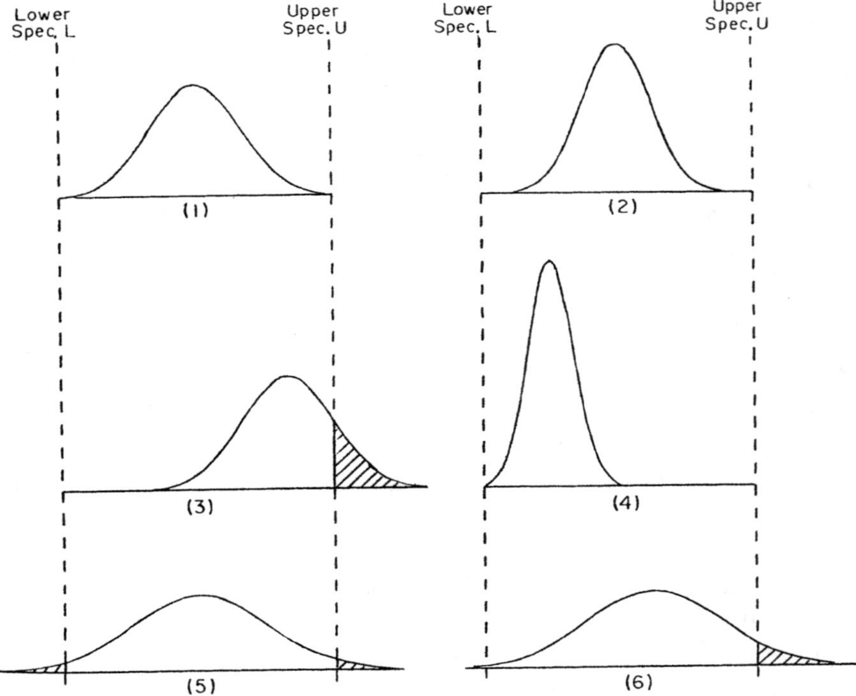

Figure 7.2 Six cases of distributions of *individual* x's in relation to specification limits L and U for individuals. See text for a description of actions that need to be taken.

(Of course, another more unlikely possibility would be a drastic cut in σ somehow, while still maintaining the high value of μ.) In (4), σ is small relative to the tolerance, U–L, and μ may have been purposely set as close as comfortable to L in order to save on material and at the same time manufacturing good product.

In (5), the process standard deviation, σ, is just too big for the tolerance. But by good centering, the percentage of pieces outside of specifications has been minimized. If the percentage outside is still too big, then steps must be taken to somehow reduce the variability, i.e., σ. In (6), the variability is the same as for case (5) but the level has been so set as to virtually eliminate pieces below L. This is often done where pieces above U can be reworked as in grinding or machining, while those below L are to be scrapped. This is the expensive case of 100% sorting, rework, and sort again.

A basic point: We must emphasize here that we do not compare control limits for averages, x-bar, with specifications for individual x's. Of course if the x-bar control limits lie *outside* specification limits, L and U, for individual x's; then some averages, x-bars, will not meet specification limits, as well as *many* individual x's. But even when x-bar control limits lie between L and U, there may well be individual x's outside. See Fig. 7.3. The x-bar distribution has 3σ limits right on L and U. But the x limits are much wider with quite a few x's outside. This is because of the following fundamental fact:

$$\sigma_{\bar{x}} = \frac{\sigma_x}{\sqrt{n}} \tag{7.1}$$

Here σ_x is what we have been calling σ, i.e., a only-to-be-expected departure of x's from μ is $\sigma = \sigma_x$. Meanwhile for x-bars, $\sigma_{x\text{-bar}}$ is the only-to-be-expected departure of x-bar's from μ. Thus if we have $n = 4$, then by (7.1)

$$\sigma_{\bar{x}} = \frac{\sigma_x}{\sqrt{4}} = \frac{\sigma_x}{2}$$

just half as much scattering for x-bars as for x's, or again if $n = 9$, $\sigma_{x\text{-bar}} = \sigma_x/\sqrt{9} = \sigma_{x\text{-bar}}/3$ so the variation among the x's is only a third as great as that among the x's.

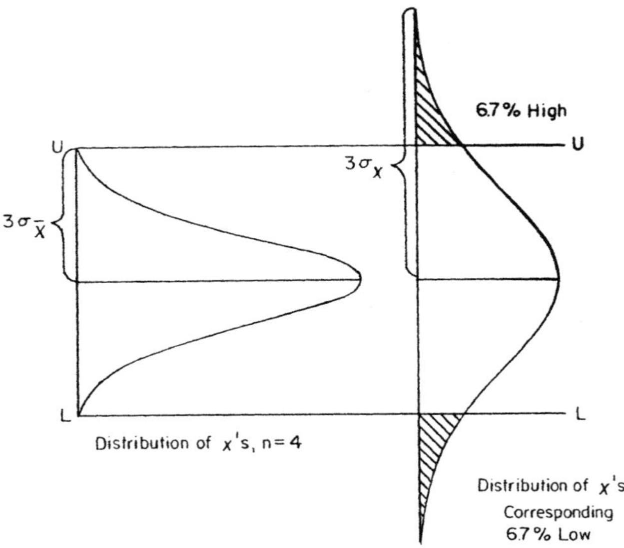

Figure 7.3 Comparison of x and $\bar{x}$ distributions with $n = 4$, where the control limits for $\bar{x}$ happen to lie right on the specification limits for individual x values, L and U. Virtually no $\bar{x}$ will lie outside L to U, but this is not important. With $n = 4$, the $6\sigma_x$ spread for x is twice as great as the $6\sigma_{\bar{x}}$ spread for $\bar{x}$. And so there are about 13.4% of the individuals x's not meeting specification limits L to U.

Thus we compare "natural" x limits with specification limits L and U for x's, namely by (6.10)

$$x\text{-double bar} \pm 3\ R\text{-bar}/d_2 \quad \text{or} \quad x\text{-double bar} \pm 3\sigma$$

If inside L to U, we are happy. If outside, we may have one of the cases in Fig. 7.1.

Better reread this section. Unless you clearly understand it, you can hardly be trusted to use measurement control charts in a plant!

7.6. PROCESS CAPABILITY INDEX

As we learned in Sec. 7.4, $\mu \pm 3\sigma$ include nearly all of the individual x's under the normal curve providing there has been no change in the process mean during the time that we obtained the data. Furthermore, we usually will use the R-chart or the s-chart to estimate the population standard deviation. This, of course, assumes that there has been no change in the process variability over the period for which the variability data were obtained. These statements lead us to make a couple of rules prior to evaluating the capability of the process to meet specifications.

Rule 1. The process must be in a state of statistical control for the period preceding the determination of process capability.

How long is a period? That probably depends on the stability of the process.

Case 1. Mature processes Some processes can go for days or weeks without having a significant departure from control. These processes are most often well understood, mature processes that have been with us for some time. The operators have been in the process for months or even years. Materials or sub-components supplies are stable and reliable. The long-term estimate of both the mean and the standard deviation is highly accurate.

Case 2. Start up In these cases, we might obtain a preliminary estimate of the variability over a relatively short period of time, e.g., the first 15–25 samples if the process appears to be stable with respect to both the mean and standard deviation. We would, of course, need to verify these estimates later on as the process settles into routine production. Here again we would require that the mean and the standard deviation are stable and in control. For the whole period of time that the process has been operating? Of course not. But what is required is stated above in Rule 1.

Rule 2. The process is only in a state of statistical control when the cause(s) of an out-of-control situation are identified and removed so that they will not recur.

All too often we find it very easy to cite the most probable cause, assume that we now know the reason and blithely go on with the process. For example, we might blame the lot of wire we purchased for the operator making too many or too few turns on a motor winding. So we change venders and, lo, the problem disappears (or seems to) only to recur later on in the process again and again. We might blame the operators and put them through a training program, the problem seems again to disappear and we have answered the problem; but once again it recurs. We might change a tool. We might implement a preventive maintenance program. Any of these could be the cause but they may not be. We must seek to find the *root cause* of the lack of control and fix that. Do this effectively and the problem disappears from the process.

Rule 3. Only after the cause(s) of the lack of control has been eliminated can we recalculate the central line and limits of the x-bar chart (estimating μ) and the central line and limits of the Range chart (estimating σ).

To illustrate this, let us pose a situation. Your process has been running for two days. You have taken 35 samples of five each during this time. One of these samples has an out-of-control range. Looking at the data, you see that one of the data points in the sample is much lower than the others, causing the range to be out of control on the high side. This occurred at point 11. You feel that it was a bad measurement so you drop it from the process control chart and recalculate the average range. What if it were not a miscalculation? How do you know? Well, you might answer, the rest of the chart is in control. What ever it was has gone away. But has it? Maybe it is still there masked by the inherent process variability! And if this is so, what is the current true R-bar? Furthermore, if we do not know what the true current R-bar is, what do the x-bar limits mean (since they are based on the R-bar that is in question).

Rule 4. From the revised R-bar (or s-bar) calculate the estimate of the variability by R-bar divided by d_2, or s-bar divided by c_4.

This estimate of the standard deviation is multiplied by six, i.e., $\pm 3\sigma$, to give the natural process limits. These natural process limits are then compared to the specification (or tolerance) limits by using the ratio

$$C_{\mathrm{p}} = \frac{\mathrm{US} - \mathrm{LS}}{6\sigma} \tag{7.2}$$

where US is upper specification limit and LS is lower specification limit.

The universally accepted notation for the process capability index is C_{p} and represents the ratio of the total specification tolerance to the natural process tolerance. When the C_{p} is less than 1.0, the tails of the normal distribution of x's are beyond the specification limits as seen in Fig. 7.2(5). Since in today's world market, this is a noncompetitive process, a program of process improvement is necessary. To illustrate this, some companies may condone or even celebrate product sent out at 1% nonconforming. Yet how many of you would go to a restaurant that published a hepatitis-C rate of 1%? How many of you would willingly use 1% nonconforming product in your process.

When C_{p} is 1.0 we have the case illustrated in Fig. 7.2(1). As the C_{p} increases to where the U–L specification limits are one standard deviation of x, σ_x, wider than the natural process limits, the C_{p} is 1.3, i.e.,

$$C_{\mathrm{p}} = \frac{\mathrm{US} - \mathrm{LS}}{6\sigma} = 8/6 = 1.3$$

This was the Q101 requirement from Ford Motor Company used by them until QS 9000 was implemented in 1994. The former stated that the process average could not be closer than four standard deviations of x to the nearest specification limit. In the latter quality management system standard issued by the automotive industry, it states that the C_{p} of 1.3 can only be applied if the process is in a state of statistical control.

The automotive industry is currently requiring a C_{p} of 1.67 if the process is *not* in a state of statistical control. This requires that the process average gets no closer than five standard deviations of x to the nearest specification.

A more ideal distribution is illustrated in Fig. 7.2(4). In this case, the specification limits are 12 standard deviations of x giving a C_p of 2.0. This is the minimum requirement for the application of the Six Sigma approach to achieving product quality, i.e., the process average shall not get closer than six standard deviations to the nearest specification limit. In practice some customers have backed off of this requirement when the supplier is using statistical process control. These have been known to essentially accept C_p of 1.67.

In some companies in the US and in the Far East, C_p's of 4.0 or greater have been obtained.

7.7. PROCESS CAPABILITY INDEX BASED ON LOCATION OF PROCESS AVERAGE

It must be immediately noticed by the reader looking at Fig. 7.2(3) that we could have a process which is in control and capable of meeting the specifications while it is making as much as 6.68% scrap! Not only do we have to have the capability but we also have to be on, or close to, nominal, i.e., in the middle of the specification interval. This apparent ambiguity in the use of C_p was recognized by original equipment manufacturers (OEMs) particularly by the automotive industry. The index developed in the 1970s was called C_{pk} which is defined as

$$C_{pk} = \frac{\bar{x} - \text{LS}}{3\sigma} \quad \text{or} \quad \frac{\text{US} - \bar{x}}{3\sigma} \quad \text{whichever is smaller} \tag{7.3}$$

In the 1970s and 1980s, some companies used the capability ratio, C_r, which is the inverse of C_p. At the time of this writing, 2003, this author has not seen this ratio used for the past 15 years. The rationale for its use seemed to be that it gave the fraction of the tolerance used up by the process. The smaller that the C_r, the better. When C_{pk} came into use, C_p had the same property of the *larger* the better and these two were built into the requirements for the automotive industry, leaving C_r in the scrap heap of time. Some authors have introduced C_m, another capability measure. This also has not been observed in use by this author over the past 10 years.

7.8. LONG-TERM PROCESS CAPABILITY INDEX—P_{PK}

Some customers, particularly those in the automotive industry, are requesting information on the long-term variability of a process.

This long-term capability index can be calculated in two different ways:

The first is to calculate the $\sigma_{x\text{-bar}}$ by inputting sequential x-bars into a tally count, histogram or computer and calculating the standard deviation of these x-bars. This value is divided by the appropriate d_2 for the sample size, n to estimate σ_x by (6.10).

The second method to calculate $\sigma_{\text{x-bar}}$ is to use all of the individual values available for the time period under study. This can be quite tedious if we do not have a computer. Still, the construction of a tally count and a simple calculator with statistical functions can be used fairly efficiently.

There are several different cases or situations that can be observed.

Case (1). The process is in a state of statistical control. In this case, the long-term variability will be only slightly greater than the short term. Given this degree of the control of the process, many customers may forgo requesting a P_{pk} particularly if the C_p is greater than 1.3.

Case (2). The process is not in a state of statistical control; however, level changes are sporadic and infrequent. Repeat causes have been essentially eliminated. Either calculation is applicable for determining the P_{pk}.

Case (3). The process average and the process variability are both out of control on a frequent basis. This could be due to a lack of understanding of the process, inadequate training of the operators, failure of corrective action to identify root causes or lack of management oversight. At any rate, the C_p and C_{pk} measures are virtually worthless for the process average and variability is not known at any given time. This leaves only the P_{pk} to be used as a measure of the process capability.

7.9. EXAMPLE

Let us go back to Figure 6.8 and Table 6.5 for an example of the application of the process capability indices.

For October 7, the range chart is out of control and the average chart is jumping all over the place. For example, what is σ? Since the range chart is out of control, we do not really know at any given time what the real variability is, e.g., what is R-bar at 6:05 pm on October 7? Is it 0.000825 as calculated or is it something else? The something else being due to some assignable cause which is operating at that time. Is it, therefore, appropriate to calculate the C_p? The authors would urge that you not do so for it might give you a false sense of security. In the same fashion we cannot calculate the C_{pk} for we know neither the σ_x calculated from the R-bar or the process average at any given time since they are out of control.

We can, however, calculate a measure of the long-term process capability, P_{pk}.

The quickest and easiest method of calculation P_{pk} without a computer would be to input the x-bar's into a hand calculator with a statistical function to calculate the x-double bar and $\sigma_{\text{x-bar}}$ using the keys identified as "x-bar" and "$\sigma_{n\text{-}1}$".

Doing this we obtain:

$$X\text{-double bar} = 0.19238 \quad \text{and} \quad \sigma_{x-\text{bar}} = 0.000756$$

$$\sigma_x = \sigma_{x\text{-bar}}\, x \text{ (square root of } n) = 0.001690$$

$$
\begin{aligned}
P_{pk} &= (\text{US} - X - \text{double bar})/3\sigma_x \\
&= (0.19400 - 0.19238)/3 \times 0.001690 \\
&= 0.319 \text{ or } (X - \text{double bar} - \text{LS})/3\sigma_x \\
&= (0.19238 - 0.19000)/3 \times 0.001690 \\
&= 0.469 \text{ whichever is lower}
\end{aligned}
$$

Note that the P_{pk} is the same calculation as is that for C_{pk}. The only difference is that the estimate of the process variability is calculated on the *long-term* variability and not from the *short-term* variability calculated using the average range of the subgroups.

Obviously there are several ways to calculate the long-term variability: (1) The method used above employing the x-bars. This may be the only method available if you have a control chart that does not record the individual measurements. (2) The second method is to use all of the individual data. This is easy when the data in electronic media and a computer is available. On the contrary, if the data must be entered manually into a calculator or a computer, the task will be time consuming and may be subject to entry errors. Obviously this method gives the most accurate estimate of the σ_x. (3) Another method available if the data are not in electronic media is to put the individual data in a grouped tally count or histogram. Using the formulas for calculating the x-bar and σ_x, we then proceed as above to calculate the P_{pk}.

In our example from October 7 from Figure 6.8 and Table 6.5, we can use the individual data in Table 6.5 to estimate σ_x and use this estimate for the calculation of P_{pk}:

$$X\text{-bar} = 0.19238 \quad \text{and} \quad \sigma_x = 0.000835$$

Thus:

$$P_{pk} = (0.19400 - 0.19238)/3 \times 0.000835 = 0.646$$

Finally for (3) above we put the data into a tally count as in Fig. 7.2. One can easily see that the distribution is not quite normal due to the lack of statistical control. It is, however, close enough to normal that a standard deviation does have some relevance. The second author strongly suggests that when estimate of σ_x from the individual values of processes that are not in a state of statistical control, the practitioner looks at the distribution. On the other hand, some would suggest that the standard deviation still be calculated and used to calculate P_{pk} even if the distribution is very different than normal. If the distribution is quite bimodal or even uniform this may be suitable; however, if the distribution is markedly skewed, the standard deviation may not represent the true situation relative to specification limits.

40–44	xxxxx xxx	8
35–39	xxxxx xx	7
30–34	xxxxx xxxxx xxx	13
25–29	xxxxx xxxxx xxxxx xx	17
20–24	xxxxx xxxxx xxxxx xxxxx xxxxx	25
15–19	xxxxx xxxxx xxxxx xxxxx xxxxx	25
10–14	xxx	3
5–9	x	1
0–4	x	1

Figure 7.2. Tally count of the individual data of October 7, in Table 6.5. The data coded are from 0.19000.

x-bar $= 0.19249$ using the class marks

$\sigma_x = 0.000853$

$P_{pk} = (0.19400 - 0.19249)/3 \times 0.000853 = 0.590$

The second author is continually impressed by the relative closeness of the statistics calculated from grouped data to those calculated from the individual data. Furthermore, there is less likelihood of input errors.

7.10. SUMMARY

The intent of this chapter has been to reinforce the material earlier presented. That is, statistical process control does not tell the whole story about a process. The process must also be able to provide a product that meets the specifications which are realistically chosen so that they reflect the distinction between good and bad quality. The measurement process used must have a variability which allows the practitioner to make unambiguous decisions on quality relative to specifications. The measures of process capability make intuitive sense and are now widely accepted in many sectors of the economy.

7.11. NOTE

A parenthetical note from the second author. It might be an interesting exercise to plot a control chart for Case 3 on page 196 using a transform of C_{pk} for each sample of the chart. The sample average and the sample standard deviation are used to calculate the C_{pk} for each sample. The major drawback of this method of control is that a computer is required to calculate the sample statistic; but as more processes are controlled or control-aided with computers, this method could be useful. The second author has plotted both simulated and real production data using this and other statistical transforms. It does appear to provide information on the process and how well it is performing against specifications.

7.12. PROBLEMS

7.1. For Problem 6.1 from the previous chapter, you are asked to estimate the process capability. Is it appropriate to use the process capability indices? Why? Calculate the C_p, C_{pk} and P_{pk} for the process. Compare these and discuss any differences among them.

7.2. For Problem 6.5 using the samples 41–60 from the previous chapter, you are asked to estimate the process capability. Discuss how you would estimate the short-term standard deviation of the process, e.g., would you exclude the out-of-control ranges? Why or why not? Calculate the C_p, C_{pk} and P_{pk} for the process. Compare these and discuss any differences among them.

7.3. For Problem 6.8 from the previous chapter, you are asked to estimate the process capability. Calculate the C_p, C_{pk} and P_{pk} for the process. Compare these and discuss any differences among them.

7.4. For Problem 6.9 from the previous chapter, you are asked to estimate the process capability? Calculate the C_p, C_{pk} and P_{pk}. What do find?

8

Further Topics in Control Charts and Applications

We now take up some additional concepts and techniques of control charts. They are useful in appropriate situations, and of interest in their own right too, in consolidating your understanding. The chapter concludes with a listing of applications.

8.1. TYPES OF SAMPLING

This section considers several different ways of drawing samples of pieces for measurement or inspection and analyzing the results. Success in quality control depends to a large extent upon how samples for analysis are drawn.

8.1.1. The Rational Sample or Subgroup

A rational sample is one in which, insofar as possible, all pieces or units within the sample were produced and measured under the same conditions. In this way, all of the variation among pieces within the sample is due to chance causes alone. The usual way to insure that conditions are substantially constant is to take *rational* samples on pieces produced close together in time. By the time the next sample is taken and measured, some of the conditions may have changed. In this way, we can say that when we use *rational* samples, all of the variation *within* each sample is a result of chance causes only, whereas variation *from sample to sample* is attributable not only to chance causes, but possibly to assignable causes as well.

If we let an assignable cause creep into our sample, then the range between the measurements will not reflect chance alone but will be enlarged by the assignable cause. *Do not inflate the range by assignable cause variation.*

8.1.2. The Stratified Sample

In production, we frequently have pieces produced from multi-spindle machines, multi-cavity molds, multi-orifices, and so on, where a number of pieces are produced more or less at the same time. For example, a five-spindle automatic machine produces a piece from spindle one, then spindle two, and finally spindle five, after which spindle one makes the next piece, always in this rotation. It may well seem quite logical to take one piece from each spindle as a sample. Such a sample is a "stratified sample", because the pieces come from "strata" or layers. Taking a series of such samples does do an excellent job of estimating the *average level* of all of the pieces produced, because it gives an equal opportunity for each spindle to contribute to the grand average. If specifications are being consistently met, then such a collection of samples is probably satisfactory; however, it should be noted that the sample does include the spindle to spindle variability.

If, however, specifications are not being met, or process improvement is desired, then it may well be desirable to take a small sample of pieces all from spindle one, another sample of pieces from spindle two, and so on. We thus have a *rational* sample from each spindle. Then later on another set of samples can be taken. Such an approach has two distinct advantages over stratified samples: (1) the variation within such samples is entirely due to chance causes only, whereas the variation within a stratified sample contains not only chance variation, but also any assignable cause variation occasioned by spindle-to-spindle differences, (2) the rational samples enable us to determine how each spindle is doing by itself, i.e., to measure spindle-to-spindle differences, whereas this is not possible if stratified samples are taken.

Let us now illustrate these points by an experimental collection of data. Consider the results in Table 8.1. First, we have stratified samples *in rows*: x_1 from spindle one, x_2 from spindle two, and x_3 from spindle three. The averages and ranges for the stratified samples in rows: x_1 from spindle one, x_2 from spindle two, and x_3 from spindle three. The averages and ranges for the stratified samples are listed and they are plotted as shown in Figure 8.1, the upper pair of charts. Look first at the x-bar chart. The x-bars seem to be running randomly enough. But now note the 3σ control limits calculated from the 36 row-samples. Does not control seem to be too good? There is not one x-bar so much as one $\sigma_{x\text{-bar}}$ away from the central line, let alone three $\sigma_{x\text{-bar}}$. (Of 36 x-bars, if control were perfect, the probabilities for a normal distribution of x-bars would say that about a third of the 36 x-bars should be more than one $\sigma_{x\text{-bar}}$ away from x-double bar.) Something is obviously wrong. To only a somewhat lesser extent, the same thing is true for the R chart. Only three of the 36 R values are more than one σ_R from the central line to the limit. Before further discussion, let us set down the calculations (with $n = 3$)

$$\text{Central line}_{x\text{-bar}} = x \text{ double bar} = +0.9/36 = 0.02$$

$$\text{Limits}_{x\text{-bar}} = x\text{-double bar} \pm A_2R\text{-bar}$$
$$= 0.02 \pm 1.023(8.47) = -8.64, +8.68$$

$$\text{Central line}_R = R\text{-bar } 305/36 = 8.47$$
$$\text{Limits}_R = D_3R\text{-bar}, D_4R\text{-bar} = \text{—}, 21.8$$

The trouble lies with the make-up of the stratified (row) samples. In each sample, there is one piece from each spindle. Thus such a sample's range will contain not only chance

Table 8.1 Data Illustrative of Sampling a Three-Spindle Automatic, with Specifications on Outside Diameter of 0.2010 ± 0.0012 in. Recorded in 0.0001 in. from 0.2010 in. Stratified Samples in Rows, One from Each Spindle. Rational Samples in Columns.

Sample number	x_1	x_2	x_3	$\bar{x}$	R	Sample number	x_1	x_2	x_3	$\bar{x}$	R
1	+5	+2	−3	+1.3	8	19	+4	+1	−6	−0.3	10
2	+5	−2	−1	+0.7	7	20	+1	0	−3	−0.7	4
3	+3	−4	−3	−1.3	7	21	+5	0	−7	−0.7	12
$\bar{x}$	+4.3	−1.3	−2.3			$\bar{x}$	+3.3	+0.3	−5.3		
R	2	6	2			R	4	1	4		
4	+7	0	−5	+0.7	12	22	+5	+1	−4	+0.7	9
5	+5	+1	−6	0	11	23	+3	0	−3	0	6
6	+7	0	−4	+1.0	11	24	+4	−1	−4	−0.3	8
$\bar{x}$	+6.3	+0.3	−5.0			$\bar{x}$	+4.0	0	−3.7		
R	2	1	2			R	2	2	1		
7	+4	−1	−2	+0.3	6	25	+5	+2	−5	+0.7	10
8	+3	0	−3	0	6	26	+2	0	−4	−0.7	6
9	+5	−3	−8	−2.0	13	27	+6	−3	−2	+0.3	9
$\bar{x}$	+4.0	−1.3	−4.3			$\bar{x}$	+4.3	−0.3	−3.7		
R	2	3	6			R	4	5	3		
10	+3	−1	−4	−0.7	7	28	−1	+1	−4	−1.3	5
11	+3	+2	−5	0	8	29	+2	+3	−5	0	8
12	+9	0	−5	+1.3	14	30	+6	+1	−3	+1.3	9
$\bar{x}$	+5.0	+0.3	−4.7			$\bar{x}$	+2.3	+1.7	−4.0		
R	6	3	1			R	7	2	2		
13	+1	+1	−4	−0.7	5	31	+6	−1	−4	+0.3	10
14	+5	+2	−3	+1.3	8	32	+3	−1	−2	0	5
15	+5	+3	−5	+1.0	10	33	+4	−2	−5	−1.0	9
$\bar{x}$	+3.7	+2.0	−4.0			$\bar{x}$	+4.3	−1.3	−3.7		
R	4	2	2			R	3	1	3		
16	+2	−2	−3	−1.0	5	34	+5	−2	−4	−0.3	9
17	+2	+3	−3	+0.7	6	35	+6	0	−4	+0.7	10
18	+3	+3	−8	−0.7	11	36	+6	0	−5	+0.3	11
$\bar{x}$	+2.3	+1.3	−4.7			$\bar{x}$	+5.7	−0.7	−4.3		
R	1	5	5			R	1	2	1		
						$\sum \bar{x}$	+49.5	+1.0	−49.7	+0.9	
						$\sum R$	38	33	32	103	305

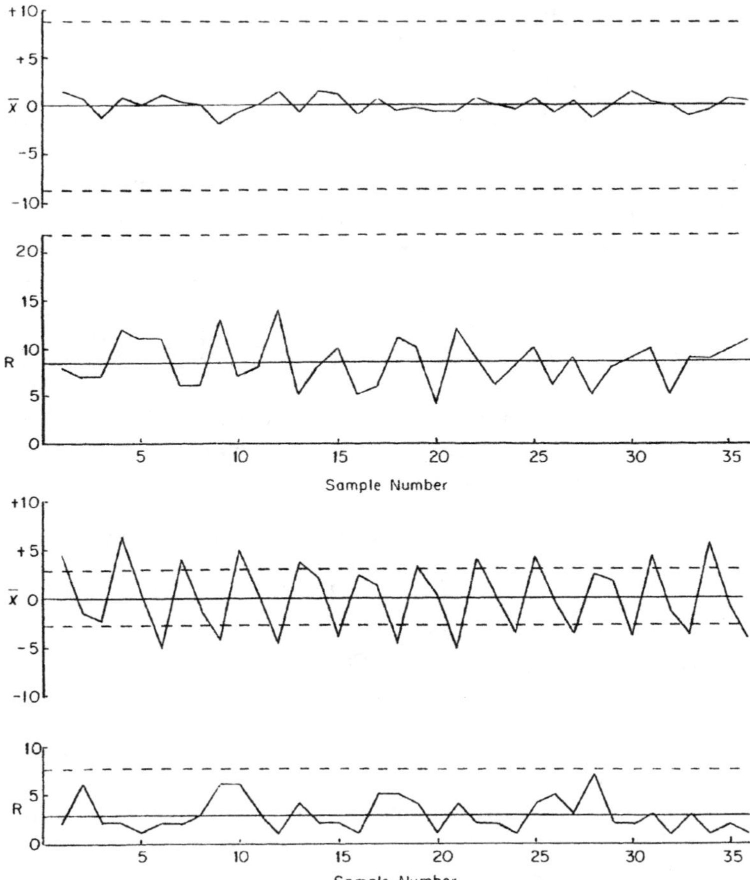

Figure 8.1 Control charts for outside diameters from a three-spindle automatic. Upper $\bar{x}$ and R charts are for stratified samples: one piece from each spindle. Lower $\bar{x}$ and R charts have $n = 3$ from spindle 1 plotted as sample 1, then three from spindle 2, and so on, in rotation. These are two different ways to plot the basic data of Table 8.1, giving radically different pictures.

variation (or causes) but also any mean differences between spindles (assignable causes). By looking at the x's in each sample, we see that with very few exceptions, spindle one contributes the maximum x and spindle three the minimum x for

$R = \max x - \min x$. Thus R-bar has been *inflated by spindle-to-spindle differences.* This gives the unreasonably wide control bands we see in the upper part of Figure 8.1.

Now consider another way of analyzing these same results. Taking the first three x_1 values as a sample to represent spindle one, the first three x_2 values to represent spindle two and so on, we have the *column* samples as shown in Table 8.1. These are plotted in rotation: spindles one, two and three then the second sample of spindle one, and so on. When plotted this gives the lower x-bar chart of Figure 8.1. We immediately note the regular cycles with all x-bar$_1$ values high, all x-bar$_3$ values low and the x-bar$_2$ values intermediate. Meanwhile the R chart looks like a reasonable random control chart, showing perfect control, but not *suspiciously* good control. The calculations for control lines follow:

$$\text{Central line}_{x\text{-bar}} = x - \text{double bar} = 0.8/36 = 0.02$$
$$\text{Limits}_{x\text{-bar}} = 0.02 \pm 1.023(2.86) = -2.91, +2.95$$

$$\text{Central line}_R = R\text{-bar} = 103/36 = 2.86$$
$$\text{Limits}_R = 0(2.86), 2.575(2.86) = \text{—}, 7.36$$

Although the two x-double bar values are identical and must be except for round-off errors, the new R-bar is only about a third of that for the stratified samples. This is because the rational samples contained only chance variation within a single spindle, whereas the stratified samples also contained spindle-to-spindle differences. Thus the R's were inflated in the stratified samples.

Again look at the x-bar chart for the rational samples. All but two of the 12 x-bar$_1$ values are above the upper control limit and all but one of the x-bar$_3$ values are below the lower control limit. Yet x-double bar is practically "on the beam" at 0.02. Clearly spindles one and three need adjustment. By how much? For this, we find

$$x\text{-double bar}_1 = +49.5/12 = +4.12,$$
$$x\text{-double bar}_3 = -49.7/12 = -4.14$$

so the two spindles need to be adjusted down and up, respectively, by about 0.0004 in. Once this is accomplished, what specifications can be met? Proceeding as usual for limits for individual x's, since the R values are in control:

$$\sigma_x = R\text{-bar}/d_2 = 2.86/1.693 = 1.69$$
$$\text{Limits}_x = 0.02 \pm 3(1.69) = 0.02 \pm 5.07 = -5.05, +5.09$$

Thus by proper adjustment of the spindles, the process can be made to meet specifications of ± 5, that is, 0.2005, 0.2015 in.

Now what really were the populations being used for this experiment? They were from Table 6.3, respectively, populations E, A, and G with averages, μ: $+4$, 0, -4, but each with $\sigma = 1.715$. These rather radically different populations were used to make the above concepts perfectly clear. We could also have used, say, B, A, and F; or, say, three of A, one each of B and F.

Another possible way to plot the rational samples for the spindle is to plot the 12 samples for spindle one in separate x-bar and R charts, then those for spindle two, and finally for spindle three. Then control lines can be figured for each spindle *separately* and compared.

Still another plotting method is called the "group control chart". It is especially useful when we have several sources, perhaps many, and do not wish to plot all values of x-bar and R. We may then take a small sample of $n = 2$, over some short time period, from each *spindle or source*, figuring x-bar and R for the sample for each source, Then instead of plotting all x-bars, only the maximum x-bar and the minimum x-bar are plotted for each time period. Then a small number indicating the source is placed by the maximum x-bar point and the number of the source contributing the minimum x-bar by its point; similarly, for the R's at each time period. Then after, say, 20 time periods, one can look along the high points to see whether any one source is contributing disproportionately to the high x-bar's or the low ones. Also the same thing is done for the R's. Control limits are set from *all* of the x-bars and R's, and, of course, out-of-control points carry their usual meaning.

A friend of the author had decided to study a 14 station machine doing 14 different jobs simultaneously on castings, such as, drilling holes and facing. Then the machine would move each casting in place for the next job. He was gathering data on one job, a facing. After completing about four rows of the 14 station measurements in a table, the foreman took notice. After explaining what he was doing, the foreman looked over the data, and at once said "Station 6 has to be jacked up and 11 lowered". Such adjustments were clearly apparent even on such meager data. This illustrates the power of gathering data in a rational manner, even without the statistical analysis.

Also, it is noted that in the authors experience, this type of stratification is often seen in the paper and film making industries where the measurement is film or paper thickness or moisture content of paper. There is often a profile of thickness or % moisture *across* the web. Samples are usually taken on a single, full width cut sheet in from the left edge, from the center and in from the right edge resulting in an x-bar control chart like that in Figure 8.1. In this case, one can establish the profile for the machine and run an x-bar and R chart *along* the machine direction at one or more positions on the web.

In the chemical industry, "stratification sampling" is often necessary. For example, if one wanted to determine the BTU (British Thermal Unit) content of coal in coal car, the sample would have to be taken at various times during the discharge of the car. The BTU content of the coal may vary with the size, e.g., fines, small pieces, and large chunks. Furthermore, some liquid chemical mixes tend to stratify during storage, e.g., surfactant solutions. More sophisticated stratification techniques are needed for sampling and analyzing fine powders, e.g., toners for copy machines.

8.1.3. *Charts for Mixed Product*

Product from multi-spindle automatic machines, multi-cavity molds, or other multiple sources is often dumped onto a conveyor or in a tote pan. Thus the pieces from the various spindles or cavities get more or less completely mixed up, and when we draw a sample, we can no longer expect one from each source in the sample. Moreover, we cannot trace back the source of each piece. What happens then? Let us see. Again we run an experiment. In order to tie in this type of sampling with the previous experiment, we took one set of each of populations A, E, and G and thoroughly mixed the 600 different-colored numbered beads. Then samples, each of $n = 3$, were drawn giving the data shown in Table 8.2. Care was taken to not look in the box while a sample of 3 was drawn, in order to avoid conscious or unconscious bias. Note that we cannot in industrial practice trace the pieces back to their source, unless some distinguishing mark is made on the piece by the process. Also it is worth mentioning again that we cannot expect one from each of the three spindles in

Table 8.2 Data Illustrative of Sampling Mixed Product from a Three-Spindle Automatic, with Specifications on Outside Diameter of 0.2010 ± 0.0012 in. Recorded in 0.0001 in. from 0.2010 in.

Sample	x_1	x_2	x_3	$\bar{x}$	R	Sample	x_1	x_2	x_3	$\bar{x}$	R
1	−8	+1	−4	−3.7	9	19	−3	+5	0	+0.7	8
2	+4	−3	+3	+1.3	7	20	+3	−4	−4	−1.7	7
3	0	+3	0	+1.0	3	21	−3	+3	−6	−2.0	9
4	−5	−2	−1	−2.7	4	22	+5	−1	+5	+3.0	6
5	+3	−5	−1	−1.0	8	23	−2	+2	0	0	4
6	+4	+2	−4	+0.7	8	24	+2	+2	−6	−0.7	8
7	−2	+1	−5	−2.0	6	25	−2	+3	−3	−0.7	6
8	+2	−1	−3	−0.7	5	26	+3	−4	−2	−1.0	7
9	+3	+3	−2	+1.3	5	27	+2	−6	+2	−0.7	8
10	−1	−5	+3	−1.0	8	28	+2	+4	−7	−0.3	11
11	+5	−5	−2	−0.7	10	29	+1	−5	+7	+1.0	12
12	−3	+3	−6	−2.0	9	30	+5	−5	+4	+1.3	10
13	−1	+7	−5	+0.3	12	31	−4	+4	+1	+0.3	8
14	−6	−3	−1	−3.3	5	32	−1	0	−4	−1.7	4
15	−5	+1	+1	−1.0	6	33	+1	−4	−4	−2.3	5
16	−4	+3	+4	+1.0	8	34	−5	0	+1	−1.3	6
17	+5	+4	+1	+3.3	4	35	−5	+1	−6	−3.3	7
18	+1	−1	+6	+2.0	7	36	+3	−3	+5	+1.7	8
										−14.9	258

a sample. (In fact the probability for such a sample here turns out to be only 0.223).

Thirty-six samples of mixed product are given in Table 8.2, and X and R are plotted in Figure 8.2. The calculation of control lines with $n = 3$ follows as usual.

$$\text{Central Line}_{x\text{-bar}} = x\text{-double bar} = -14.9/36 = -0.41$$
$$\text{Limits}_{x\text{-bar}} = -0.41 \pm 1.023(7.17) = -7.74, +6.92$$

$$\text{Central Line}_R = R\text{-bar} = 258/36 = 7.17$$
$$\text{Limits}_R = 0(7.17), 2.575(7.17) = \text{—}, 18.5$$

Plotting these control lines, we find both charts in perfect control. And this is despite the fact that the pieces from which the samples were drawn came from three quite distinct populations. The reason is that we were drawing in reality from just one population, namely, the compound population made

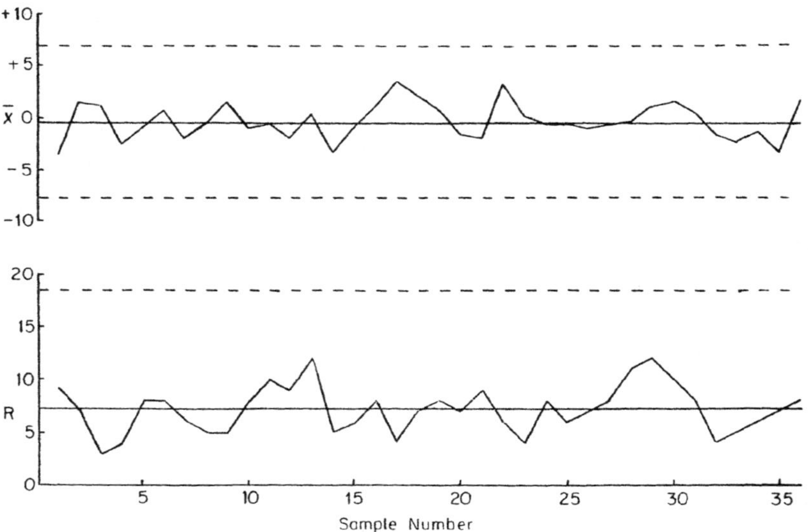

Figure 8.2 Samples from mixed product from three spindles. Outside diameter to meet specifications of 0.2010 ± 0.0012 in. Data from Table 8.2 and recorded in 0.0001 in. from 0.2010 in.

up of the three separate populations (one population from each spindle).

Now what do these control charts of Figure 8.2 tell us about meeting specifications of ±12 (in the coded units)? As usual, when we have good control, we set limits for the individual *x*'s:

$$\sigma_x = R\text{-bar}/d_2 = 7.17/1.693 = 4.24$$

$$\text{Limits}_x = x\text{-double bar} \pm 3\sigma_x = -0.41 \pm 3(r.24)$$
$$= -13.13, +12.31$$

Since these limits for *x* lie approximately on the specification limits ± 12, it would appear that we are meeting specifications quite well. However, if we look at the process capability indices, we get a very different picture.

$$C_\text{p} = \frac{\text{UL} - \text{LL}}{6\sigma} = \frac{0.0024}{6(0.000424)} = 0.943$$

$$C_\text{pk} = \frac{\bar{x} - \text{LL}}{3\sigma} \quad \text{or} \quad \frac{\text{UL} - \bar{x}}{3\sigma}$$

$$= \frac{0.200959 - 0.1998}{3(0.000424)} \quad \text{or} \quad \frac{0.2020 - 0.200959}{3(0.000424)}$$

$$= 0.910, 0.818 = 0.818$$

Even though we have such lenient limits, it would still be necessary to examine more deeply for spindle-to-spindle differences.

The population from which we drew the samples of Table 8.2 was the following composite of populations A, E, and G of Table 6.3:

x	−9	−8	−7	−6	−5	−4	−3	−2	−1	0
f	1	3	10	23	40	51	49	46	50	54
x	+1	+2	+3	+4	+5	+6	+7	+8	+9	
f	50	46	49	51	40	23	10	3	1	600

This is not a "normal" population, being (irregularly) flat-topped, but it is the population we drew from. Its non-normality caused the x limits (x-bar $\pm$ 3 s $= -13.13$, $+12.31$) calculated to be wider than the *actual* population limits of ± 9.

But now suppose that specifications were not ± 12, but instead ± 5. Then we would examine the process in much more detail for differences between spindles by taking rational samples *separately*, from each spindle. Then, as we have seen, the process could meet limits of ± 5, provided each spindle is set up properly to have an average close to the nominal 0 (0.2010 in.).

Let us give a couple of examples to touch up this section. One was a problem worked on by Holbrook Working, who was one of those who developed and taught the World War II short courses in statistical quality control. In a plant visit, the foreman said that they were having trouble with a production process. Almost everyone blamed the raw material (an at all-too-common tendency). Working examined the operation and then said, “It can’t be the raw material”. His reason was that the way this material came into the shop was such that it was very well mixed. Thus any charts would show good control because of the mixing, even from quite an out-of-control process producing the raw material. But they *were* finding charts out of control on the product, following the two supposedly identical production machines. Then the foreman admitted that they did have some misgivings on one of them. Additional carefully taken samples from each machine separately enabled them to find the trouble. The whole affair was a matter of an hour or so, whereas the trouble had been occurring for quite some time.

In another case, a man came back from a short course and made a lot of control charts in a sheet and tin mill associated with a steel plant. His charts all showed good control! (It is very rare in an initial study to find good control immediately.) The superintendent, foreman, and others were very skeptical. About this time, the man was taken in the draft! Examining the charts and data indicated that the man had taken the cards for the process in random order. This of course causes the charts to show control, despite non-controlled production conditions.

8.2. TOOL WEAR, SLANTING LIMITS

In this section, we describe a variation of control charting which can prove of much use when the tolerance $T = U - L$ for individual x values is considerably greater than $6\sigma_x$ where σ_x is the short-term standard deviation of pieces or product produced over a short time interval. Under these conditions, there is some latitude within which we may let the process average μ vary and still be meeting the specifications. The techniques we are about to describe are especially useful when μ tends to shift gradually. For example, we may find that because of tool wear, the average *outside* diameter of the parts gradually increases, or the average *inside* diameter gradually decreases. Other examples of drifts are: (1) grinder wear, (2) polisher wear, (3) viscosity, (4) temperature as of room conditions, (5) human fatigue, (6) moisture in lumber, (7) surface finish, and (8) die wear.

Let us consider for an example the data on an outside diameter of a spacer as given in Table 8.3. The quality control man was considerably confused by this operation when he applied the usual analysis of a preliminary run to the data, just preceding those in Table 8.3. The x-bar chart showed very bad control, but there seemed to be cycles, and the spacers were meeting specifications. Then he began to realize that the process mean could be permitted to drift somewhat. Suppose we analyze the data of Table 8.3. Look first at the R chart in Figure 8.3.

$$R\text{-bar} = 133/44 = 3.02$$

$$\text{UCL}_R = D_4 R\text{-bar} = 2.115(3.02) = 6.39$$

Drawing in these control lines shows two points clearly out of control. What is the cause? Looking at the make-up of these two samples of five x values, we can see the presence of two quite distinct levels of x. Looking now also at the x graph, it is seen that the high ranges occur at or just after a peak on the x's. This then looks like a case of resetting the process average downward because it is getting too close to the upper specification. Moreover, the high-diameter spacers produced

Table 8.3 Department: Automatic; Sampling Interval: 30 min; Sample Size: 5; Part: Spacer; Specifications: Outside Diameter = 0.1250 + 0.0000, −0.0015 in.; Recorded in 0.0001 in. from 0.1250 in.

Sample number	x_1	x_2	x_3	x_4	x_5	$\bar{x}$	R
1	−10	−3	−9	−2	−3	−5.4	8
2	−12	−13	−13	−11	−10	−11.8	3
3	−12	−8	−13	−13	−11	−11.4	5
4	−11	−10	−8	−12	−11	−10.4	4
5	−11	−10	−13	−10	−12	−11.2	3
6	−11	−9	−13	−11	−13	−11.4	4
7	−10	−12	−11	−10	−9	−10.4	3
8	−12	−12	−11	−8	−11	−10.8	4
9	−8	−10	−9	−10	−11	−9.6	3
10	−11	−10	−8	−9	−8	−9.2	3
11	−10	−8	−10	−9	−9	−9.2	2
12	−9	−10	−9	−8	−10	−9.2	2
13	−10	−10	−8	−9	−8	−9.0	2
14	−8	−7	−9	−9	−8	−8.2	2
15	−7	−7	−6	−8	−10	−7.6	4
16	−7	−6	−9	−8	−7	−7.4	3
17	−6	−6	−7	−7	−8	−6.8	2
18	−4	−5	−6	−7	−4	−5.2	3
19	−4	−2	−4	−3	−3	−3.2	2
20	−11	−2	−10	−9	−10	−8.4	9
21	−10	−9	−8	−11	−9	−9.4	3
22	−11	−10	−11	−9	−9	−10.0	2
23	−9	−7	−11	−11	−9	−9.4	4
24	−9	−7	−9	−10	−8	−8.6	3
25	−8	−10	−9	−8	−7	−8.4	3
26	−10	−11	−8	−8	−7	−−8.8	4
27	−7	−8	−8	−7	−6	−7.2	2
28	−6	−8	−7	−9	−9	−7.8	3
29	−6	−7	−7	−8	−8	−7.2	2
30	−7	−9	−5	−7	−7	−7.0	4
31	−5	−7	−7	−6	−5	−6.0	2
32	−5	−8	−7	−6	−7	−6.6	3
33	−7	−6	−8	−8	−7	−7.2	2
34	−8	−8	−7	−9	−7	−7.8	2
35	−7	−8	−7	−6	−7	−7.0	2
36	−6	−5	−6	−5	−5	−5.4	1
37	−5	−8	−5	−6	−4	−5.6	4
38	−5	−6	−4	−6	−6	−5.4	2

(Continued)

Table 8.3 (*Continued*)

Sample number	x_1	x_2	x_3	x_4	x_5	$\bar{x}$	R
39	−4	−4	−5	−5	−4	−4.4	1
40	−11	−12	−10	−11	−12	−11.2	2
41	−9	−8	−11	−10	−10	−9.6	3
42	−9	−9	−8	−10	−8	−8.8	2
43	−7	−8	−9	−10	−8	−8.4	3
44	−8	−9	−8	−7	−10	−8.4	3
						−361.4	133

before the resetting and the smaller diameter pieces produced after the resetting were both in the tote pan, and some of each were included in each of these two samples. This will give a high range inflated by the adjustment. Such a range does not measure chance variation only. The appropriate action is obvious, namely, to arrange to have kept separate the pieces produced before a resetting from those produced afterward. If this action is carried out, then the two high ranges become no longer typical of the process, and can be discarded. Hence we now revise R-bar:

$$R\text{-bar} = (133 - 8 - 9)/42 = 2.76$$
$$\text{UCL}_R = 2.115(2.76) = 5.84$$

Now the remaining 42 R values are all below this limit, and thus control is perfect. We are, therefore, in a good position to estimate σ_x for the process by (6.10).

$$\sigma_x = R\text{-bar}/d_2 = 2.76/2.326 = 1.19$$

Comparing this with the tolerance, $U - L = 15$ we find that the tolerance is $15/1.19 = 12.6$ standard deviations and the $C_p = 15/6(1.19) = 1.68$. Thus, there is a considerable amount of latitude or permissible variation within which we can permit the process average μ to vary.

Now suppose we were to set the *usual* x-bar limits. We have (omitting the two x-bar values for samples with the high R's):

$$x\text{-double bar} = (-361.4 + 5.4 + 8.4)/42 = 8.28$$
$$\text{Limits}_{x\text{-bar}} = -8.28 \pm 577(2.76) = -9.87, -6.69$$

Looking at Figure 8.3, we can see that had we used these control limits, a great many x-bars would be outside of these limits, mostly at the beginning of a run upwards and also at

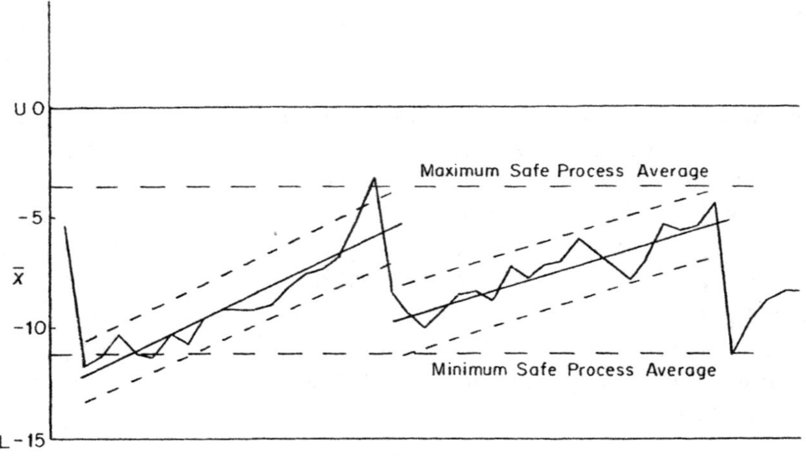

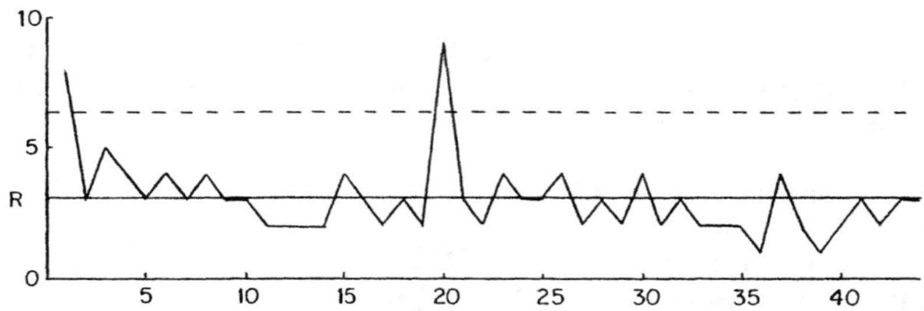

Figure 8.3 Charts for $\bar{x}$ and R from data on spacers in Table 8.3. The x's show tool wear and were analyzed by trend lines for μ and parallel control limit lines. Also shown are maximum and minimum safe process average lines for guidance as to where to start and end a run.

the end. Of course, the acid test is whether spacers *outside* of the specification limits have been or are being produced. A good way to analyze the process is to draw by eye a "trend line" following the x-bar points. That is, we try to picture the way the true process average grows when we disregard the chance variations of the x-bars around this trend. We have drawn in two such trend lines for μ in Figure 8.3 (This was done by eye, but there are objective mathematical methods. Also the trend could be a curve.)

Now how much variation of the x-bar's around this trend for μ is only to be expected? As usual, we use $3\sigma_{x\text{-bar}}$ limits, *measured, vertically* from the estimated trend line. This is found by

$$\pm A_2\ R\text{-bar} = \pm 0.577(2.76) = \pm 1.59$$

We therefore draw slanting control limit lines, above and below the trend lines, by a *vertical* distance of 1.59. In the first run, the last point indicates an assignable cause of rapid rise in μ and should be investigated. It was noticed, however, and resetting was done promptly. The tool wear in the second run was in control, and at a slightly slower rate than for the first run.

We now ask the useful questions as to what level μ we should aim at when beginning a run, and to what level we should let μ rise before resetting. For these, we find the following two levels from our estimated standard deviation σ_x for individual x's.

$$\text{Maximum safe process average} = U - 3\sigma_x \tag{8.1}$$

$$\text{Minimum safe process average} = L + 3\sigma_x \tag{8.2}$$

(Use of $3\sigma_x$ is quite conservative, since only about 0.1% would be out at these levels. We could use $2.5\sigma_x$ if we can permit 0.5% to be out for a short time.) In the example since $\sigma_x = 1.19$, these give

$$\text{Maximum safe process average} = 0 - 3(1.19) = -3.57$$
$$\text{Minimum safe process average} = -15 + 3(1.19) = -11.43$$

Then our objective, in order to maximize the length of a run, is to aim the initial setting at -11.4, and then to try to reset when the *trend line* for μ hits -3.6. Note carefully that when μ is at -11.4, an x-bar point can well be below -11.4, in fact half of them will be.

And similarly when μ is at -3.6, half of the x-bars will lie above -3.6. Thus we watch the *trend* for μ, much more than the separate x-bar points. Reset when it reaches -3.6.

Using these as guidelines, the first run seems to have been started about right or perhaps a little low. But for the out-of-control run-out point, the run might have been permitted to continue longer. The second run seems to have been started high and ended too soon.

In one division of a corporation manufacturing farm implements, the management said that die life had been doubled by use of trend charts. In one example cited, the time between downtimes was increased from 4 to 11 days, by investigating points out of the slanting control lines.

Commonly the trend line for μ is not available until several x-bar points are plotted. Then the line can be sketched in. However, if the slopes of the trends appear to be quite constant, it may become possible to draw in a tentative trend and slanting control limits quite early.

8.3. CHARTS FOR INDIVIDUAL *X'*S AND MOVING RANGES

Normally when we wish to analyze a process by control charts for measurements, we take a preliminary run of some 100 x values divided into rational samples, within which conditions are supposed to be basically identical. But between samples, conditions may vary. The 100 x values may yield 20 samples of $n = 5$, or 25 of $n = 4$. However, if the individual measurements occur at some sizable time gap, perhaps 1 hr, 4 hr, a shift, or a day, then it is more hazardous to assume that conditions have remained constant, and moreover it may take a very long time to accumulate 100 measurements. Another point too is that the measurements may be quite costly, and we cannot afford to take very many. In the manufacture of chemicals where a batch or run of product is homogeneous, multiple samples only reflect the measurement error that is (hopefully) very much smaller than the variability from batch to batch. It then becomes desirable to obtain as much information as we can from rather meager data in a time or serial order of production. The methods of this section provide one approach to analyzing such data. This is the method of control charts for individual x's and for "moving ranges".

Let us consider the data of Table 8.4. These are daily percent of steel rejections in a plate mill. They are daily figures for the first 21 days of June, at which time the mill was shut down for general repairs. The moving ranges are the numerical differences between each set of two *consecutive* x's. Thus, using vertical bars for the absolute value,

$$R_1 = |x_1 - x_2|,\ R_2 = |x_2 - x_3|,\ R_3 = |x_3 - x_4|, \ldots, R_{n-1} = |x_{n-1} - x_n| \qquad (8.3)$$

This yields only $n - 1$ moving ranges out of n x-values. In Table 8.4,

$$R_1 = |3.69 - 3.53| = 0.16 \quad \text{and} \quad R_{26} = |3.80 - 7.65| = 3.85$$

One must be very careful in the subtraction. Also note that the ranges are recorded between the x's from which they were found.

We next plot the 27 x values as shown in Figure 8.4, one for each date. But in plotting the moving ranges, we plot R_1 half-way between June 1 and 2, because, in time, it is related to both dates. This gives the 26 R points as shown in the figure.

For central lines, we use as usual the averages

$$x\text{-bar} = \sum x/n = 97.89/27 = 3.63$$
$$R\text{-bar} = \sum R/(n-1) = 39.22/26 = 1.51$$

For the moving range chart

$$\text{UCL}_R = D_4 R\text{-bar} = 3.267(1.51) = 4.93$$

Table 8.4 Percent Steel Rejections in a Plate Mill for 27 Days in June. Measurements x and Moving Ranges Shown

Date	x	R	Date	x	R
1	3.69		14	2.30	
		0.16			0.79
2	3.53		15	3.09	
		0.92			0.63
3	2.61		16	3.72	
		0.25			0.71
4	2.36		17	4.43	
		2.00			1.75
5	4.36		18	2.68	
		2.78			0.04
6	1.58		19	2.64	
		2.45			2.19
7	4.03		20	4.83	
		0.99			0.47
8	3.04		21	4.36	
		1.09			1.25
9	4.13		22	5.61	
		2.31			0.46
10	1.82		23	6.07	
		0.75			4.02
11	1.07		24	2.05	
		2.08			3.14
12	3.15		25	5.19	
		0.95			1.39
13	4.10		26	3.80	
		1.80			3.85
			27	7.65	
				97.89	39.22

All ranges lie below this limit, and so no assignable cause is indicated. (Such an assignable cause if found on the moving range chart would be for an exceptionally large jump or drop between x's.)

Next the limits for the individual x's are taken from σ_x as estimated by R-bar$/d_2$. (See (6.10)). Here using d_2 for $n = 2$

$$\sigma_x = R\text{-bar}/d_2 = 1.51/1.128 = 1.34$$

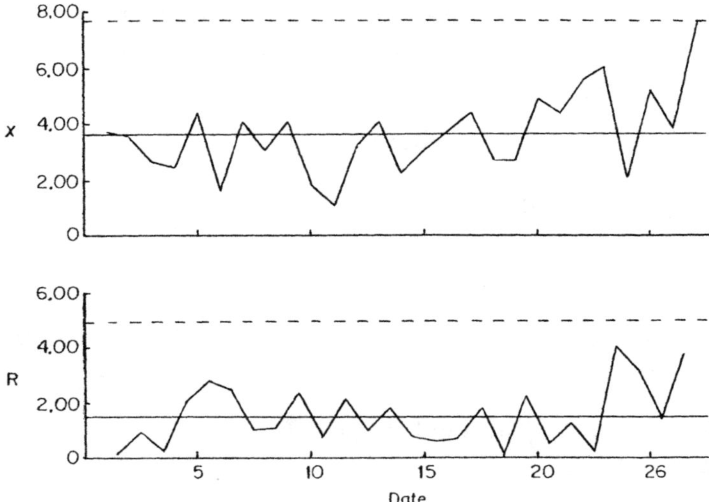

Figure 8.4 Control charts for individual x values and moving ranges R for daily percent steel rejections. Data from Table 8.4. Moving ranges plotted between the two associated x values.

Then 3σ control limits for x's are

$$x\text{-bar} \pm 3\sigma_x = x\text{-bar} \pm 3R\text{-bar}/d_2 \ (3\sigma \text{ limits for } x) \qquad (8.4)$$

In our case, we obtain from this

$$\text{Limits}_x = 3.63 \pm 3(1.34) = \text{—}, 7.65$$

the lower limit being negative, and thus meaningless for percent rejection. The last x is on the control limit. This high point is probably associated with the mill being shut down after the 27th.

Notice particularly that we did not use x-bar $\pm A_2R$-bar, which are limits for x-bar's of $n = 2$ each, which we do not have here.

Now there is one variation some workers use, the authors being among them. This variation uses 2σ limits for x. The reason is that with such meager data available, we must try to get "something" out of it. So we deliberately increase the risk of thinking an assignable cause is present when *none* is, so as to *decrease* the *chance* of missing a signal when an assignable cause *is* present.

Then we have

$$x\text{-bar} \pm 2\sigma_x = x\text{-bar} \pm 2R\text{-bar}/d_2 \qquad (8.5)$$

and in our example this gives

$$\text{Limits}_x = 3.63 \pm 2(1.34) = .95, 6.31$$

It happens here that these limits do not exclude any additional x points.

8.3.1. Risks of Making Wrong Interpretations of Control Charts

Control charts of measurements, e.g., x-bar and R chart, have limits that are plus and minus three standard deviations of the mean. It then follows that 99.73% of all means from that population of process means will fall within the '3-sigma' limits. It also follows that 0.027% or about 3 in every 1000 means will fall outside these limits. The rules for interpreting control charts state that we had better look for a cause of these out-of-control points (even though there has been no change in the mean of the process). This error is called a Type I error or more simply, "concluding that a change in the process has taken place when, in fact, there has been no change". The symbol for the probability of this type of error is alpha, α, and is, in this case would be 0.027%.

Control charts also have a probability that we will not detect a change in the process average when the process has, indeed, changed. This error is called a Type II error and is related to sample size and the amount of the shift of the process parameter, e.g., mean. For example, a shift of one standard deviation of x in the process mean would be likely caught by an out-of-control point usually within three to four samples in a control chart using a sample size of five. (Freund, 1968) This also means that the change was missed two or three times. The symbol for the probability of this type of error is beta, β.

In charts of individuals, x and MR, there are two significant differences that distinguish them from x-bar and R charts.

First, there is no benefit from using a sample mean as the measure of the process. The sample size is 1. If three standard deviation limits are used, the probability of finding an out-of-control point when the process mean has shifted one standard deviation is only 2.28%. The process might go on operating at this new level for some time before we catch it. The beta risk would be 97.72%! In order to make a more favorable beta risk, two standard deviation limits are often used. Paul Clifford calculated limits to give similar beta risks to that of the x-bar and R charts. Since the α and β risks are related to one another, we have much greater risk of looking for a change in the process when, in fact, none has taken place. When two standard deviation limits are used, alpha, α, is 2.28% at either limit or 4.56% for the chart. Many practitioners overcome this difficulty in decision by immediately taking another sample if an out-of-control point is observed. If this second sample is near the limit or out-of-control then the process is assumed to be out of control. Other practitioners suggest that it would be a good thing to look for a reason for the out-of-control point anyway. Using the two sigma limits results in a much improved chance at finding a change in the process when there has been a change in the process average of one standard deviation. The probability of detection on each sample becomes 15.87% in each direction. This would give the chart a better than 50% chance of detecting a one standard deviation change within the first four samples.

Secondly in x and MR charts, there is no protection from the application of the Central Limit Theorem. The distribution of the x's may be nonnormal, e.g., skewed. This means that decisions made using symmetrical limits may have very different α and β probabilities than those assumed. Computers can now apply a number of transforms, e.g., Johnson or log normal distributions to make the charts look symmetrical but interpretation of the standard deviation becomes difficult since it varies with the value of x. The latest author suggests that it is imperative that when using a chart of individuals, that the distribution of x's be summarized in a tally count or a histogram to see if it is somewhat symmetrical. In severe cases of asymmetry, why not calculate separate limits based on an estimated standard deviation for each half of the distribution, i.e., asymmetric two standard deviation limits?

8.4. PERCENT NONCONFORMING OF BULK PRODUCT

There are some types of percent or fraction nonconforming data which should not be analyzed by a fraction nonconforming, i.e., p or d chart. For example, consider percent spoilage of bulk product, say, paint. We might have

Batch	Volume (gal.)	Spoilage	Percent
1	20,000	400	2.00
2	18,000	390	2.17
3	20,000	420	2.10
4	21,000	380	1.81

Now why cannot the last column be analyzed as ordinary percent nonconforming data (or converted to fraction nonconforming data)? The reason is that we have not taken 20,000 1-gal. cans of paint and declared each one either good or nonconforming. If this were the case, then n would be 20,000. But instead the defective paint might be skimmed off the top of a vat or strained out. (The authors are not paint manufacturers.) Thus there is no natural sample size. Furthermore, if we used pints instead of gallons, the first sample size would seem to be 160,000 instead of 20,000. Thus we cannot use a p chart for analysis.

But we can treat the series of percents of spoilage as measurements x. Then, if we have quite a few in relatively short time periods, we can use x-bar and R charts. Or if the percents available are few in number and/or relatively far apart in time, we can use x and moving range charts, as in the preceding section.

Another case where we do not use the p chart approach on fraction nonconforming data is that in which we have overall data on total production. Thus for an organization making piston rings, a week's production may be 4,000,000. Suppose that at final inspection, the average fraction defective is about 0.02. Then what would be the standard control limits?

$$\bar{p} \pm 3\sqrt{\frac{\bar{p}(1-\bar{p})}{n}} = 0.02 \pm 3\sqrt{\frac{0.02(0.98)}{4,000,000}} = 0.02 \pm 0.0002$$
$$= 0.0198, 0.0202$$

Probably, no weekly figure will lie between such exceedingly narrow limits! The reason is that so many things can happen over all lines and types of rings in a week that fluctuations in p will be much greater. In this case again, we can treat such fractions as measurements x, and use an x chart.

One other case where fraction defective data are not appropriately analyzed by a p chart is that in which we make moldings, stampings, plastic parts, rubber molded parts, or stacks of small metal castings. The reason is that the occurrence of nonconforming pieces is not independent. Thus if one piece is nonconforming, it is more likely that others near it are nonconforming too, because of nonconformity-producing conditions. Or if a good one occurs, this tends to mean that conditions were adequate and we have a bit *lower* probability of nonconformities near it. The binomial distribution is not applicable. Again we can take such a fraction defective and treat as an x measurement.

8.5. AVERAGE RUN LENGTH FOR A POINT OUT

An important technique for comparing control charts as to their efficiency is to determine the *average run length* of pieces measured or gauged until a point outside of the limits is found. In other words, we are interested in the number of samples, on the average, which we must observe before a point goes outside of the control limits. There are two sides to this matter. On the one hand, if the process remains in control relative to the standard distribution assumed, we would like to have a long average run length (ARL); the longer the better. But if the standard distribution changes by a significant amount, we would like to have a quick warning of this change, that is, a short ARL. To be truly comparable, two charts should have the same ARL when the process is in control. Then we can meaningfully compare the ARLs when the standard distribution changes by some amount of interest, that is, goes out of control by some amount. The shorter the ARL the better, under the changed condition.

To determine the ARL for a chart, we need to know the probability of a point, representing a sample of n, to be outside of the control band. Then calling this probability P, we have

$$\text{ARL} = n/p \tag{8.6}$$

for the average *number of pieces* until a warning occurs.

Thus for example if we have a c chart with $c_o = 3$, the control limits are

$$c_o \pm 3\sqrt{c_o} = 3 \pm 3\sqrt{3} = \text{—}, 8.2$$

Now in Table B given $c_o = 3$, $P(8 \text{ or less}) = 0.996$, so that $P(\text{point out}) = P(9 \text{ or more}) = 0.004$. Then by (8.6)

$$\text{ARL} = 1/0.004 = 250$$

So, *on the average*, there will be 250 samples before a false alarm occurs by a point outside. This is quite a comfortable ARL.

Now suppose that we continue to use the upper control limit of 8.2 but suddenly production conditions change from the standard $c_o = 3$ to $c_o = 6$. Again using Table B with $c_o = 6$, we find

$$\begin{aligned} P(\text{point out}) &= P(9 \text{ or more}) \\ &= 1 - P(8 \text{ or less}) \\ &= 1 - 0.847 \\ &= 0.153 \end{aligned}$$

Then using (8.6)

$$\text{ARL} = 1/0.153 = 6.5$$

This may or may not be short enough to suit us. If not, then we might try taking three pieces of the product upon which we will count the total number nonconforming. Now however, $c_o = 9$ (for the count on three) which give control limits of $9 \pm 3\sqrt{9} = 0,\ 18$. Given $c_o = 9$,

$$\begin{aligned} P(\text{point out}) &= P(0) + 1 - P(17 \text{ or less}) \\ &= 0.000 + 1 - 0.995 = 0.005 \end{aligned}$$

$$\text{ARL} = 3/0.005 = 600$$

Comparable to a jump to $c_o = 6$ on one piece is a jump to 18 nonconforming on *three*.

Given $c_o = 18$,

$$\begin{aligned} P(\text{point out}) &= P(0) + 1 - P(17 \text{ or less}) \\ &= 0.000 + 1 - 0.469 \\ &= 0.531 \end{aligned}$$

$$\text{ARL} = 3/0.531 = \text{—}, 5.6$$

Comparing, we see that we have considerably improved the length until a false alarm when in control, but have not obtained much improvement of an ARL when the one-piece c_o goes from 3 to 6. To obtain a more comparable ARL under control, we can try an upper control limit, for the count on three pieces, of 17 or 16. (These would not be 3σ limits, but might be desirable in balancing risks.)

Moreover, we can consider the option of x-bar and R charts vs. a p chart. The former requires numerical measurements, whereas the latter merely requires a determination as to whether the measurement for the piece does, or does not, lie between the specification limits. For example, this can be done by a "go no-go" gauge. Thus, suppose that a production process is just comfortably able to meet the limits L to U with a normal distribution. That is, $L = \mu - 3\sigma_{x\text{-bar}}$, $U = \mu + 3\sigma_{x\text{-bar}}$. Then by Table A, since L corresponds to $z = -3$, and U to $z = +3$, the fraction nonconforming when in control is 0.0026. Now for 3σ limits on the x-bar chart, the probability for a false alarm of a point out while still in control is 0.0026 for all sample sizes. Thus the ARL is

$$\text{ARL} = n/0.0026 = 385n$$

Now suppose that the process mean should jump up to $\mu + 1\sigma$. Then the z values for L and U for individual x's are

$$z_L = \frac{\mu - 3\sigma - (\mu + \sigma)}{\sigma} = -4$$
$$z_U = \frac{\mu + 3\sigma - (\mu + \sigma)}{\sigma} = +2$$

Using Table A,

$$P(-4 < z \leq +2) = P(z \leq +2) - P(z \leq -4)$$
$$= 0.9772 - 0.0000 = 0.9772$$

and so the fraction defective under the changed condition is $1 - 0.9112 = 0.0228$.

We may now try a sample size for x and a sample size for a p chart. For the latter, given n, we can use $p_o = 0.0026$ and find the probability of a false-alarm point out. Then do the same when p_o jumps to 0.0228. These will give ARLs by (8.6).

Likewise with a chosen n for x-bar values, the ARL for a false alarm is $385n$, as given above. And for the ARL when μ jumps up by σ_x, we have the control limits $\Sigma \pm (3\sigma_x/\sqrt{n})$

$$z = \frac{\mu + 3\sigma + (3\sigma/\sqrt{n}) - (\mu + \sigma)}{\sigma/\sqrt{n}} = \frac{3/\sqrt{n} - 1}{1/\sqrt{n}} = 3 - \sqrt{n}$$
$$z = \frac{\mu - 3\sigma + (3\sigma/\sqrt{n}) - (\mu + \sigma)}{\sigma/\sqrt{n}} = \frac{-3/\sqrt{n} - 1}{1/\sqrt{n}} = -3 - \sqrt{n}$$

If $n = 4$, then $z = 1$, $z = -5$. Neglecting $z = -5$, the probability above $z = 1$ for the upper control limit for x is 0.1587 from Table A. This is for a warning when μ jumps up by $1\sigma_x$. Then

(8.6) gives

$$4/0.1587 = 25$$

So for x-bar ($n = 4$), the two desired ARLs are 385 (4) = 1540 and 25 for a false alarm and an alarm when the μ has changed by 1σ, respectively.

Now compare with a p chart and $n = 250$. When in control, $p_o = 0.0026$ and $np_o = 250(0.0026) = 0.65$ for the central line, and by (6.7)

$$\text{Limits}_{np} = 0.65 \pm 3\sqrt{0.65(0.9974)} = 0.65 \pm 2.42 = \text{—}, 3.07$$

So 4 is out of control.

Now if the process is still in control, we have $np_o = 0.65$ and use Table B for

$$P(\text{point out}) = P(4 \text{ or more, given } 0.65) = 1 - P(3 \text{ or less})$$
$$= 1 - 0.996 = 0.004$$

This gives

$$\text{ARL} = 250/0.004 = 62,500$$

Now if p_o jumps to 0.0228, by the assumed process change, we have $np_o = 250(0.0228) = 5.7$. Now when we *expect* 5.7, we have

$$P(\text{point out}) = P(4 \text{ or more, given } 5.7)$$
$$= 1 - P(3 \text{ or less}) = 1 - 0.180 = 0.820$$

Thus by (8.6)

$$\text{ARL} = 0.250/0.820 = 305$$

The working out of specific plans to maximize efficiency at some particular out-of-control condition does not seem to the author to be very fruitful, because we cannot expect a process to be either in control, or else out by the one specified amount. Nevertheless, the concept of average run length is of much importance.

One example of p charts vs. measurement charts was one where the head-to-shoulder length of cartridge cases was the characteristic in question. Originally snap gauges were used giving an attribute decision on each piece: either within or outside of limits. Samples of $n = 10{,}000$ were used because even in this many, only 1 or 2 would be rejects. But even this few was undesirable. Little progress was made because too much could happen while 10,000 were being produced. So a method of *measurement* was developed. The with samples of $n = 4$, x-bar and R charts permitted much better control, and causes for off-length were found and eliminated. And fewer nonconformities were *produced* than formerly had been slipping past 100% inspection.

8.6. CHART FOR DEMERITS, RATING QUALITY

A method of rating product quality for more or less complicated product, such as electronic assemblies, was developed by Dodge(1928) (see also Dodge and Torrey, 1956). The method can, however, be used for any product which may contain nonconformities of varying degrees of seriousness.

The plan involves giving demerit points for each nonconformity observed in a unit of product. All possible defects are listed in classes, according to seriousness, and assigned demerit values appropriate. In the original operation of the plan in the Bell System, four classes were used:

—Very serious (demerit value 100 points).
—Serious (demerit value 50 points).
—Moderately serious (demerit value 10 points).
—Not serious (demerit value 1 point).

Nonconformities included in each class must be carefully defined. Then inspection of a unit of product is done, yielding counts c_1, c_2, c_3, and c_4 of nonconformities within each class, where the c's may be 0, 1, 2, ...Now calling the weights w_1, w_2, w_3, and w_4, the demerit points D for a unit of product are defined by

$$D = c_1w_1 + c_2w_2 + c_3w_3 + c_4w_4 \tag{8.7}$$

This D score provides a demerit rating for each unit of product. A control chart can be run for such individual D

values. But probably the more commonly used application is for describing the quality for a shift, day, week, or a month. Then for one such time period we will have, say, n units and thus demerit ratings D_1, $D_2, \ldots,$ D_n. Thus the total demerits for the time period are

$$\sum D_i = w_1 \sum c_1 + w_2 \sum c_2 + w_3 \sum c_3 + w_4 \sum c_4 \quad (8.8)$$

Such an aggregate of demerit points is of some use, but since the number of units may vary a bit, it is usually desirable to use as the measure, the *demerits per unit,* called U. Thus for n units, we define

$$\begin{aligned} U = \sum D_i/n &= \Big(w_1 \sum c_1 + w_2 \sum c_2 + w_3 \sum c_3 \\ &\quad + w_4 \sum c_4\Big)/n \\ &= w_1 c_1\text{-bar} + w_2 c_2\text{-bar} \\ &\quad + w_3 c_3\text{-bar} + w_4 c_4\text{-bar} \end{aligned} \quad (8.9)$$

Such a U value then describes the quality over the time period in terms of average demerits per unit. As such it is quite readily interpreted.

For a series of k time periods, we may wish to run a control chart, and thus we need a central line and control limits. We will have k U values, such as in (8.9). This will provide

$$U\text{-bar} = \sum U_i/k \quad (8.10)$$

Or to find this we could average the last expression of (8.9) over k time periods finding

$$\begin{aligned} U &= w_1 c_1\text{-double bar} + w_2 c_2\text{-double bar} \\ &\quad + w_3 c_3\text{-double bar} + w_4 c_4\text{-double bar} \end{aligned} \quad (8.11)$$

this being the central line for U's and the c-double bar values being grand averages over all nk units in question. We then have the formulas for analysis of past data:

$$\sigma_U = \sqrt{\frac{w_1 c_1^2 + w_2 c_2^2 + w_3 c_3^2 + w_4 c_4^2}{n}} \tag{8.12}$$

$$\text{Limits}_U = U\text{-bar} \pm 3\sigma_U \tag{8.13}$$

If standard values c_1, c_2, c_3, and c_4 had been decided upon, they would be used to replace the respective c-double bars in (8.11) and (8.12).

Trends are especially to be looked for. Also when a U value goes above the upper control limit, we look in the records for the particular defect or defects which were responsible, and seek the causes so as to control the trouble.

The assignment of demerit points is somewhat arbitrary. For example, Hill (1952) uses demerit points 50, 20, 5, and 1. One author has used 10, 5, 2, and 1 for the demerit points in order to simplify the arithmetic. It might also be mentioned in passing that we could instead run a c or u chart for the nonconformities in each class, thus giving four c charts instead of one U chart.

8.7. SOME TYPICAL APPLICATIONS

(1) **Saving time in looking for assignable causes when none are present**: In one case, a process was producing about 40% of the pieces nonconforming. One thing after another was tried and sometimes improvement seemed to be obtained; but then a reversal would be noted. But the sample sizes were quite small. The quality control person said, "I don't think any of our experimenting has had any effect. I think I can duplicate these results out of one box." So he got beads of two colors, 40% being red and repeatedly scooped up the sample size being used. Wild fluctuations in the observed p values occurred from the one box, much like the variations noted during the trying of different remedies. So after "brainstorming" they agreed to try something entirely new. This dropped p-bar to 5%. Then another new idea dropped it to 1%. The point was that during the early

experimentation, the process had remained nearly in control but with a very poor p.

An extremely widespread evil in manufacturing is to reset processes on insufficient evidence. When a process is in reasonable control (even though not meeting specification limits), then tampering with the setting when a sample gives a seemingly off-result, but which is not out of control, only serves to *make matters worse*, by increasing the variability above that natural to the process. In one actual case, a process was not meeting specifications well. It was reset 68 times in an 8-hr shift. Two quality control persons got permission to run it for a whole shift, without adjustment. They had the set-up person do a careful job, and then let it run. The result was much fewer pieces out of limits. Such excessive tampering with the setting is called "hunting."

(2) **Help in finding causes of trouble by telling when to look, and by rational sampling, where**: Many cases could be cited for this basic aim of a control chart program. In one plant making vacuum tubes, excellent results were obtained in refining and improving the manufacture and assembly of the tubes. In one instance, an extremely difficult but important tube was developed. The engineers said, "We are only getting five percent good tubes and you will probably never get over a ten percent yield. But don't worry. We need these tubes." Manufacturing with help from quality control people using charts to control all component characteristics worked up to a 95% yield.

In another instance in wool processing, trouble occurred, and experimentation for months was unsuccessful. Then someone ran a control chart on all the wool sample results and found the first point, 6 weeks *before* trouble really hit. Checking the records, they found a process change at that time. Restoring the earlier condition solved the problem.

(3) **Knowing when a machine or process is doing the best that can be expected of it**: If a process is not in control, whether or not meeting specifications, we really do not know what it is capable of. Only when the process is brought into control can we say what it can do. In one case

a person returning from a course on quality control asked permission to experiment with a couple of old machine tools which were not being used, because everyone said that they would not hold useful tolerances. Armed with control charts and with help, he was able to find quite a few assignable causes. After getting good control at last, they found that these machine tools could do almost as close work as any others (although at a slower rate than the newest ones). So they were able to save the two and also learned a great deal about variability and control.

It is very useful to obtain good control of all one's manufacturing processes and thus to know the process capabilities. Then it is possible to efficiently allocate jobs to the various lines, as well as to maintain their known capabilities.

(4) **Determining the repeatability of a measuring technique and its error**: Dimensional or weight measurements and chemical analyses are, of course, subject to variability, i.e., when the same material or piece is repeatedly measured, the results will differ at least slightly, even if we try to work closely. A series of repeated measurement made on identical (as nearly as possible) material without the measurer knowing it is identical, should show good control on x-bar and R or x and moving R charts. If not, steps should be taken to secure control. Then R-bar/d_2 estimates the repeatability standard deviation. If there is no bias in the measurement, this is the standard deviation of measurement error. However, if x-double bar or x-bar is substantially off from a known true measurement or analysis μ, then there is a bias error. Steps should be taken to eliminate the bias or else to obtain a calibration curve.

Sometimes a proposed measurement technique or gauge proves incapable of controlled results and is, therefore, to be abandoned. Moreover, such (unfortunately not infrequently) proves to be the case with some time-honored measuring techniques.

(5) **Decreasing product variability**: The very act of eliminating undesirable assignable causes commonly reduces the process variability, often drastically, to perhaps a half or less.

(6) **Saving on scrap and rework costs**: In an actual case in automobile manufacture, the division management of the lower-priced automobile had been urging the management of the top quality automobile division to give statistical quality control a try. Finally they agreed and asked that someone be sent over. They asked that the person look into the crankshaft production. They said to themselves, "No bad crankshaft has gone out in a car in two years. Let's see what the person can do with that record." The visitor got out the blueprints, borrowed gauges and followed the processes through. Within a month the visitor found that this was an ideal spot for the quality control methods. For one thing, 14% of the crankshafts went out the back door "grade A scrap iron"! Then too there was a huge amount of unnecessary rework and excess machining. The charts clearly showed that in the first rough work too much metal was removed, necessitating scrapping. At others, too little was removed calling for excessive metal removal on subsequent operations. Presenting the case within about 6 weeks, action was taken reducing scrappage to 1% and greatly reducing machining and grinding time by an unknown amount!

Often when processes are incapable of meeting limits and reworking is necessary, it is possible to run at a level so as to balance scrap and rework costs for a minimum loss overall.

(7) **Increasing tool and die life**: This is a "natural" for the methods discussed in Sec. 8.2, using slanting control limits. The idea is to aim at starting the process as close to one specification as is safe and let the average increase or decrease watching control until approaching as near as is still safe to the other limit. Thus maximum length runs are obtainable. Some chemical processes can use the same approach.

(8) **Decreasing inspection for processes in control at a satisfactory level**: When the appropriate charts show the process to be in control at a satisfactory quality performance, we can in general decrease the frequency of taking control chart samples. Or possible, the size n may be cut. In one famous case for a molded part in World War II, after showing the process to be in control and incapable of meeting

specification limits, the latter were examined and parts out of limits assembled. They worked satisfactorily; so the tolerance was much increased. No longer was 100% sorting necessary. Samples of $n = 5$ for an x-bar and R chart set-up were taken only every hour. Then only one sample each 4 hr and finally only on in each 8 hr.

Note that some follow-through of an occasional sample was still maintained. This is highly recommended otherwise production may become careless. Moreover, a record of some surveillance is desirable in case of some legal action. One must be able to defend their methods.

(9) **Safer guaranteeing of product, reducing customer complaints**: The basis of sound guarantees and customer satisfaction is controlled processes operating at a satisfactory level. If control is uncertain, we are in a much weaker position. Sales managers need to know our process capabilities, i.e., both statistical capability and what can or cannot do. Then they are in a position to promise only what can be achieved and delivered.

(10) **Improving production–inspection relations**: Control charting of production can make a real contribution by objective treatment taking variation into account. It is the variability in results which can cause much argument unless allowance is made for discrepancies explainable by chance. Differing results may or may not be compatible.

(11) **Improving producer–consumer relations**: Many problems in this relationship are similar to those in (10). In addition, there are problems of sampling-acceptance results. See Chapters 9 and 10. One commonly occurring problem is that of compatibility of measuring technique, e.g., gauges, and of definitions of nonconformances. All such problems can be aided by statistical methods.

(12) **Setting specifications more realistically, sounder relations between engineering and production**: The objective is to obtain good control of processes so as to know the production *capabilities*. Often by obtaining full control of processes enough decrease in variability is obtained to meet the specified tolerances. Also not infrequently the process, even when in control, is not capable of meeting specifications

or maintaining a requisite level of C_{pk}, e.g., 1.33. The choices are then: (1) making fundamental changes in the process, (2) sorting 100% to the specification limits or (3) widening the specification limits. Solution (2) is very expensive, time consuming, and only 85% effective. It also requires scrapping and reworking. Solution (1) is often very expensive too. Solution (3) is often feasible. If the customer can be convinced that production is getting all it can out of the process, then the customer may well relax the tolerance if the parts made can be shown to work satisfactorily.

In one case, 30-in rubber tubes (extruded) were cured on mandrills, then each tube was cut into 300 gaskets for assembly in metal caps for food jars. Thickness of rubber was important for proper sealing. Specified limits for doubled thickness of rubber from single gaskets were so tight that no one tube could yield gaskets all within limits, let alone a whole lot of tubes. The reason was that in curing on a mandrill, the rubber at each end of a tube would pull together a bit, giving thicker gaskets at the ends than in the middle. Thus some of the 300 gaskets from *single* tubes would always lie outside *both* limits. When this was clearly shown, engineering and the customer were willing to set much more realistic limits, within the process capabilities.

(13) **Decreasing nonconformances on subassemblies**: As we have seen, c and u charts are potent tools for attacking such problems. Posting results can be a strong inducement to improvement. In one case, a very large manufacturer of auto radiators decreased the number of leaks at the initial test following the assembly of two sides, from seven to less than two, in a few months.

(14) **Comparison of several inspectors, machines or processes**: Taking rational samples from each as in Sec. 8.1.2 and putting on one control chart permits comparison with reasonable allowance for natural variability. If the various samples are all in control, then there is no reliable evidence of any real differences. But if one or more points are out, we do have reliable evidence of real differences in quality performance. Charts could be on measurements or attributes.

(15) **Stabilization of chemical or metallurgical processes**: By running control charts on chemical additions, temperatures, pressures, timing, throughput, and then on the resulting chemical compositions and physical characteristics of the final product, the whole operation may be improved and stabilized. For example, in a large production of cast manganese steel tank shoes, spectacular results were obtained, the percentage of nonconforming castings dropping from 13.9% to 2.0% in 10 months. And this was at a benefit-to-cost ratio of 20 to 1.

(16) **Justifying a " pat on the back", fostering quality mindedness**: If carefully approached and sold to personnel, control charts can easily become a source of pride, even friendly competition among workers. It is a help to be complimented by a foreman, when an objective chart shows that a fine job is being done by an operator. Moreover, under some incentive systems or profit sharing programs, statistical quality control methods may actually put more money into a worker's pocket.

(17) **Determining the stability and quality level of a producer**: Control charts run on samples from a supplier form an excellent record of their performance. If in control at a satisfactory level, receiving inspection may well be decreased or eliminated. And if not in control and/or the quality level is inadequate, appropriate steps can be taken objectively. Suppliers may also be objectively compared. Regular reports to suppliers are facilitated. Moreover, if the supplier uses statistical control charting, they may submit the results to the user as is required by the automotive companies in lieu of incoming inspection. This can also be the basis for vender certification.

(18) **Convenient and meaningful records**: Control charts form a compact and objective record of product performance throughout the production process. Such records can be vital in the case of customer complaints, or in the extreme case of a liability lawsuit. In such a situation, our methods are likely to be under close scrutiny and we must be able to show that due care was exercised and objective methods used. Even with such sound methods, judges and

juries may be difficult to convince because of ignorance of probability and statistics.

(19) **Saving on weight control**: There are enormous savings available in all types of manufacturing. The first thing you may think of is the package or container content weight. The general approach is to secure better and better control of weight of fill in two respects, i.e., (1) freedom from assignable causes of erratic weights, and (2) decreasing the variability. As improvement is made in both respects, it becomes possible to run the process average closer to the specified average or the specified minimum, whichever is applicable. But this subject is much broader than container weights. It literally applies to nearly all products. Furthermore, it applies to factors of safety which some, not without reason, call “coefficients of ignorance”.

(20) **Interpretation of the many management figures**: One often sees in manager’s offices a series of charts with figures plotted, e.g., production, unit costs, late shipments, absenteeism, man-hours per ton, power consumption, etc. But unfortunately such running records frequently have no central line nor control limits. The placement of control lines can aid decision making so that matters which deserve attention receive it and those apparently off values (which are really in control) do not receive unjustified attention and concern. Blaming people for what is really only a random fluctuation with nothing different being done, does not improve morale. Quite the contrary. Moreover action may be taken which is harmful rather than beneficial.

(21) **Improving visual inspection**: The first step, of course, is to obtain clear, objective definitions of each nonconformity. A set of “limit samples” can help. These are a collection of pieces with the borderline example of the nonconformity. More severe examples would be “nonconforming” and less noticeable examples would be considered as okay and disregarded. Such can be a great help to inspectors or operators. But this may not altogether solve the problem. If no trouble seems to be coming along, inspectors may become lax and then if trouble hits, the foreman may get “tough” and

clamp down. Then inspectors may start rejecting pieces with nonconformities less noticeable than the borderline example.

In a piston-ring corporation, this precise problem was encountered on holes in completed rings. For each inspector, *two* audit charts of the np-type were run. On an ordinary vertical scale, 0, 1, 2, . . . , was plotted for the day, all of the *good* rings *rejected* by the inspector found among those rejected by the inspector. These were found by a lead inspector auditing the rejected rings. Then immediately beneath the foregoing chart was another np chart, with scale 0, 1, 2, . . . , *increasing downward* (back to back with the first one). On this chart was plotted the number of *nonconforming* rings found by the lead inspector from among a sample of 200 rings *accepted* by the inspector. Now if both points were higher (many good one rejected and few nonconforming ones missed), this tended to mean that the inspector was too lenient. Out-of-control points on either chart were, of course, most significant. The savings on good rings which were no longer being rejected amounted to 25% of the corporation's net income for one year.

(22) **Decreasing the number of clerical errors**: The *c* chart is a natural approach to cutting the number of clerical errors. Clerical work can be audited by a top clerk or inspector. All sorts of clerical jobs may thus be controlled, including typographical errors and accuracy of proofreading. Yes, even in the age of computers and spell checking, we can make these, types of errors. It came to the attention of one of the authors that one employee was inputting his own misspellings into the spell checker's dictionary! A mail order company achieved excellent results by using a *c* chart.

(23) **Decreasing accidents**: One company found a relation between hospital calls and serious or lost-time accidents. By watching a chart on hospital calls and putting on a more intensive program of accident prevention when the call significantly increased, the accident rate was cut.

(24) **Classifying product to be processed**: One highly lucrative application was made in a hardwood veneer company. Lumber to be dried is measured with respect to moisture content. Prior to the application of a control chart method, lumber of may different moisture contents would be

put in a kiln and dried as the wettest lumber. The following approach was devised by a person returning from a short course in quality control. For each stack of lumber, the person would take six readings, x, find x-bar and R. Then they would find $x\text{-bar} + 3R/d_2$ as an estimate of the wettest lumber in the stack which was called "moisture number". Categories for moisture numbers were chosen, placing the data on different colored cards. Then they would have all the, say, blue card stacks dried together. They would be dried to the estimated wettest lumber using x-double bar $+ (2R\text{-bar}/d_2)$ from the various blue card stacks to be dried. In this way, all the lumber would be dried at about the right rate, thus avoiding spoilage ("honeycombing"). Millions more board-beet of lumber were able to be dried per kiln-month because some especially wet lumber would not be holding up each load to be dried.

(25) **Improving packaging**: In one case of packaging of cleaning powder, the metal ends were not always well crimped onto the cardboard cylinder. Rigorous tests on finished packages in connection with np charts got at the causes of trouble which were eliminated. The quality control people were subsequently called "the boys with the iron thumbs" by production. Control charts can be used in a wide variety of ways in packaging operations.

(26) **Facilitating random assembly**: In a corporation, piston ring castings were being ground three or four times for edge width (thickness) by disc grinders. An efficient materials-handling method put all the rings in order onto stakes. After the final fine grind, the rings would be made up into pots 15 or 20 for machining grooves and the like. Unfortunately, as a result of poor control, shims had to be used in making up the pots. A complete graphing of 6000 consecutive rings gauged for edge width led to the discovery that the thicknesses showed runs of thicker, then thinner rings. By the materials-handling method, theses runs would be accentuated in the next grinds. Finally, it was decided to dump the rings from each grind helter-skelter into a box. This broke up the cycling. Also it facilitated much better control of edge width so that 100% sorting after each grind was eliminated in favor of sampling inspection thereby decreasing the cost

in the grinding department by 80%. Another bid gain was that with a random order of rings and better control, the pots of rings for machining could be assembled randomly without resorting to the use of shims of various thicknesses.

(27) **Getting at the causes of trouble rather than expecting the nonconforming pieces to be sorted out and not missed by inspection**: This is the basic aim in all process control and is greatly aided by the control chart approach.

(28) **Reducing the number of adjustments needed in making chemical mixes**: In a chemical mixing operation, a preliminary mixture of components is made so that the analysis will be lower in the critical component. Over addition of this component would result in large rework costs. Once this preliminary mixture is analyzed, a calculated addition is made. Usually the operator will put in less than the calculated amount "just in case". This cycle may be repeated several more times. The average number of *adjustments* was 3–5 per batch which resulted in total mix preparation times in excess of 1–2 days. A quality engineer ran an x and moving range chart on the results of the preliminary mix. It was found that the process was in a state of good statistical control and that the standard deviation was such that the process capability, C_p, was about 1.6. The engineer and the operator then moved the target for the preliminary run to the center of the specification range. The operator continued the control chart and began making 100% good batches with no adjustments thereby reducing the mix time from over 20 hr down to 4 hr.

This list should give you some ideas of how to apply control charts on your job in whatever work environment you experience. A bibliography of early applicational references in journals is given in Burr (1976).

8.8. PROBLEMS

8.1. Obtain 30 stratified samples, each of three x values, consisting of the total points in a throw of two dice, then three dice and finally four dice. Tabulate the three x values always in the same order as in Table 8.1. (a) Analyze the stratified samples as in rows: using the usual control chart formulas and drawing the charts. Comment. (b) Similarly analyze the rational samples in turn. Comment. (For the totals on two dice $\mu = 7$, $\sigma = 2.42$; three dice $\mu = 10.5$, $\sigma = 2.96$; four dice $\mu = 14$, $\sigma = 3.42$; and in general for k dice $\mu = 3.5k$, $\sigma = \sqrt{(35k/12)}$.)

8.2. Simulate a three-spindle automatic being run for production at specification limits of ± 5 (coded). Use 30 samples of $n = 3$ each consisting of one from each of populations A, B and F of Table 7.3. Always take in the same order as in Table 8.1. (a) Analyze the stratified samples in rows using the usual control chart formulas and drawing the charts. Comment. (b) Similarly analyze the rational samples in turn. Comment.

8.3. This problem involves measurements on the outside diameter at the base of the stem of an exhaust-valve bridge. Specifications were 1.1550–1.1560 in. All pieces from the machining operation were measured and samples of $n = 5$ in succession were formed. Three vendors were supplying forgings. There were constant tool changes and tool resets. As a result of these changes, tool life was short and pieces were often out of specifications. It was the general opinion that differences among vendors were responsible for pieces being out of specifications. However, a study was made on each vendor's pieces and there was no significant difference in the trends of any of them. A chart was placed on the machine and the operator instructed as to where to start a run and at what level to reset. The chart indicated that approximately 75 pieces could be produced before resetting. By using this type of trend chart, the process was made to run without nonconforming pieces and 100% inspection was reduced to five pieces every 30 min. Tool life was considerably increased.

The data given below were in 0.0001 in above 1.1550 in with $n = 5$.

Sample 1	2	3	4	5	6	7	8	9	10	
x-bar	3.0	4.0	4.4	4.6	5.0	5.6	6.8	7.0	7.6	7.8
R	2	2	2	2	2	2	1	2	1	2
Sample 11	12	13	14	15	16	17	18	19	20	
x-bar	8.0	8.6	9.0	1.8	2.4	2.6	2.8	3.0	3.4	3.6
R	2	1	2	2	1	2	1	2	1	2

(a) Plot the points for x-bar and R. Check the R chart for control. Estimate σ if justified. (b) Draw by eye the trend lines for μ and show on the chart the minimum and maximum safe process averages. Also draw slanting control limits around the trend lines. (c) Were the runs started and ended at about the right levels? Was the tool wear running in control?

8.4. The data shown below constituted the original run preceding that of Table 8.3. As in that table, specifications were 0.1235 to 0.1250 in. with data in 0.0001 in. units below 0.1250 in. and $n = 5$.

Time July 27	8:00	8:30	9:00	9:30	10:00	11:00	12:30
x-bar	−8.0	−7.4	−6.8	−5.4	−5.6	−4.0	−3.4
R	2	3	2	2	3	4	3
Time July 27	1:15	2:30	2:45	3:15	4:00	4:30	
x-bar	−10.8	−11.6	−10.4	−9.6	−9.8	−9.4	
R	2	3	2	3	2	4	
Time July 30	9:00	9:30	10:00	11:00	12:15	12:45	1:15
x-bar	−7.4	−7.8	−7.4	−5.4	−7.0	−5.8	−3.8
R	4	4	2	3	2	3	2
Time July 30	1:45	2:15	2:45	3:30	4:00	4:30	
x-bar	−4.6	−4.2	−2.8	−6.2	−5.6	−6.2	
R	4	4	3	5	4	2	
Time July 31	8:00						
x-bar	−9.4						
R	4						

(a) Plot the x-bar and R points, check the R chart for control. Estimate σ if justified. (b) Would the x-bar chart with ordinary limits be in control? (c) Draw by eye the trend lines for μ. Show the minimum and maximum safe process averages, and the slanting control limits for x-bar. (d) Were the trend lines for μ started and ended at about the right level?

8.5. The following analyses, x, give the carbon content of 20 consecutive heats of 1045 steel. Three heats are made in an open hearth furnace per day. Check the control by an x chart (with 2σ limits) and a moving range chart. Specified limits are 0.45% to 0.50%. The following are given in 0.001% units: 480, 470, 470, 455, 515, 495, 460, 465, 520, 530, 485, 495, 455, 475,

515, 455, 500, 470, 480, 505. (a) How is control? (b) Can you estimate the proportion of heats meeting specifications?

8.6. A steel plant was producing hot-rolled bars which were shipped to a customer who cold drew them. After cold drawing, the bars were inspected and certain ones rejected for defects purporting to originate in the steel plant. Rejections were totaled for each month and expressed as a percentage of the total drawn. Twenty-four monthly figures in percent rejection follow: 4.4, 2.9, 4.2, 5.4, 6.6, 3.6, 1.0, 1.3, 1.2, 1.2, 1.0, 2.8, 2.0, 3.0, 1.8, 2.3, 1.3, 5.2, 4.1, 4.3, 5.3, 5.3, 0.6, 0.8. Make an x chart (2σ limits) and a moving range chart and comment.

8.7. Give two examples from your own experience of data on fraction nonconforming which cannot appropriately be analyzed by a p chart. State why not in each case.

8.8. Find the average run lengths in terms of individual pieces, when using an upper control limit of 17 (17 nonconformities are out of control) for the total number of nonconformities on three units: (a) when c_o for one piece is 3 and (b) when c_o for one piece is 6.

8.9. Suppose that on 100 units of equipment, $\sum c_1 = 1$, $\sum c_2 = 3$, $\sum c_3 = 32$, $\sum c_4 = 350$ with corresponding weights $w_1 = 100$, $w_2 = 50$, $w_3 = 10$, $w_4 = 1$. Standard values for c_o are 0.001, 0.005, 0.3, and 3, respectively for 1, 2, 3 and 4. Find U and compare with control lines.

REFERENCES

HF Dodge. A method of rating manufactured product. Bell system Tech J 7:350–368 Bell Telephone Laboratories Reprint B315 1928.

HF Dodge, MN Torrey. A check inspection and demerit rating plan. Ind Quality Control 1956; 13(1):5–12.

DA Hill. Control of complicated product. Ind Quality Control:18–22 1952; 8(4).

IW Burr. Statistical Quality Control. New York: Dekker, 1976.

9

Acceptance Sampling for Attributes

For the moment, let us consider the general problem of deciding whether to accept a lot from a vendor, subcontractor, or another department or line. This can of course be done by inspecting or testing every item or piece in the lot. But this can be expensive and time consuming. Moreover, it is quite often unnecessary, as we shall see. Sound decisions can often be made from a sample from the lot. The criteria of quality of pieces and of the lot may be on the basis of measurements or the presence or absence of defects, that is, by attributes. It is the latter class of criteria, which we will take up in this chapter and the next. Sampling acceptance by measurements is discussed in Chapter 10.

9.1. WHY USE A SAMPLE FOR A DECISION ON A LOT?

There are several reasons why we may wish to use a sample from a lot or process for decision making as a basis for action, rather than inspecting or testing all pieces:

1. **To save money and time**: Unless the characteristic in question is extremely critical, sampling acceptance plans are often satisfactory for a decision. They can be set up so at to *reliably* distinguish between "good" lots and "rejectable" lots. The degree of reliability of such discrimination can be specified in advance and to any desired degree of confidence.
2. Very often the inspection of a sample of pieces can be and is done much more carefully than 100% inspection of the entire lot. This is because of fatigue and psychological factors. It may even be possible to find out the lot quality more accurately from a sample than from inspection of the entire lot! One hundred percent inspection can be notoriously poor. Some mechanical inspection can, of course, be highly accurate.
3. When the test is *destructive*, 100% testing of pieces is impossible, for then there would be no pieces left to use even if the lot were shown to have been perfect. In this case, the only possibility is to base the decision on the lot upon a sample.

At this point in the discussion of acceptance sampling, a word of caution is necessary. All of plans of this chapter and the next are based on a sample that is *representative* of the population from which it was withdrawn. In many, if not most, situations it may be difficult to obtain a truly representative sample. One auditor was faced with getting a sample of 55 baseball (tear gas) hand grenades from a storage area consisting of several hundred cases containing 50 hand grenades each. To obtain a truly representative sample, would usually require randomly choosing 55 cases and then randomly selecting one grenade from each of these cases. Since this would take a great deal of time, the sample of 55 was taken by selecting one of the cases on top of the pile and taking another five from an adjacent case. Was this sample representative? In nearly every sampling situation that this author has observed, samples were taken conveniently, i.e., from the front, top, or side of the product.

Convenience sampling is also seen in the selection of a sample from a lot of a powder chemical that is being dried in an oven. The sampling protocol states that the sample is to consist of portions of the chemical selected throughout the shelves of the drying oven. The process operator was observed scooping a bottle of the chemical from the front of the oven and taking it to the laboratory for analysis. Was it representative?

It is essential that the persons taking the sample be educated on the need for obtaining a representative sample, the techniques for obtaining such a sample on the specific products to be sampled and feedback on their performance.

9.2. LEVELS OF INSPECTING OR TESTING A LOT

Consider the problem of assuring that a lot of pieces is of adequate quality in all characteristics. How many characteristics are there on one piece? The author once saw a rather small piece part for the electronics industry that had some 60 dimensions of varying importance! In another case, there were 20 pages of specifications on insulated wire. It is simply not possible to give full attention to all characteristics. As a consequence, it can be said that inspection and testing of the characteristics will run from (1) none, (2) spot-checking, (3) sampling inspection, (4) 100% inspection, to (5) several hundred percent inspection. Which level to use will depend upon (1) the practical importance of the characteristic in the product, (2) the ease of controlling the characteristic, and (3) the producer's record or history on it.

In sampling inspection, as in this chapter and the next two, we are assuming that it is feasible and desirable to base the decision about a lot upon a sample. Because of the great flexibility in the setting up of sampling plans, such acceptance sampling can be used for incidental nonconformities of little significance, minor nonconformities, major nonconformities and even quite critical ones. We can set the risks or probabilities of wrong decisions at whatever levels we wish, keeping in mind that the smaller the risks are which we specify, the larger the sample size must be. One way to cover a multiplicity of different characteristics is to include them in classes of nonconformities. Thus, "major nonconformities" may include, say, 12 nonconformities (some measurable and some visual).

You can have almost any desired reliability built into your acceptance sampling plan provided you are willing to pay the price of the required sample size, and provided you will take steps to see that samples are randomly chosen, i.e., is representative of the population, and that inspection is done soundly.

9.3. THE OPERATING CHARACTERISTIC OF A PLAN

Probably the most important characteristic of a sampling plan is its operating characteristic (OC). This is the probability that a lot or process will be approved, called P_a for "probability of acceptance". Now such a probability for any given sampling plan is a function of the submitted lot or process quality, that is, the better this quality, the higher the probability of acceptance is. Of course, P_a also depends upon the criteria of the sampling plan itself, such as the sample size or sizes and the acceptance numbers given in the plan, as we shall be seeing. But always keep in mind the "if/then" character of any plan: if the lot is of such and such quality, then P_a is thus and so. In mathematical terms, P_a is a conditional probability, conditioned upon the submitted lot quality.

Furthermore in practice, P_a, as calculated, will assume random sampling and accurate inspecting or testing. If one samples only out of one corner of a lot where good pieces happen to be, but there are many nonconforming ones in other parts of the lot, it is possible to accept a very bad lot.

9.4. ATTRIBUTE SAMPLING INSPECTION

Up to now what we have said applies to both sampling inspection by attributes and by variables (or measurements). We will now concentrate on the former, namely where the lot quality of interest is concerned with attributes. As we have seen in Chapter 6, there are two ways to inspect:

1. By counting *nonconforming pieces* wherein each piece is "good", i.e., free of nonconformities, or else it is nonconforming, i.e., it possesses one or more nonconformities.
2. By counting the number of *nonconformities* on one or more pieces.

For the moment, we shall be primarily concerned with the case of nonconforming pieces.

Let us again emphasize that although the word "nonconformity" has a "bad" meaning to many and is undesirable to all, there are nonconformities of all sorts of severity and importance, from quite unimportant discrepancies all the way to very critical nonconformities. It is because of recent litigation that the use of "nonconforming" pieces, rather than "defective" pieces is being used by the American Society for Quality and the quality technologists.

9.5. CHARACTERISTICS OF SINGLE SAMPLING PLANS

A "single sampling" plan is one in which we take a random sample of n pieces from the lot or process, inspect or test them, at the end of which a decision is made to either accept or reject the lot (or process). On the other hand for "double sampling", we select a random sample of n_1 pieces and inspect or test them. The results lead to a decision to accept or to reject, or to request that another sample of n_2 pieces be drawn randomly from the remainder of the lot. These are then inspected or tested, after which a firm decision of acceptance or rejection is always reached. We may also have "multiple sampling" in which several samples may be required before reaching a decision.

Let us now set down a few notations for acceptance sampling based upon nonconforming pieces.

$$N = \text{number of pieces in the lot} \tag{9.1}$$

$$n = \text{number of pieces in the sample} \tag{9.2}$$

$$Ac = \mathrm{c} = \text{acceptance number for nonconforming pieces in sample} \tag{9.3}$$

$$\mathrm{Re} = \mathrm{c} + 1 = \text{acceptance number for nonconforming pieces in sample} \tag{9.4}$$

$$D = \text{number of nonconforming pieces in lot} \tag{9.5}$$

$$D = \text{number of nonconforming pieces in sample} \tag{9.6}$$

$$P_0 = D/N = \text{fraction nonconforming in the lot} \tag{9.7}$$

$$p_0 = d/n = \text{fraction nonconforming in the sample} \tag{9.8}$$

$$P_\mathrm{a} = \text{probability that lot will be accepted by plan} = P_\mathrm{a}\ (\text{given } p_0) = \text{probability of acceptance, if } p_0 \text{ in lot} \tag{9.9}$$

The acceptance–rejection decision process for single sampling may be diagrammed as follows:

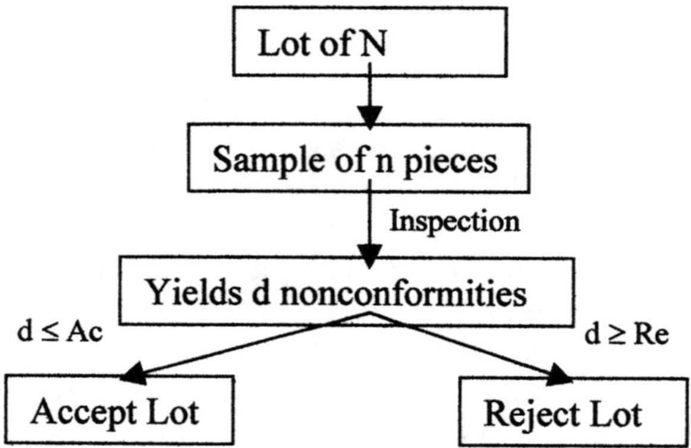

This is simple and exactly what you the reader would come up with, given a little thought.

Thus, a single sampling plan consists of three numbers: N for lot size, n for sample size, and Ac for acceptance number. Now how do we find out what any given plan does for us, for example, how well does it distinguish between lots of various qualities or p_o values?

9.5.1. The Operating Characteristic Curve

It is basic to know for lots of any given quality p_o how often they stand to be accepted by the sampling plan. Naturally if p_o is relatively small, then P_a will be high, i.e., close to 1. And the larger p_o becomes, the lower we can expect P_a to be. How do we obtain specific answers? This is accomplished through use of the appropriate attribute distribution from Chapter 4. We have the following two cases:

1. **Type A OC curve:** Probability of accepting a lot of N pieces, where account is taken of the lot size, N. The exact calculation involves using the "hypergeometric distribution" of Section 3.6, which you may not have studied. (But do not worry, it can be readily "approximated!")
2. **Type B OC curve:** Probability of acceptance of a lot of N chosen at random from a *process* with true fraction nonconforming, p. The exact calculation involves using the binomial distribution of Section 3.4. But again, we can in general obtain entirely useful approximations by using the Poisson distribution of Section 3.5 and thus Table B.

Type A OC curves are correct when we have a series of lots of N pieces, in each of which there are exactly D nonconforming pieces. This seems to be and is quite artificial. But if we should consider that we have just one lot of N, containing D nonconforming pieces, we likely would wish to know the probability of such an isolated lot being accepted by our sampling plan.

On the other hand, a Type B OC curve is, in effect, a series of lots of N chosen from a *process* with fraction nonconforming, p_o. This is usually what we are interested in if we have a series of lots. Given a process at p_o, what proportion of lots will be accepted in the long run? Here the binomial distribution is exactly correct (and the size of the lot, N, is immaterial.) If n is, say, at least 20 and p_o is 0.05 or less, the Poisson distribution is an excellent approximation to the binomial by letting $c_o = np_o$. Even when $n = 10$ and $p_o = 0.10$, the approximation is quite good. Therefore, we will use the Poisson distribution. (It is much easier to use too.)

Now for an example, let us find the Type B OC curve for the sampling plan

$$N = 1000 \quad n = 80 \quad \text{Ac} = 3$$

In the first column of Table 9.1 are listed some average lot or process fraction nonconforming of interest, and in the second column, the expected or theoretical average number of nonconforming pieces per sample, i.e., np_o, from (3.8). Here since $n = 80$, this is $80p_o$. Then using this to set the row in Table B, we use the column for $c = 3$ to give $P(d \leq 3) = P(d \leq \text{Ac}) = P_a$. Thus, the entries in Table B give directly the probabilities of acceptance. The last column gives the P_a's as found by the formally correct binomial distribution using an available table. Comparing the last

Table 9.1 Type B OC Curve Calculations for Plan: $n = 80$, $\text{Ac} = 3$, $\text{Re} = 4$

Process, p'	Expected defectiveness in sample, $80p'$	Poisson approximation for Pa from Table B	Exact binomial for P_a from Harvard Univ., Computer Lab (1955)
0.00	0	1.000	1.000
0.01	0.8	0.991	0.991
0.02	1.6	0.921	0.923
0.03	2.4	0.779	0.781
0.04	3.2	0.603	0.602
0.05	4.0	0.433	0.428
0.06	4.8	0.294	0.286
0.07	5.6	0.191	0.181
0.08	6.4	0.119	0.109
0.09	7.2	0.072	0.063
0.10	8.0	0.042	0.035

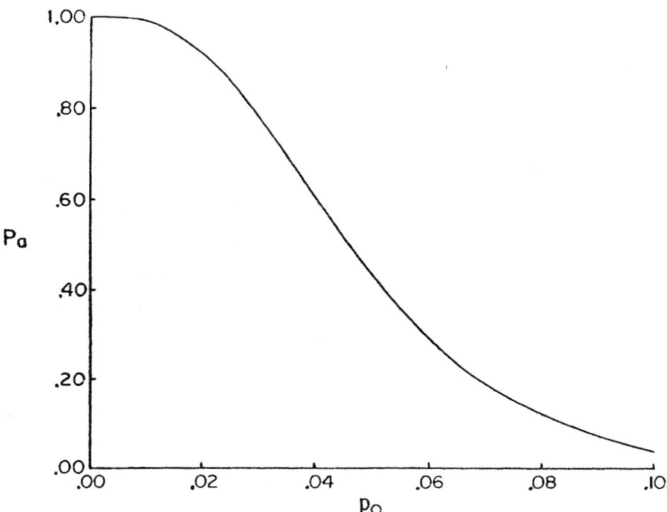

Figure 9.1 The type B OC curve for the single plan: $N = 1000$, $n = 80$, Ac = 3, Re = 4.

two columns of Table 9.1, we can see that the Poisson distribution provided an excellent approximation to the binomial.

Figure 9.1 shows the OC curve for the plan. If the full grid were shown, we could easily read off P_a for any given quality p_o. Or we can reverse the question and ask for what p_o is P_a 0.95? This appears on the graph to be about 0.017. Of course, we can answer such questions from Table B. In this instance, we look in the $c = 3$ column of Table B for P(3 or less) to be 0.950. We obtain

np_o	P(3 or less)
1.3	0.957
	7
	0.950
	4
1.4	0.946

The desired np_o is interpolated by $1.3 + (7/11)(0.1) = 1.364$. This being np_o we have but to divide by $n = 80$

for the desired p_o, that is, 0.017. Such a p_o being accepted 0.95 of the time is called $p_{0.95}$.

9.5.2. The Average Sample Number Curve

For a single sampling plan, we always inspect (or test) precisely n pieces in order to reach a decision. Thus, the average sample number (or ASN) for a single sampling plan is n. The graph of the ASN is merely a horizontal line at a height of n for the ASN curve for the whole collection of p's. However, as we shall see in Section 9.6, the ASN curve for a double sampling plan really is a curve. Thus, in order to reach a decision on a lot which contains no nonconformities, we will always accept on n_1 pieces. But as p_o increases above zero, we will sometimes require a second sample of n_2 additional pieces and it, therefore, takes $n_1 + n_2$ pieces for a decision. If p_o continues to increase, the chance of *rejecting* the lot on the first sample comes into play and the n_2 additional pieces may not be needed for a decision. The ASN begins to decrease toward n_1 again. The ASN is a *cost* characteristic of a sampling plan.

9.5.3. The Average Outgoing Quality Curve

This characteristic of a sampling plan is another index to the *protection* it supplies, the OC curve being the first such protection characteristic which we considered. The average outgoing quality is the average fraction defective for *all* lots submitted, after those lots which were rejected by the sampling plan have been sorted 100% and cleared of defectives. Thus, for example, suppose a series of lots comes in at about 2%, that is, $p_o = 0.02$, and we sample inspect each lot. Some lots will be accepted as they stand, still at about 2%. Meanwhile other lots, even though they also had $p_o = 0.02$, will be rejected and screened of defectives so that for them p_o has become zero. Therefore, the average outgoing quality is a weighted average of $p_o = 0.02$ and $p_o = 0.00$. In particular, suppose $P_a = 0.80$. This means that 80% of the lots are passed and remain at $p_o = 0.02$, whereas 20% of the lots are rejected and rectified by screening so that $p_o = 0.00$. Therefore, using the respective weights 0.80 and 0.20 (which add to one), we have

$$\text{AOQ} = 0.80(0.02) + 0.20(0.00) = 0.016$$

In general, it can be seen that we have the following formula, since the second term, $(1 - P_a)(0.00)$, always drops out:

$$\text{AOQ} = (P_a)p_o \tag{9.10}$$

Now there are in practice various ways of handling the nonconforming pieces found in the samples and the sorted remainder of the lots. That is, we may consider that we replace by good pieces each nonconforming piece found in the sampling and/or in the sorted remainder. If replaced, this gives more good pieces in the outgoing lots than if not replaced. Also there is the question as to whether we are using Type A calculations (all lots at exactly p_o) or Type B (lots chosen randomly from a process at p_o). However, for nearly all situations in practice, it makes comparatively little difference on the AOQ which assumptions we use. In general we use () for AOQ's. For our plan of $n = 80$, Ac = 3, we have, using the Poisson column of Table 9.1:

p_o	0.00	0.01	0.02	0.03	0.04	0.05
AOQ	0.0000	0.0099	0.0184	0.0234	0.0241	0.0217
p_o	0.06	0.07	0.08	0.09	0.10	
AOQ	0.0176	0.0134	0.0095	0.0065	0.0042	

The AOQ curve is drawn in Figure 9.2. We see there that the curve rises initially along the 45° line AOQ $= p_o$, then

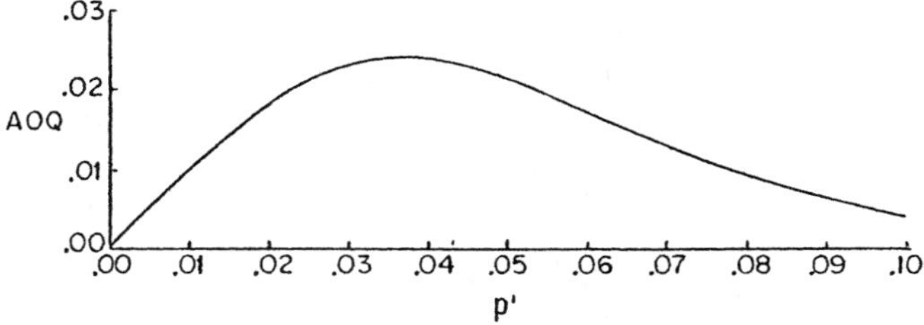

Figure 9.2 Average outgoing quality curve for the single plan $N = 1000$, $n = 80$, Ac = 3, Re = 4, AOQL = 0.024 (maximum AOQ).

begins to fall away from this line, reaching a maximum AOQ of about 0.0242 when p_o is about 0.036.

Definition. The maximum of the AOQ's for all p_o values is called the *average outgoing quality limit* (AOQL) of the plan.

Thus, the AOQL tells us what is the worst long-run *average* fraction nonconforming we will have to work with, no matter what comes in. But p_o has to be at just one level for the average outgoing quality to approach the AOQL. Thus, for the sampling plan being analyzed, p_o must be close to 0.036 in order to have an AOQ close to the limit AOQL of 0.0242. See the curve. For p_o on either side of 0.036, the AOQ will be less.

9.5.4. The Average Total Inspection Curve

The fourth characteristic of a sampling plan considered here is also a cost curve as was the ASN curve.

Definition. *The average total inspection* (ATI) for a sampling plan is the average number of pieces inspected per lot for a series of lots, including those pieces in samples and the pieces inspected in the remainder of rejected lots.

Thus, the ATI measures the total inspection load to maintain an AOQ for given p_o. The ATI depends upon the lot size, the sampling plan, and, of course, the incoming fraction defective p_o.

In order to calculate the ATI for any p_o, we note that for the accepted lots, we only inspect n pieces. A proportion, P_a, of all the lots are accepted. Meanwhile for the rejected lots, we look at n pieces for the decision, and, having rejected the lot we now must look at $N-n$ *additional* pieces. Among the lots, the proportion of rejected lots is $1-P_a$. Since we look at n pieces in all lots, we have as an average inspection load:

$$\text{ATI} = n + (N-n)(1-P_a) \quad \text{(single sampling)} \tag{9.11}$$

The ATI for a single sampling plan starts at n for $p_o = 0$, since then the $P_a = 1$ and the second term on the right- h and side of (9.11) drops out. As p_o increases, P_a starts to decrease, and the second term begins to increase. The ATI

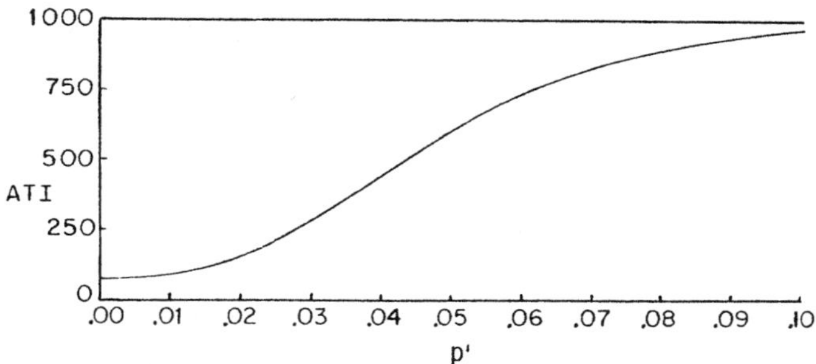

Figure 9.3 The average total inspection curve for the single sampling plan $N = 1000$, $n = 80$, $\text{Ac} = 3$, $\text{Re} = 4$.

rises above n and eventually reaches the lot size N, when the probability of acceptance P_a becomes zero.

The ATI curve is shown in Figure 9.3. The calculations are simple. For our example, we have $n = 80$, $N - n = 920$, and we use P_a from the Poisson approximation column of Table 9.1. Thus, we have the following if $p_o = 0.04$: $P_a = 0.603$, $1 - P_a = 0.397$,

$\text{ATI} = 80 + 920(0.397) = 80 + 365 = 445$. Continuing we obtain

p_o	0.00	0.01	0.02	0.03	0.04	0.05
$1 - P_a$	0.000	0.009	0.079	0.221	0.397	0.567
$(N - n)(1 - P_a)$	0	8	73	203	365	522
ATI	80	88	153	283	445	602
p_o	0.06	0.07	0.08	0.09	0.10	
$1 - P_a$	0.706	0.809	0.881	0.928	0.958	
$(N - n)(1 - P_a)$	650	744	811	854	881	
ATI	730	824	891	934	961	

The four curves, two protection—OC and AOQ—and two cost—ASN and ATI—give a wealth of information about a sampling plan and how it operates. In passing, we may mention that none of these curves is of much practical importance beyond a p_o value where P_a is 0.6. If such a p_o quality level should continue, there would be so many rejected lots that sampling inspection would be discontinued and action taken to have the producer drastically improve his process. In our example, this would mean that if p_o goes much above 0.04, we would drop sampling inspection and begin 100% sorting each lot until the quality level improves.

9.6. DOUBLE SAMPLING PLANS AND THEIR CHARACTERISTICS

We now take up the somewhat more complicated but highly useful subject of double sampling. The objective of double sampling is to obtain a more favorable (lower) ASN curve while still supplying the same power of discrimination, that is, the same OC curve. Two other advantages of double sampling are (1) the psychological appeal to giving a lot a second chance and (2) the fact that no lot is ever rejected on less than two nonconforming pieces under double sampling.

The following notations will be used for double sampling, acceptance–rejection criteria:

$$n_1 = \text{number of pieces in first sample} \tag{9.12}$$

$$n_2 = \text{number of pieces in second sample (when required)} \tag{9.13}$$

$$d_1 = \text{number of nonconforming pieces in first sample} \tag{9.14}$$

$$d_2 = \text{number of nonconforming pieces in second sample} \tag{9.15}$$

$$\text{Ac}_1 = \text{acceptance number for } d_1 \tag{9.16}$$

$$\text{Re}_1 = \text{rejection number for } d_1 \tag{9.17}$$

$$\text{Ac}_2 = \text{acceptance number for } d_1 + d_2 \tag{9.18}$$

$$\text{Re}_2 = \text{rejection number for } d_1 + d_2 \tag{9.19}$$

$$\text{Re}_2 = \text{Ac}_2 + 1 \tag{9.20}$$

These notations can perhaps better be described by an example than by a flow chart. As our example we shall take a double sampling plan, which has an OC curve quite similar to that of the single plan we have already analyzed. This is the plan $N = 1000$, $n_1 = 50$, $n_2 = 50$, $\mathrm{Ac}_1 = 1$, $\mathrm{Re}_1 = 4$, $\mathrm{Ac}_2 = 4$, and $\mathrm{Re}_2 = 5$.

For the operation of this plan, we take a random sample of 50 from the lot of 1000 and inspect the pieces, finding d_1 nonconforming pieces. Now if d_1 is 0 or 1, that is, less than or equal to Ac_1 we accept the lot at once. Or if d_1 is 4 or more we reject the lot at once. But if d is 2 or 3, we are in between the acceptance and rejection numbers for the first sample, and cannot make a decision until after we have drawn a second sample of 50, randomly, from the 950 pieces as yet uninspected in the lot. Then after this inspection yields d_2 additional nonconforming pieces, we find the total nonconforming $d_1 + d_2$ in the two samples. Now if $d_1 + d_2$ is less than or equal to $\mathrm{Ac}_2 = 4$, we accept the lot, but if $d_1 + d_2$ is equal to or greater than $\mathrm{Re}_2 = 5$, we reject the lot. Since there is no gap between Ac_2 and Re_2 there is always a firm decision after completing inspection of the second sample.

Let us summarize the foregoing by a general flow chart.

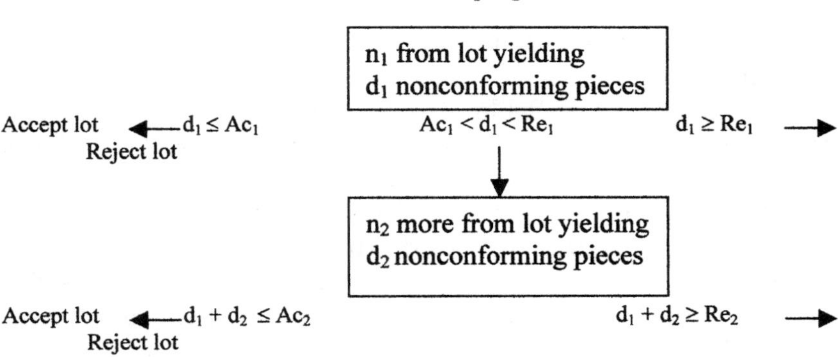

9.6.1. The OC Curve, Double Sampling

For the OC curve, we need to find the probability for a lot of given incoming fraction nonconforming, p_o, to be accepted by a given double sampling plan. This probability of acceptance, P_{a}, is made up of several distinct ways in which acceptance can occur. Thus, for the double sampling plan which we are using as an example, these distinct routes for acceptance are

On first sample of 50: $d_1 = 0,\ d_1 = 1$

On second sample with total pieces inspected of 100, the acceptable combinations are

$d_1 = 2 \quad d_2 = 0$

$d_1 = 2 \quad d_2 = 1$

$d_1 = 2 \quad d_2 = 2$

$d_1 = 3 \quad d_2 = 0$

$d_1 = 3 \quad d_2 = 1$

These may be grouped as follows in order to use Table B—the Poisson distribution:

On first sample: $d_1 \leq 1$

On second sample: $d_1 = 2$ with $d_2 \leq 2$

$d_1 = 3$ with $d_2 \leq 1$

Then we have only to calculate the probabilities for each way and add.

We recall that Table B—the Poisson distribution—has cumulative or additive probabilities such as $P(d \leq 3)$ for a given np_o or c_o. Therefore, to find $P(d = 2)$, we use $P(d \leq 2) - P(d \leq 1)$. We now work out P_a for our sampling plan, for the case where $p_o = 0.04$. We can do this calculation in a simple table form as follows. A little explanation may be useful. On the first sample, as explained in general terms, $P(2) = P(2 \text{ or less}) - P(1 \text{ or less}) = 0.667 - 0.406 = 0.271$. Similarly, $P(3) = 0.857 - 0.677 = 0.180$. Now suppose that we had 1000 lots coming in from a process with $p_o = 0.04$. Of these, we would expect (on the average) to have 406 accepted on the first sample by having only 0 or 1 nonconforming pieces. Then also we would expect to have

$p_o = 0.04 \ n_1 = 50 \ n_2 = 50$

Expectations: $n_1 p_o = 2. \quad n_2 p_o = 2$

$p_o = 0.04$	$n_1 = 50$	$n_2 = 50$		Contrib to P_a
Expectations:	$n_1 p_o = 2.$	$n_2 p_o = 2$		
$P(1 \text{ or less}) = 0.406$				0.406
$P(2) = 0.271$	$P(2 \text{ or less}) = 0.677$		$0.677(0.271)$	$= 0.183$
	$P(2 \text{ or less}) = 0.677$			
$P(3) = 0.180$	$P(1 \text{ or less}) = 0.406$		$0.406(0.180)$	$= 0.073$
$P(3 \text{ or less}) = 0.857$				$P_a = 0.662$

Two hundred and seventy-one lots go into a second sample with two nonconforming pieces against them. Of these, 0.617 will be passed by the second sample yielding 2 or less nonconforming pieces (total 4 or less). Thus of the original 1000 lots, we expect to pass 0.611 of the 271 lots yielding $d = 2$, or 0.677(0.271) as a probability. Similarly, we expect to have 180 lots of the 1000 with three nonconforming pieces against them from the first sample. To pass they can afford not over 1 nonconforming piece on the second sample. The probability of this latter event is 0.406. So the probability for this route to acceptance is 0.406(0.180) or 0.073. Adding the probabilities for the three distinct routes to acceptance yields $P_a = 0.662$. Let us do one more for comparison:

$p_o = 0.08$ $n_1 = 50$ $n_2 = 50$

Expectations: $n_1 p_0 = 4.0$ $n_2 p_o = 4.0$

Contribution to P_a

$P(1 \text{ or less}) = 0.092$

0.092

$P(2) = 0.146$ $P(2 \text{ or less}) = 0.238$

$0.238(0.146) = 0.035$

$P(2 \text{ or less}) = 0.238$

$P(3) = 0.195$ $P(1 \text{ or less}) = 0.092$

$$0.092(0.195) = \underline{0.018} \qquad P(3 \text{ or less}) = 0.435$$
$$P_a = 0.145$$

In these calculations, we have used the Poisson distribution (Table B) as an approximation. The correct distribution for a series of lots chosen from a process at p_o is the binomial. As an illustration of the accuracy, the author used binomial tables (Harvard Univ., Comput. Lab., 1955) on the first calculation. This yielded $P_a = 0.66116$ instead of 0.662.

Doing other similar calculations to those we have shown yields

p_o	0.00	0.01	0.02	0.03	0.04	0.05	0.06	0.07	0.08	0.09	0.10
P_a	1.000	0.996	0.950	0.831	0.662	0.488	0.399	0.224	0.145	0.091	0.057

OC curves are shown in Figure 9.4 for our single and double sample plan examples. It is easily seen that the double

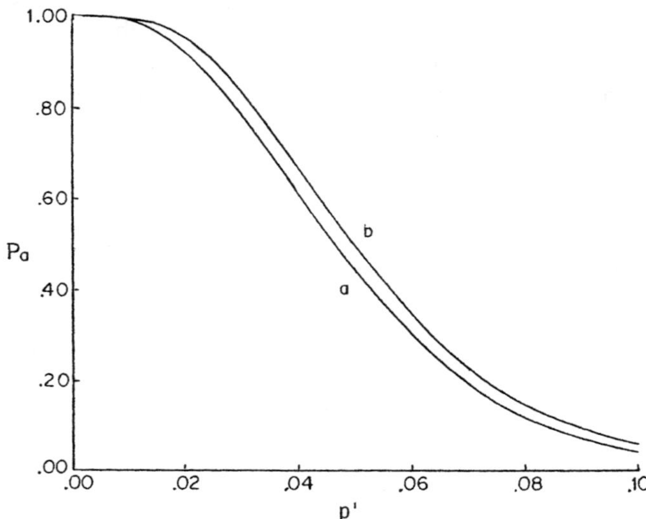

Figure 9.4 Operating characteristic curves (Type B) for (a) the single plan $n = 80$, $\mathrm{Ac} = 3$, $\mathrm{Re} = 4$ and (b) the double plan $n_1 = n_2 = 50$, $\mathrm{Ac}_1 = 1$, $\mathrm{Re}_1 = 4$, $\mathrm{Ac}_2 = 4$, $\mathrm{Re}_2 = 5$.

sampling plan is everywhere more lenient than the single plan, but that they give rather similar protection against making wrong decisions on lots. Thus, lots at $p_o = 0.00$–0.02 are accepted the great majority of the time, and those at $p_o = 0.08$ or more are seldom accepted whenever offered.

9.6.2. The Average Sample Number Curve, Double

Let us now consider the determination of the average sample number (ASN) curve, that is, the *average* number of pieces to be inspected per lot to reach a decision. For any double sampling plan we always inspect a first sample of n_1 pieces, but then go on to inspect a second sample of n_2 pieces only some of the time. How often does this occur, i.e., what is its probability? Now if

$$d_1 = \mathrm{Ac}_1 \quad \text{or} \quad d_1 \geq \mathrm{Re}_1$$

we have a decision on the first sample, respectively, acceptance and rejection. But if d_1 lies *between* Ac_1 and Re_1 then a second sample is required, i.e., when $\mathrm{Ac}_1 < d_1 < \mathrm{Re}_1$. What is the probability? Let us see by studying the calculation table for P_a when p_o was 0.04. There we see that when $d_1 = 2$ or 3 a second sample is needed, so we want $P(2) + P(3) = 0.271 + 0.180 = 0.451$. Another way to obtain this is by P(3 or less) $-$ P(1 *or* less). Hence at $p_o = 0.04$

$$\mathrm{ASN} = 50 + 0.451(50) = 72.6$$

since of 1000 lots at 0.04, all need 50 pieces inspected, and we expect that 451 lots of the 1000 will require a second sample for a decision. So, on the *average*, the number inspected is as given. Generalizing

$$\mathrm{ASN} = n + nP(\mathrm{Ac} < d_1 < \mathrm{Re})(\text{doubling sampling}) \tag{9.21}$$

Notice especially that there are no equal signs in the probability, only inequality signs.

Similarly for $p_o = 0.08$, we have $P\ (1 < d_1 < 4) = P$ (3 or less) $-$ P (1 or less) $= 0.433 - 0.092 = 0.341$. Therefore,

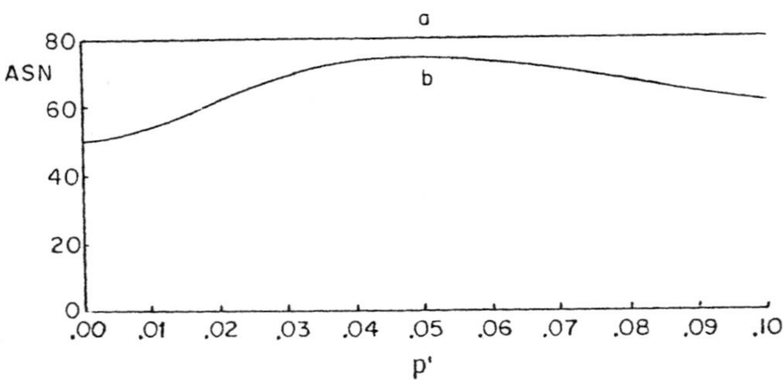

Figure 9.5 Average sample number curves for two quite comparable sampling plans: (a) single, $n = 80$, Ac = 3, Re = 4; (b) double, $n_1 = n_2 = 50$, $\text{Ac}_1 = 1$, $\text{Re}_1 = 4$, $\text{Ac}_2 = 4$, $\text{Re}_2 = 5$.

$$\text{ASN} = 50 + 50(0.341) = 67.0$$

Proceeding we may find

p_o	0.00	0.01	0.02	0.03	0.04	0.05
ASN	50.0	54.4	62.2	68.8	72.6	73.6
p_o	0.06	0.07	0.08	0.09	0.10	
ASN	72.4	70.0	67.0	64.0	61.2	

These results are plotted in Figure 9.5, giving a curve starting at ASN = 50, increasing to a maximum of around 73.6, then dropping back gradually toward 50. Also shown is the ASN graph for the single plan, a straight line. The ASN for the double plan is all times below that for the single plan. Remember, however, that for the double plan the actual number of pieces to be inspected in a lot will be either 50 or 100, depending upon whether or not a second sample is required for a decision.

9.6.3. The Average Outgoing Quality Curve

As we have seen, this characteristic of a sampling plan gives the average fraction nonconforming, outgoing, when lots come in at fraction defective p_o. The average outgoing quality (AOQ) is made up of the accepted lots, still at about p_o, and the rejected lots which have been screened of nonconforming pieces and are now substantially free of them, giving a zero fraction nonconforming. We could have either of two cases (1) all lots exactly at some fraction nonconforming $p_o = D/N$, and (2) lots of N chosen at random from a process in control at p_o. Then too, there are cases according to whether we replace with good pieces all defective pieces found in the samples and/or the sorting. However, unless the lot size is quite small, these various cases all give quite closely the same AOQ. Thus as a practical matter we may again use the simple formula (9.10) , that is,

$$\text{AOQ} = (P_a)p_o \tag{9.10}$$

Using this formula we find the following AOQ's for the double sampling plan being studied.

p_o	0.00	0.01	0.02	0.03	0.04	0.05
AOQ	0.0000	0.0100	0.0190	0.0249	0.0265	0.0244
p_o	0.06	0.07	0.08	0.09	0.10	
AOQ	0.0203	0.0157	0.0116	0.0082	0.0057	

These are plotted in Figure 9.6 along with the AOQ curve for the single sampling plan for comparison. Since P_a for the double sampling plan was everywhere above that for the single plan (except at $p_o = 0$) by (9.10) the AOQ curve for the double sampling plan lies everywhere above that for the single plan (except at $p_o = 0$).

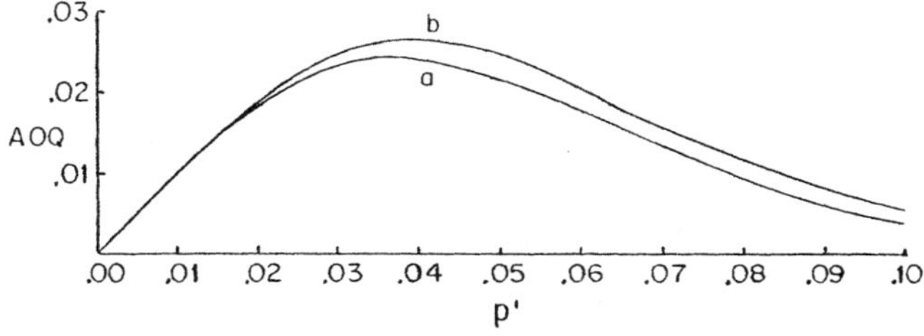

Figure 9.6 Average outgoing quality curves for the two sampling plans (a) single, $n = 80$, $\text{Ac} = 3$, $\text{Re} = 4$; (b) double, $n_1 = n_2 = 50$, $\text{Ac}_1 = 1$, $\text{Re}_1 = 4$, $\text{Ac}_2 = 4$, $\text{Re}_2 = 5$.

For the double sampling plan, the maximum AOQ, i.e., the AOQL, is about 0.0265, whereas for the single plan it is about 0.0242. Thus, no matter what the incoming lot fractions nonconforming may be, the average *outgoing* fractions nonconforming will be no worse than these respective AOQL values. In practice, the actual outgoing fraction nonconforming is much less than the AOQL because to reach this level, p_o would have to remain at the level giving a peak on the AOQ curve. And this occurs when P_a is low, say at 0.60. Should p_o persist at such a level, sampling acceptance would soon be abandoned in favor of 100% sorting.

9.6.4. The Average Total Inspection (ATI) Curve, Double Sampling

This total number inspected per lot, made up of sampled pieces for a decision and of 100% screening of pieces in rejected lots make up the ATI. There are many possible formulas for ATI in double sampling. But it seems easiest to understand by using three categories: (1) lots accepted on the first sample; (2) lots accepted after a second sample; and (3) lots rejected. The respective inspection loads are (1) n_1, (2) $n_1 + n_2$ and (3) N. Meanwhile, the probabilities of such cases are, respectively, (1) $P(d_1 \leq \mathrm{Ac}_1)$, (2) $\mathrm{P_a} - P(d_1 \leq \mathrm{Ac}_1)$ and (3) 1 – $\mathrm{P_a} = P(\text{reject})$. The first is for immediate acceptance on the first sample, while the second covers the remainder of the *accepted* lots. Then the third covers all of those lots which are *rejected.* Using the probabilities as weights of the respective inspection amounts, we have

$$\begin{aligned}\text{ATI} = n_1 \cdot P(d_1 \leq \text{A}]c_1) + (n_1 + n_2) \cdot [P_\text{a} - P(d \leq \text{A}c_1)] \\ + N \cdot [1 - P_\text{a}](\text{double sampling}) \qquad (9.22)\end{aligned}$$

The calculation of such an ATI looks complicated to one not accustomed to using formulas. But it is really quite easy to do, using the tables made in finding P_a for a double sampling plan. Let us see.

For our double sampling example, we have at $p_o = 0.04$, $P_a = 0.662$, and also $P(d_1 \leq \mathrm{Ac}_1) = P(d_1 \leq 1) = 0.406$. Therefore,

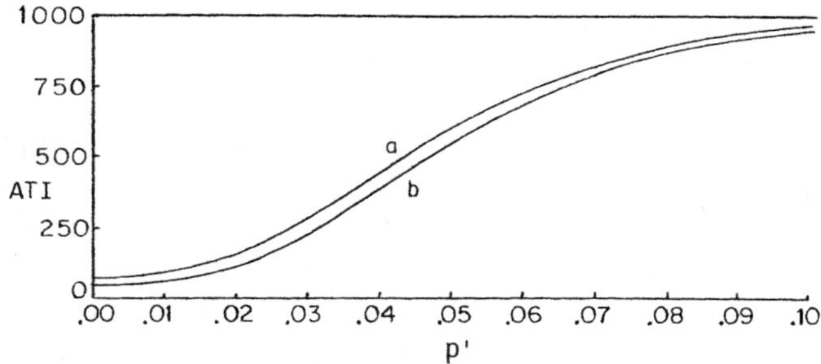

Figure 9.7 Average total inspection curves for lots of $N = 1000$, for (a) single plan, $n = 80$, $\text{Ac} = 3$, $\text{Re} = 4$, and (b) double plan, $n_1 = n_2 = 50$, $\text{Ac}_1 = 1$, $\text{Re}_1 = 4$, $\text{Ac}_2 = 4$, $\text{Re}_2 = 5$.

using (9.22),

$$\begin{aligned} \text{ATI} &= 50(0.406) + 100(0.662 - 0.406) \\ &\quad + 1000(1 - 0.662) = 20 + 26 + 338 = 384 \\ \text{At } p_o &= 0.08, \text{we have } P_a = 0.145, P(d_1 \leq 1) \\ &= 0.092 \text{ and hence} \\ \text{ATI} &= 50(0.092) + 100(0.145 - 0.092) \\ &\quad + 1000(1 - 0.145) = 5 + 5 + 855 = 865 \end{aligned}$$

Proceeding in this way we obtain

p_o	0.00	0.01	0.02	0.03	0.04	0.05	0.06	0.07	0.08	0.09	0.10
ATI	50	59	109	224	384	546	685	192	865	915	947

This ATI curve is shown in Figure 9.7, along with the ATI curve for the single sampling plan. The latter is above the former, in part because it has a uniformly higher ASN curve.

9.6.5. Truncated Inspection and the ASN Curve

Truncated or curtailed inspection is quite often used in practice. What this means is that for decision making it may be possible to stop sampling inspection on a lot before completing the sample size(s) called for in the plan. For example, in our single sampling plan $n = 80$, Ac $= 3$, Re $= 4$, it could happen that the fourth nonconforming piece occurs on the 37th piece inspected. It has now become certain that the lot will be rejected, because even if the remaining 43 pieces to complete $n = 80$ were all good ones we would still have $d \geq \text{Re} = 4$. This could be quite a saving in the case of a rejection. On the other hand, in the case of lot acceptance, much smaller savings might occur. The maximum saving would be if all of the first 77 pieces inspected were good. Then, even if all of the remaining three pieces should be nonconforming, we would still have $d < \text{Re} = 4$. Hence, we would know what we could accept. But 77 is the smallest number to which we could ever truncate under *acceptance*. Thus, with such small potential saving *under acceptance*, we do not bother with truncation.

But even with lot *rejection* under *single* sampling, we ordinarily do not truncate. Instead we complete the inspection of the sample for the record, so as to have the same sample size for each lot; thus we can give each lot the same weight in determining p. In fact, this also makes easier the construction of a control chart record of performance, p or np, which is an excellent thing to maintain in acceptance sampling.

Now with double sampling the situation is a bit different. In general, we will always complete the first sample for the record. Since many lots do not call for a second sample, only the results of the first samples will be used for the control chart record. Thus, we are free to truncate the second sample. But for the reasons given above, it only pays to truncate under early rejection. But this does provide substantial savings and gives a lower ASN curve than such a one as in Figure 9.5, especially at relatively high p_0. (See Burr, 1976, for an explanation of how to make the calculations.)

9.6.6. Use of Calculations by Others

Sections 9.5 and 9.6 on the characteristics of single and double sampling plans have been given so that you may be able to understand the meanings of them and how the curves are developed. That is, this material provides a broad orientation to acceptance sampling plans. But you may never need to calculate any plans yourself. The reason is that there are so many available which have been calculated by others. (See the published plans in Chapter 10.) Then too, if you do need some curves calculated, there is likely to be available in your organization calculators and programmers who can do the job for you. But you should know what is going on and what the results mean.

9.7. ACCEPTANCE SAMPLING FOR NONCONFORMITIES

Up to now in this chapter we have been studying plans where the acceptance–rejection criteria were in terms of counts of *nonconforming pieces* in samples, and determining how the characteristics depended upon lot or process quality, p_o. Let us now consider criteria in terms of counts of *nonconformities*. We are quite likely to use such an approach to quality when we are concerned with a variety of possible nonconformities, any one of which makes a piece nonconforming. Moreover, two or three nonconformities on a single piece still gives only one nonconforming piece. Now if c_o is so small that we have on the average only five *nonconformities* per hundred pieces, it is unlikely that two or more of the five occur on a single piece. But if c_o is great enough to yield 10 nonconformities per 100 pieces, duplication becomes more likely and a c_o of 10 per 100 is about equivalent to a p_o of 0.09. So with such a c_o or a higher one, we tend to use criteria in terms of counts of nonconforming pieces rather then nonconformities, although either could be used. Then instead of criteria such as $d \leq \text{Ac}$ or $d \geq \text{Re}$, we merely change to counts, c, on samples of n pieces and use $c \leq \text{Ac}$, $c \geq \text{Re}$. And interest centers on P_a, AOQ, ASN, or ATI vs. lot or process *nonconformities per 100 units* rather than nonconforming units per 100 pieces, that is, percent nonconforming. The basic distribution for such calculations on defects is the Poisson (Table B) rather than the binomial.

9.8. FINDING A SINGLE SAMPLING PLAN TO MATCH TWO POINTS ON THE OC CURVE

An approach to acceptance sampling plans which has rather wide appeal is to set two fractions nonconforming, p_o, and *appropriate* risks or probabilities of wrong decisions. This makes especially good sense when we are concerned with a single isolated lot, or with lots coming along at quite infrequent intervals. Then we want to be given good assurance that we will accept a lot if $p_o = D/N$ is at some acceptably low p_o, and also good assurance that we will reject a lot if p_o is at some rejectably high p_o. For example, we might ask for P_a to be 0.95, if $p_o = 0.01$. Thus, the producer of a lot with $p_o = 0.01$ is subjected to a risk of only 0.05 of his lot being rejected. On the other hand, a lot at $p_o = 0.04$ might be considered quite undesirable, and we might be willing to accept such a lot only 10% of the times offered, that is, $P_a = 0.10$. Thus, we desire to find a plan with the following two points on the OC curve:

$$\text{For } p_1 = 0.01, \quad P_a = 0.95 \quad \text{and for } p_2 = 0.04, \quad P_a = 0.10$$

Now with such a plan what are the odds on a lot at say $p_o = 0.02$? This can be found from the complete OC curve for the plan which was found. But obviously its P_a would lie between 0.95 and 0.10.

Here, let us say a word about "risks". At $p_o = 0.01 = p_1$, we might be willing to accept all such lots. But decision making is always subject to some risks of wrong decisions. Here the risk of *rejecting* a lot at $p_o = 0.01$ is set to be 0.05. Thus, about one time in twenty such an *acceptable* lot will be rejected. This risk or probability is often called the producer's risk, PR, and is sometimes known by the small Greek alpha, α. On the other hand, the consumer regards lots at $p_o = 0.04$ as undesirable and would like to never accept such lots. But in order to avoid having to sort every lot 100%, the consumer must be willing to take some risk of erroneously *accepting* such a *rejectable* lot, should one be offered. This risk was set at 0.10 and is called the consumer's risk, CR, and is sometimes known by the small Greek beta, β. Thus the consumer is willing to erroneously accept such lots 10% of the times they are offered. The 0.05 value for PR and 0.10 value for CR have somehow gotten quite well entrenched in the literature, but they do make quite good sense.

Now if you are willing to use a 5% PR and a 10% CR, the problem of finding a plan for nonconforming pieces which matches as closely as possible the two points on the OC curve

$$p_1, P_\mathrm{a} = 0.95 \text{ and } p_2, P_\mathrm{a} = 0.10$$

becomes very easy using Table 9.2. This follows the plan in Peach and Littauer (1946).

Let us use Table 9.2 on our problem. We first take the desired fractions nonconforming, $p_2 = 0.04$ and $p_1 = 0.01$ and divide for the ratio $p_2/p_1 = 4$. (The smaller such a ratio is the more discriminating power we are asking of our plan.) Then we look for 4.00 in columns 1 and 4. Our ratio 4 lies between 4.06 and 3.55; so we take the row with 3.55 and find Ac = 5. Now we need a sample size n. The third entry in this row is $np_2 = 9.27$. But we started with $p_2 = 0.04$. Thus we have $n(0.04) = 9.27$, giving $n = 9.27/0.04 = 232$. Therefore, the desired plan is $n = 232$, Ac = 5, Re = 6. This plan matches the 0.04, $P_\mathrm{a} = 0.10$ point on the OC curve as well as the

Table 9.2 Operating Ratios, p_2/p_1, and Acceptance Numbers for Finding a Single Attribute Plan Matching Two Points on the OC Curve, with PR = α = 0.05 and CR = β = 0.10

p_2/p_1	Ac	np_2	p_2/p_1	Ac	np_2
44.7	0	2.30	2.50	10	15.41
10.9	1	3.89	2.40	11	16.60
6.51	2	5.32	2.31	12	17.78
4.89	3	6.68	2.24	13	18.96
4.06	4	7.99	2.18	14	20.13
3.55	5	9.27	2.12	15	21.29
3.21	6	10.53	2.07	16	22.45
2.96	7	11.77	2.03	17	23.61
2.77	8	12.99	1.99	18	24.76
2.62		9	14.21		

Poisson approximation allows. Meanwhile the PR at $p_1 = 0.01$ is less than 0.05. This is because we did not hit p_2/p_1 exactly, but instead took an operating ratio of 3.55. We arrived at a slightly more discriminating OC curve than we originally asked for. We can find our actual PR at $p_1 = 0.01$ by using Table B. Thus, for $np_1 = 2.32$, $P(5 \text{ or less}) = 0.968$, $PR = 1 - 0.968 = 0.032$. The full OC curve can be drawn as in Section 9.5.1.

A set of single sampling plans following this approach is given in Kirpatrick (1965).

9.9. SOME PRINCIPLES AND CONCEPTS IN SAMPLING BY ATTRIBUTES

By way of review and summary, we now give some general material in acceptance sampling for attributes.

9.9.1. Prerequisites to Sound Acceptance Sampling for Attributes

For sound decision making on a lot or process, we should take care to provide the following prerequisites:

1. Each type of nonconformity must be clearly defined, so that insofar as possible everyone involved will call any given imperfection a "nonconformity", or else everyone will call it "no nonconformity". This is especially important in all measurable characteristics which can make a piece good or nonconforming. There must be a good way to measure the piece or an accurate go-not-go gauge. Gauges and measuring instruments need frequent checking or calibration.
2. Samples must be taken in an unbiased way, preferably by a random procedure. At the very least, the sample or samples should be drawn from all portions of the lot. If there are only a few pieces in the lot and these can be numbered, then we may use random numbers of Table D, starting at a random spot, to select the pieces for our sample. Or if there are quite

a few cartons in the lot, we may use Table D to select a sampling of cartons to be opened, each of which is then sampled to obtain the desired total of n pieces. For example, suppose we have 60 cartons and decide to select 20, and from each carton four pieces to give $n = 80$. We number the cartons 1–60, bring down a pencil point in Table D to give a two-digit pair of numbers. Say it is 37, then carton 37 is taken. Next, 06 is next below 37, so carton 6 is taken. Next is 73, so go to the next. Suppose it is 37 again. Then go to the next pair of digits until 20 cartons are chosen. *Results are unreliable unless samples are chosen in an unbiased manner.*

3. Sound inspection must be somehow *obtained* and *maintained* through careful training and supervision. Also there must not be undue pressure for speed in inspecting samples. (In one case, it was found that inspectors were making bets as to who could find the fewest nonconformities!)
4. The rules for decision making on lots from samples must be exactly adhered to. No flinching. Harold Dodge related an amusing case to the author. He and a foreman saw an inspector take a piece, glance at it, set it back and take another. (d was already equal to Ac.) They immediately asked the inspector about it. The reply was "Oh, this is random sampling so it does not matter which piece I take".

9.9.2. The Conditional Character of P_a

Let us emphasize again that P_a is an if-then probability. For example, in Section 9.8, if $p_o = p_1 = 0.04$, P_a is 0.10. That is, there is a 10% risk of accepting a lot at 0.04, *if offered.* But if no such lot (or a worse one) ever comes along, then there would be no risk at all of accepting a lot with $p_o = 0.04$. Thus, we must guard against statements such as "This plan gives a 10% risk of accepting a lot at 0.04". We should add "if offered," or "should one come along". Or even worse, we must not say "10% of the lots passed by this plan will be at $p_o = 0.04$".

9.9.3. What is a Nonconformity?

As has already been pointed out, nonconformities come in all degrees of severity and importance. Many are comparatively unimportant, such as mild blemishes, or a little grease not wiped off from a machine whereas others may be very serious indeed. Then there is the question as to whether the nonconformity may be caught in subsequent operations or in assembly, in which case its seriousness may be lessened. It may be wise to use the term "nonconforming piece" if it fails to meet exactly at all requirements.

9.9.4. The Proportion of the Lot Samples, n/N is Relatively Unimportant

This concept seems to be difficult for many to grasp. First, let us suppose that we are talking about Type A operating characteristic curves, i.e., we want the probability of acceptance for a series of lots, every one of which has exactly the same number of nonconformities, D, in the given constant lot size N. Thus, $p_o = D/N$ is constant from lot to lot. Let us now use the same sample size n and acceptance number Ac for various lot sizes N and see what effect the lot size has. For the curves in Figure 9.8, we used $n = 10$, $\mathrm{Ac} = 0$ for each. But for the bottom curve $N = 20$, so that this was 50% sampling.

Actually D can only be 0, 1, 2, and so on, so the only possible p_o values are 0.00, 0.05, 0.10, and so on. Thus, the "curve" should be just a series of dots. The same is true for the middle curve, for which $N = 50$, giving 20% sampling. (The exact calculation is done by the hypergeometric distribution, Section 4.4, but we used a table Lieberman and Owen, 1961). Finally, the top curve was for an infinite lot size N; in practice a very large lot size. Thus, for the top curve we are sampling at 0%!

Now for practical purposes, are not the top two OC curves practically identical, especially for quite low p_o where we are likely to use such a sampling plan? Even the bottom curve is not greatly different. Thus, the factor giving these OC curves is n and Ac, far more than n/N.

We also mention that the comparison was for a Type A OC curve. If our interest lies in Type B OC curves, that is,

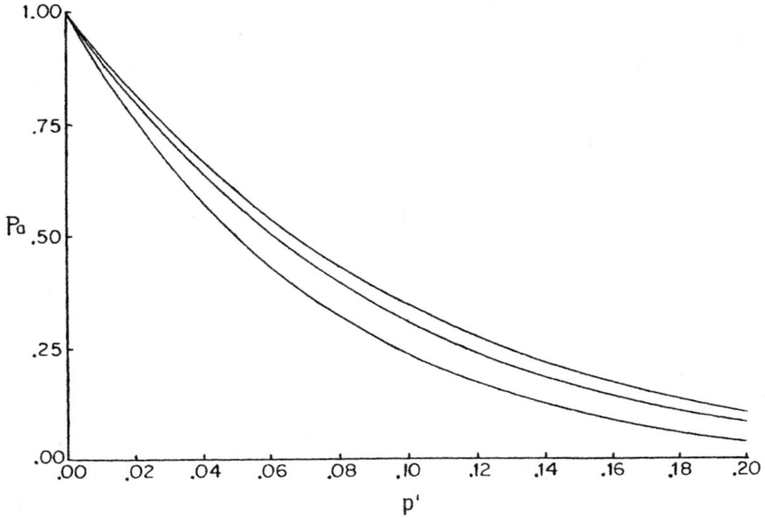

Figure 9.8 Three operating characteristic curves, all with $n = 10$, Ac= 0, Re = 1. They take account of lot size, that is, Type A OC curves. The bottom curve has $N = 20$, middle curve $N = 50$, top curve N infinite, that is, a Type B OC curve.

if we are concerned with P_a for a series of lots, each chosen from a process at p_o, then the OC curve for all N's *is the same*. In particular for $n = 10$, Ac = 0, the top curve of Figure 9.8 holds for all N's, and n/N has no influence at all!

9.9.5. Why not Sample a Fixed Percentage of Each Lot?

As we have just seen, the thing which basically determines the shape and discriminating power of a sampling plan is the sample size and acceptance–rejection numbers, rather than the proportion n/N of the lot we sample. Thus, provided we draw our sample randomly, a sample of 100, say, will do just as well on a lot of 1,000,000 pieces as it will on a lot of 500 or 1000.

To illustrate this further, let us consider what sampling 10% of the lot does for us. We will draw the usual type of OC curve (Type B) for 10% samples from lots of $N = 100$, 500,

2400, and 10,000. For each, we will use an acceptance number so that P_a is about 0.90 when $p_o = 0.01$. Thus, if $N = 100$, $n = 0.10$, $N = 10$, and using Ac = 0 gives $P_a = 0.904$. Similarly, if $N = 500$, $n = 50$, and letting Ac = 1, gives $P_a = 0.911$ at $p_o = 0.01$. The other plans are

$$N = 2400 \quad n = 240 \quad \text{Ac} = 4 \quad P_a = 0.905 \text{ at } p_o = 0.01$$
$$N = 10{,}000 \quad n = 1000 \quad \text{Ac} = 14 \quad P_a = 0.918 \text{ at } p_o = 0.01$$

We cannot hit $P_a = 0.90$ exactly because we have to make a choice between whole numbers for our acceptance number.

In Figure 9.9 we show the four OC curves. Although each plan samples 10% of the lot and each has P_a about 0.90 when p_o is at 0.01, the curves are markedly dissimilar. The OC curve for $n = 10$, Ac = 0 is very lazy and undiscriminating because the sample size is so small. Would this be an

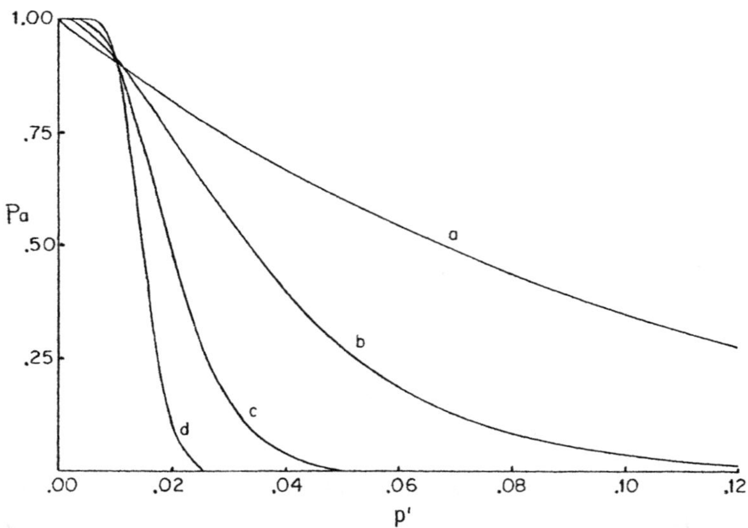

Figure 9.9 Four operating characteristic curves, based on 10% samples, giving P_a at about 0.90, when $p' = 0.01$. (a) $N = 100$, $n = 10$, Ac = 0, (b) $N = 500$, $n = 50$, Ac = 1, (c) $N = 2400$, $n = 240$, Ac = 4, (d) $N = 10{,}000$, $n = 1000$, Ac = 14.

acceptable plan? On the other hand, the plan for $n = 1000$, $\mathrm{Ac} = 14$ is very sharply discriminating. If $p_o = 0.008$, $P_a = 0.983$, whereas if $p_o = 0.02$, $P_a = 0.103$. Is such sharp discrimination with such a steep, square-shouldered OC curve really needed? Very probably not.

The moral is that for small lots we likely need more than 10% of the lot to be in the sample, whereas with large lots, 10% is more than enough. Therefore, pay attention to the sample size and acceptance–rejection criteria rather than the lot size. Thus, we can gain by random sampling of relatively large lots, instead of sampling many small lots.

9.9.6. There are Risks Even with 100% Inspection

A good many people seem to have great confidence in 100% inspection of entire lots of articles. But risks are still present. Studies have often been made on the efficiency of 100% sorting. They have shown that seldom does sorting efficiency reach 80%, that is, will find 80% of the nonconforming pieces present in a lot. Efficiency, of course, depends upon the inspector's training, skill, and motivation. But it also depends upon the number of characteristics being checked, the objectivity of the definition of a nonconformity, and the fraction nonconforming present. Efficiency tends to be best when the definitions are clear, there are few characteristics to be checked, and many nonconformities are present. But even then fatigue and lack of perfect attention cause nonconformities to be missed, and also, good pieces to be rejected. Even with a gauge and only one dimension being checked, inspection is not perfect. An assistant plant manager noticed one of his *better* inspectors sorting piston rings for thickness by a dial gauge. It had a green sector for good thicknesses. The inspector's eyes were open but seemed a little glassy. The manager put his hand over the dial while the inspector checked 50 rings! In checking 100% of food jar caps, an inspector is asked to watch about six lines of caps moving along on parallel conveyors, for some 15 different nonconformities. The hypnotic effect is such that 100% efficiency is impossible. One more example may be enough. A book of the Bible was to be reproduced as the "perfect book". After six expert proofreaders each read the proof 100 times, the proof was reproduced and sent to college campuses, with a challenge to find any errors, and a reward as stimulus. Later the book was published in full confidence. Subsequently, six typographical errors came to light, including one on the first line of the first page!

It has also been shown that sample inspection may possibly give a more accurate figure for the lot fraction nonconforming than 100% sorting, because of the better job of inspection in the sampling work.

The only solution is continually to try to improve all inspection through motivation, objectivity, good tools and not requiring undue speed.

9.9.7. Why not Reject All Lots Yielding any Nonconforming Pieces in the Sample?

This suggestion is sometimes made, but will it prevent nonconforming pieces from ever occurring in the lots? Obviously, the answer is no. The OC curve for sampling plans with $\mathrm{Ac} = 0$, are always of the general shape of those in Figure 9.8, that is, have the concave side of the curve always upward. But always, even with a large sample size n and $\mathrm{Ac} = 0$, there will sometimes be lots passed having p_o at some undesirable value, if any such lots reach the sampling inspection station. Meanwhile with such a shape of an OC curve, there may well be lots rejected which have p_o at a satisfactorily low level. Of course, if the nonconformities are so critical that none be tolerated, then the only things to do are (1) to try to perfect the production processes so that no defects are *made* and (2) to do the best you can with 100, 200, 300 or more percent inspection. On the other hand, if a suitably small percentage of nonconformities or nonconforming pieces can be tolerated, for example, (1) where subsequent operations will find them or (2) where the defects are not all that critical, then sampling inspection can be used. The discriminating power (OC curve) can be set as desired. Of course, the most basic step is the perfecting of the production process. You cannot pass any bad lots, if no such lots are ever offered.

9.9.8. Will a Sampling Plan Give the Same Decision Every Time?

Some persons rather thoughtlessly think that if a sampling plan is any good it will give the same decision every time on a given lot. To "test the sampling plan, they may try it several times on a lot and are disillusioned when the decision is not always the same". A glance at any OC curve will show that repeated use of the *same* sampling plan on any lot can well lead to both kinds of decisions. In fact for given p_o, only P_a of the time will we expect the lot to be passed. For example, for an intermediate p_o, P_a can be 0.50, so that half of the decisions will be to accept and half to reject. But this is not so bad because such a p_o is of "indifference" quality.

9.9.9. Misconception: Sampling Acceptance Plans can Only be Used if the Production Process is in Control

This idea is occasionally heard. As a matter of fact almost the opposite is true. For if we find that the production process is in control with satisfactory p_o, then we can even begin to think of eliminating most of our sampling inspection and using the producer's records. On the other hand, if the producer is in control but p_o is too high, then all lots should be 100% inspected to rectify the high p_o, and thus we might well eliminate sampling inspection until p_o is improved.

But if the producer's process is not in control, so that there are both acceptable and rejectable lots being produced, then sampling inspection is an excellent tool for locating the latter lots for rejection or rectification, and for passing the former.

9.9.10. Some Notations on OC Curves

Let us now show some old and new notations on a typical OC curve in Figure 9.10, namely for the plan: $n = 100$, $\text{Ac} = 3$. The

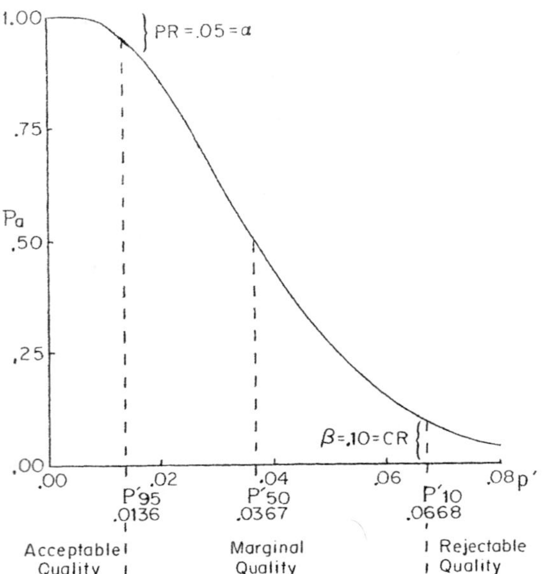

Figure 9.10 A typical operating characteristic (OC) curve (Type B), calculated from the Poisson distribution, Table B. Notations shown for the sampling plan $n = 100$, Ac $= 3$.

Poisson distribution of Table B was used for the calculations, the binomial distribution being the formally correct distribution for this Type B OC curve.

The vertical dashed lines were drawn to the points on the OC curve where P_a is 0.95, 0.50, 0.10. The points on the p_o axis where these lines intersect are designated as follows:

$$p_{95} = p_o \text{ having } Pa = 0.95 \tag{9.23}$$

$$p_{50} = \text{ indifference quality } = p_o \text{ having } Pa = 0.50 \tag{9.24}$$

$$p_{10} = p_o \text{ having } Pa = 0.10 \tag{9.25}$$

Usually we think of p_o values from zero up to p_{95} as being regarded by the sampling plan as fully acceptable. Likewise, those p_o values above p_{10} are regarded by the sampling plan as clearly rejectable. Meanwhile, those p_o values between represent marginal quality of varying degrees. For these, it is not too great an error to accept or to reject.

Note that in Section 9.8, p_1 was the notation for p_{95} and p_2 for p_{10}. Furthermore, in some areas p_{95} is called the "acceptable quality level" (AQL) and p_{10} the "lot tolerance proportion defective" (LTPD) or the "rejectable quality level" (RQL). But it is more common practice as in the ANSI/ASQ Standard Z1.4 (see Chapter 10) to use AQL and RQL for nominal figures not specifically tied to any P_a's.

In the vertical line above p_{95} we find a brace of length 0.05 which is the probability of *rejection* of a lot at p_{95}. This is designated the producer's risk or PR since it is the probability that a "good" lot will be *rejected* by the sampling plan. The PR is also often designated by the small Greek alpha, α. Meanwhile on the vertical line above p_{10} is a brace for the 0.10 probability of a lot of quality p_{10} being erroneously *accepted* by the plan. This is called the consumer's risk, CR, since it is the risk the consumer assumes of accepting a "bad" lot if one should be offered. The CR is often also designated by the small Greek beta, β.

Now unless some such risks α and β can be assumed, respectively, for p_{95} and p_{10} then *no* sampling plan can be used. But these four: α, β, $p_1 < p_2$ can be set arbitrarily for a sampling plan. Similarly, for nonconformities we could set α, β, $c_1 < c_2$.

9.10. SUMMARY

In Section 9.9 we have been reviewing principles, concepts, and notations. We complete the review by emphasizing the four curves which describe and characterize any sampling plan based upon attributes, either nonconforming pieces or nonconformities.

1. Protection curves

   a. OC curve: P_a vs. p_o or c_o.
   b. AOQ curve: Average outgoing quality vs. p_o or c_o.

2. Cost related curves

   a. ASN curve: Average sample number vs. p_o or c_o.
   b. ATI curve: Average total inspection vs. p_o or c_o.

If willing to pay the price, you can have just about any degree of protection you desire. Scientific acceptance sampling consists of obtaining a good balance between protection and cost, while actively ensuring that the prerequisites of Section 9.9.1 are carried out.

9.11. PROBLEMS

9.1. For the example of single sampling given in Section 9.5 check the results given for P_a, AOQ, and ATI at $p_o = 0.02$ and at $p_o = 0.06$. Find p_{95} and p_{10} for the plan by reversing the use of Table B, that is, finding np_o when P(3 or less) is 0.95 and 0.10.

9.2. For the example of double sampling given in Section 9.6, check the results given for P_a, AOQ, ASN, and ATI at $p_o = 0.02$ and at $p_o = 0.06$.

9.3. Sketch the OC, AOQ, ASN, and ATI curves labeling units on the axes, for the following two acceptance sampling plans for lots of 1000: (a) $n = 80$, $\text{Ac} = 2$, $\text{Re} = 3$ and (b) $n_1 = n_2 = 50$, $\text{Ac}_1 = 0$, $\text{Re}_1 = 3$, $\text{Ac}_2 = 3$, $\text{Re}_2 = 4$. Points for $p_o = 0.01, 0.02, \ldots, 0.06$ should be enough.

9.4. Sketch the OC, AOQ, ASN, and ATI curves, labeling units on the axes for the single sampling plan: $N = 2000$, $n = 100$, $\text{Ac} = 2$, $\text{Re} = 3$. Also find the AOQL, and p_{95} and p_{10}. Use points for $p_o = 0.01, 0.02, \ldots, 0.05$ initially.

9.5. Sketch the OC, AOQ, ASN, and ATI curves, labeling units on the axes for the double sampling plan: $N = 2000$, $n_1 = 50$, $n_2 = 100$, $\text{Ac}_1 = 1$, $\text{Re}_1 = 4$, $\text{Ac}_2 = 3$, $\text{Re}_2 = 4$. Use points for $p_o = 0.01, 0.02, \ldots, 0.06$ which may be enough.

9.6. Find a single sampling plan (using Section 9.8) having $p_{95} = p_1 = 0.01$ and $p_{10} = p_2 = 0.05$. Similarly, find one for $p_1 = 0.002$, $p_2 = 0.01$.

9.7. Find a single sampling plan (using Section 9.8) having $p_{95} = p_1 = 0.01$ and $p_{10} = p_2 = 0.03$. Also find one for $p_1 = 0.001$, $p_2 = 0.003$.

REFERENCES

IW Burr. Statistical Quality Control Methods. New York: Dekker, 1976.

Harvard Univ., Comput. Lab., Tables of the Cumulative Binomial Probability D'istribution. Cambridge, MA: Harvard Univ. Press, 1955.

RL Kirpatrick. Binomial sampling plans indexed by AQL and LTPD. Indust. Quality Control 22:290–292, 1965.

GJ Lieberman, DB Owen. Tables of the Hypergeometric Probability Distribution. Stanford, CA: Stanford Univ. Press, 1961.

P Peach, SB Littauer. A note on sampling inspection by attributes. Ann. Math. Statist. 27:81–84, 1946.

10

Some Standard Sampling Plans for Attributes

In this chapter, we shall concentrate upon the ABC Standard sampling plan for attributes. We shall also describe briefly some other sampling plans which are available for their intended purposes. Interesting as it is, we shall not go into the history of acceptance sampling plans. An authoritative and readable summary is available in Dodge (1969, 1970). The Military Standard MIL-STD 105D (or E) has been replaced by the American National Standards Institute/ American Society for Quality, ANSI/ASQ Zl.4. This latter standard has come by the work of a joint committee of the two specified bodies in the Z1 Committee. At the time of the revision of this book, the new ISO standard for Acceptance Sampling by Attributes is being approved. This will replace the Z1.4; however, it is very close to it.

10.1. ANSI/ASQ Z1.4

The ANSI/ASQ Z1.4 Sampling Procedures and Tables for inspection by Attributes is widely used throughout the world. It is used for receiving inspection, within-house inspection between departments, and final inspection. It is used in attribute inspection, on either nonconforming units or nonconformities. For the case of acceptance sampling by measurements, there is a companion standard, ANSI/ASQ Z1.9 which replaced the MIL-STD 414, will be briefly described in Chapter 11. The ANSI/ASQ Z1.4 is highly useful, but its objectives of quality control may only be obtained if *all* of the instructions are precisely followed.

It is appropriate to initiate the discussion of this standard for acceptance sampling by attributes by noting a little history. During the Second World War, the United States industrial base was producing far more airplanes (for example) than could be inspected by the current staff. A high level commission was established under Dr. Wallace to develop a set of sampling plans that would allow shipment of specific lots of airplanes by the inspection of only some of them. This, of course, did not make for happy pilots who had to ditch in the Atlantic Ocean; but it did overcome the blockage caused by the lengthy inspection process. It also gave them assurance that the high quality manufacturers would get their product lots accepted in a minimum time.

The assumptions behind the early standards were primarily: (1) The manufacturing process providing the product is a continuous process from which consecutive lots are drawn, and (2) the purpose of the set of plans was to identify good quality lots. In most applications encountered today both of these assumptions are violated. Even when a lot of product is manufactured by a continuous process any particular customer purchases discontinuous lots while other customers purchase the alternate lots. In the experience of the second author over the past 30 years, the sampling plans are used to identify lots of bad product, i.e., lots that do not meet the quality level identified by management, usually by setting the AQL. As will be seen this can lead to some pretty bad quality creeping into the plant.

For example, in the 1970s some companies had chosen a sampling plan which required 10 pieces to be inspected and if any of these were nonconforming, the lot was rejected. It was touted as a cost effective plan and assures good quality. As the reader will see if they construct an OC curve, a result of this plan is that a lot having 7.5% nonconforming parts will be accepted half of the time.

The ANSI/ASQ Z1.4:2003 states that this standard can be used for isolated lots but the user should study the OC curves to select a plan that meets the real quality needs of the using company.

10.1.1. Aim

The basic aim of the ANSI/ASQ Z1.4 is the maintenance of the outgoing quality level at a given 'acceptable quality level" or better. There are two criteria of quality level, either of which may be specified, namely, percent nonconforming, that is the number of *nonconforming units* per hundred units, or secondly, the number of *nonconformities* per hundred units. Then the objective is to insure that the outgoing quality is at the AQL or better.

Moreover, the standard is designed so that if the producer runs consistently at precisely the AQL (average quality level), then the great majority of the lots can be expected to pass. But if his process is a little worse than the AQL, they can expect to have trouble and may face "tightened" inspection. Thus, the AQL is the *minimum* quality performance at which the producer may safely run, rather than a perfectly safe level at which to run, as some have erroneously thought of the AQL. The producer is, therefore, advised to run at or better than the AQL.

Let us here point out that the reader should carefully distinguish between the two quite similar abbreviations: AQL and AOQL. AQL stands for acceptable quality level, whereas AOQL represents average outgoing quality limit, that is, the worst long-run average outgoing quality, no matter what comes in.

10.1.2. How ANSI/ASQ Z1.4 Operates to Fulfill the Aim

Let us look at a typical OC curve, for example, Figure 9.10 There we see that for the plan $n = 100$, $\mathrm{Ac} = 3$, $p_{.95} = 0.0136$. This means that if the producer is running the process at $p_o = 0.0136\%$ or 1.36%, then he can expect to have 95% of the lots passed by the plan, that is, 19 out of 20. Now suppose that the consumer, in using this plan, has set 0.0136 as the AQL. The "greatest majority of lots" from a production process at $p_o = 0.0136$ will be accepted. So far so good. Now to obtain such a high Pa when $p_o = 0.0136$, they must let Pa be also quite high when p_o is a bit above 0.0136. Specifically, if p_o should deteriorate to 0.020, the Pa is still high, namely 0.857. Therefore, if the consumer continues to use the same plan $n = 100$, $\mathrm{Ac} = 3$, they will be passing most of the lots at around $p_o = 0.02$, which they do not want to do.

What is the remedy? As we shall see, if a producer runs much worse than the AQL set, then quite soon the plan will invoke "tightened inspection". Specifically, it might well tighten to the plan $n = 100$, $\mathrm{Ac} = 2$. Now, if $p_o = 0.02$, then $\mathrm{Pa} = 0.677$ so that only two lots out of three at 0.02 will be accepted. This is quite a drop from $\mathrm{Pa} = 0.857$ for the original plan, at $p_o = 0.02$. But also note that for this tightened plan, $n = 100$, $\mathrm{Ac} = 2$, the AOQL is at 0.0136, it happens, so the consumer will have long-run quality at 0.0136 or better, that is, at his specified AQL. So he will have the quality requested even though the producer was not cooperating. Moreover, according to the standard, if the producer does not improve from $p_o = 0.02$, the plan will soon call for abandonment of sampling altogether, going to 100% inspection.

Thus, in order to give a high probability of acceptance (low producer's risk) when the process is at the AQL or better, and yet to give the consumer adequate protection against worse process averages than the AQL, we must be sure to use tightened inspection whenever it is called for. The appropriate criteria are discussed further on.

10.1.3. Saving Inspection by "Reduced Sampling"

There is also provision for using smaller sample sizes than normal when quality is consistently better than the AQL and production is continuous. Under these conditions, bad quality lots are very unlikely and a less steep OC curve, coming from a smaller sample size, can be tolerated. There are, as we shall see, adequate safeguards which can call for reverting to the normal inspection plan. Reduced sampling, when available, is still optional and only used to save on inspection, not to achieve the basic aim of maintenance of outgoing quality at $p_o = \text{AQL}$.

10.1.4. Multiple Sampling

The ANSI/ASQ Z1.4 standard makes available single, double, and multiple sampling plans. A typical multiple sampling plan having an AQL of 4.0 percent is the following one:

Sample size, n	20	20	20	20	20	20	20
Cumulative sample size	20	40	60	80	100	120	140
Acceptance number, Ac	0	1	3	5	7	10	13
Rejection number, Re	4	6	8	10	11	12	14

This table is almost self-explanatory. If in the first sample of 20 pieces, there are no nonconforming units, we can accept the lot at once. Or, if there are four or more nonconforming units, we reject the lot at once. But if there are 1, 2, or 3 nonconforming units in the first 20, we take another random sample of 20. Then we compare the cumulative number of nonconforming units $d_1 + d_2$ against $\text{Ac}_2 = 1$ and $\text{Re}_2 = 6$. If

$d_1 + d_2 \le 1$	Accept lot
$d_1 + d_2 \ge 6$	Reject lot
$1 < d_1 + d_2 < 6$	Take a third sample of 20

In the last case, we find d_3 nonconforming pieces and then similarly compare $d_1 + d_2 + d_3$ against $\text{Ac}_3 = 3$ and $\text{Re}_3 = 8$. This process might possibly continue for seven samples at which point a decision becomes certain because there is no gap between $Ac_7 = 13$ and $Re_7 = 14$. Now, although it is possible to require all seven samples totaling 140, it is quite unlikely. In football language, "It would take a very good broken field runner to reach the end".

The standard provides two companion plans:

$$\begin{array}{lllll} \text{Single} & n = 80, & \text{Ac} = 7, & \text{Re} = 8 & \\ \text{Double} & n_1 = 50, & \sum n = 50, & \text{Ac}_1 = 3, & \text{Re}_1 = 7 \\ & n_2 = 50, & \sum n = 100, & \text{Ac}_2 = 8, & \text{Re}_2 = 9 \end{array}$$

These two plans and the multiple plan all have practically the same OC curve and, therefore, AOQ curve. But the multiple plan will save inspection time by having a more favorable (lower) ASN curve than either of the companion plans. Also the double plan has a more favorable ASN curve than does the single plan. Of course, the price you pay for these more favorable ASN curves is greater complexity in the plans, requiring more training and supervision time. But all three plans are practically identical in fulfilling the aims of the standard; so it is a management decision on which plan to use.

10.1.5 General Description of the ABC Standard

There are nine basic tables of specific sampling plans. Tables (II) (A), (B), and (C) are all single sampling plans, respectively, for normal, tightened, and reduced sampling. Likewise Tables (III) (A), (B), and (C) are for double sampling, and (IV) (A), (B), and (C) for multiple sampling. To find a specific plan in any one of these nine tables, one needs a "sample size code letter", A, B, ..., R, and an AQL, which may be in terms of nonconforming units per 100 units, i.e., percent nonconforming or nonconformities per 100 units. Either may be used for AQL's of 0.010 up to 10, but beyond 10, the basis will be nonconformities per 100 units. The sample size code letter is found from Table I, using the lot size N to determine the row and the "inspection level" to give the column, from which the letter is found. Inspection levels I, II, and III are the commonly used ones, the other four S-1, ...,S-4 being for special purposes. Level II is used in general unless for some reason smaller samples than normal are desired and level I specified, or larger samples and level III specified.

Let us now summarize the steps for finding a specific sampling acceptance plan in the Standard.

1. Decide on the size of lot N, which is to be sample inspected. This need not be a production lot size.
2. Decide upon an inspection level, in general II.
3. Using 1 and 2, enter Table I to find the corresponding sample size code letter A, B, . . . ,R, the last calling for largest sample sizes.
4. Decide upon single, double, or multiple sampling.
5. Decide whether to start with normal (almost always), tightened, or reduced sampling.
6. Table IIA, IIIA, or IVA will thus be determined by 4 and 5 (if normal inspection is to be used).
7. Decide upon the inspection basis of nonconforming units or nonconformities.
8. Decide upon the desired AQL: for percent nonconforming units (10.0 or less) or nonconformities per 100 units; using only what is available in the tables.
9. Enter the table determined in 6, using the AQL for the column from 8 and the row from the sample size code letter in 3. This commonly gives the acceptance–rejection numbers in the block, the sample size(s) being given to the left of this block.
10. Following the above, we may reach an asterisk or an arrow. An asterisk means to use the single sampling plan for the desired AQL and code letter, instead of double or multiple sampling. If an arrow is encountered, follow it to the first block with acceptance–rejection numbers, using sample sizes to the left of *this block*, *not* to the left of the *original block*.

For example, suppose that we have lot sizes of 500 units and will use inspection level II, for normal sampling with an AQL of 0.65% nonconforming. The first two items specified, on entering Table I tell us to use sample size code letter H. For normal double sampling, we use Table III (A). In this table in the 0.65 column and the H row, we find in the block a downward arrow. Following it down to the first block containing numbers (that is, the next lower block here), we find $Ac_1 = 0$, $Re_1 = 2$, $Ac_2 = 1$, $Re_2 = 2$. Now going over to the left of this block, we find $n_1 = n_2 = 50$. Note very particularly that we do not use the sample sizes $n_1 = n_2 = 32$, which are normally used for code letter H. Thus our plan is to take a random sample of 50 from the 500 in the lot, inspect, and finding d_1 nonconforming units (1) accept if $d_1 = 0$, (2) reject if $d_1 \geq 2$, or (3) take another sample if $d_1 = 1$. Now this second sample of 50 from the 450 remaining in the lot yields d_2 nonconforming units. Then (1) accept the lot if $d_1 + d_2 = 1$ or (2) reject if $d_1 + d_2 \geq 2$.

Under the same conditions but for tightened inspection, we use Table III (B), and find $n_1 = n_2 = 80$, $Ac_1 = 0$, $Re_1 = 2$, $Ac_2 = 1$, $Re_2 = 2$. This is indeed a tighter plan making for a lowered Pa at any p_o. Further, under the same conditions, except using reduced inspection, we go to Table 10.3 (C). Code letter H calls for $n_1 = n_2 = 13$, but in the column for AQL = 0.65 we again find an arrow pointing downward. It yields a block with $Ac_1 = 0$, $Re_1 = 2$, $Ac_2 = 0$, $Re_2 = 2$, but these are for $n_1 = n_2 = 20$ (the reduced double sample sizes for code letter J, not H). The author hopes that you have wondered at the gap between Ac_2 and Re_2. This seems to mean that no decision is reached if $d_1 + d_2$ happens to be 1. This is covered in Section 10.1.4 of the ANSI/ASQ Z1.4. There it says that if such an event occurs, we still accept this lot (because the previous quality has been excellent relative to the AQL), but we are now alerted to the possibility that quality has slipped from its previous excellence. Therefore we abandon reduced sampling with its quite lenient OC curve and go back to normal sampling.

Now let us have a look at Figure X-J which we have included as a sample. There is in the standard such a pair of pages for each code letter A–R. They give a wealth of information on the respective sampling plans for each code letter. All sampling plans—single, double, and multiple—are listed on the second pages as in Table X-J-2. At the top of Table X-J-1 are given the OC curves for all AQLs for single normal inspection (most being calculated using the Poisson distribution as in our Table B; but the binomial is used if the single plan has $n \leq 180$ and also the AQL ≤ 10.0). As usual, Pa is the vertical scale, the bottom scale being percent nonconforming or nonconformities per 100 units for the process. The OC curve for the normal single sampling plan for code letter J and AQL = 0.65%, namely, $n = 80$, Ac = 1, Re = 2 is shown in the graph next to the number "0.65". Thus, for example, if $p_o = 3\%$ nonconforming, the Pa = 0.30 or if $p_o = 2\%$ nonconforming, the Pa = 0.55.

This table in the lower half of the first page gives certain convenient points on the OC curve, namely in the percents p_{99}, p_{95}, p_{90} and so on to p_{01}. These figures are for single normal sampling plans, as are the OC curves. The OC curves for double and multiple normal sampling "are matched as closely as practicable". Our examples of Section 9.5 and 9.6 were such a pair of matched plans and appear in Table X-J-2, for an AQL of 1.5%. Note that the table at the bottom of the first page of Figure X-J is entered from the top for normal sampling AQLs and from the bottom for tightened sampling AQLs. Some of the plans duplicate; others do not. For example, the normal plan with an AQL of 1.5% is the same plan as that for the tightened plan with an AQL of 2.5%. OC curves for some tightened plans are not shown. There are two sides to this table, one for nonconforming units (on the left) and the other for nonconformities.

With this description of the tables in mind, carefully read the first nineteen pages of the standard. Section 7.1 specifies random sampling. Section 7.2 provides for proportional sampling from boxes on containers within the lot. Section 8 is of much importance in using the standard, providing as it does the "switching rules". The rule 8.3.1 for going from normal to tightened inspection is clear and easily followed as long as a continuing record of action on consecutive lots is preserved in order. Note see. 8.3.2 in conjunction with 8.4. When on tightened inspection, the quality must be quite quickly improved so as to return to normal inspection by having five consecutive lots passing; or else inspection will be discontinued until action is taken to improve the quality submitted. Bear in mind that invoking tightened inspection, when called for, must be carried out. This is not an optional part of the standard. It is essential in order to provide the consumer with adequate protection, and to ensure that the outgoing quality is at the AQL or better, in the long run. If not followed, then you are not using the ABC standard!

On the other hand, use of reduced sampling when available is optional. It is not necessary to use it in order to ensure quality at the AQL or better. Its only function is to safely save inspection with adequate safeguards. All four requirements of Section 8.3 must be met in order to go onto reduced inspection. They ensure consistently excellent quality better than the AQL. Use is made of Table VIII, which gives limit numbers of nonconforming units or nonconformities found in cumulations of the single samples or in first samples for double or multiple sampling for at least 10 lots. But to go from reduced sampling back to normal, only one of the four conditions of Section 8.3.4 need occur. One of the provisions involves an acceptance of a lot under conditions which raise a question as to whether quality may have deteriorated, as noted before and covered in Section 10.1.4.

Let us now describe the remaining tables which we have reproduced. Only the normal and tightened multiple sampling plans in Tables IV(A) and IV(B) have been included omitting the corresponding reduced plans.

Table V(B) provides the average outgoing quality limits, AOQLs, for all tightened, single- sampling plans. This table thus gives a measure of the consumer protection provided when on tightened inspection. Except for those single tightened plans for which the acceptance number, Ac = 0, all of these AOQLs are at about the corresponding AQL or better. This means that when we go onto tightened inspection, we are, in general, using a plan providing outgoing quality averaging no worse than the AQL.

Table VI(A) and VI(B) gives quality levels having a 0.10 probability of acceptance if offered, respectively, in terms of percent nonconforming and nonconformities per 100 units. These are particularly useful in case we are sample inspecting isolated or infrequent lots, since these figures tell what quality levels, of which a given plan provides a 90% assurance against acceptance, if offered.

Figure IX contains very useful information on the average sample number (ASN) curves, in a convenient compact form, by means of which we may compare single, double, and multiple sampling. They are for both normal and tightened inspection, but do not include cases for which the single plan has Ac = 0. All other acceptance numbers for a single plan are included.

For example, if we wish the ASN curves for normal sampling, code letter J, AQL = 1.5% nonconforming, we will use the third square which says $c = 3$ (the letter, c, being an alternative to "Ac". Now pay close attention to the way the scales are made up. The three plans call for the following:

$$n = 80, \qquad \text{Ac} = c = 3, \qquad \text{Re} = 4$$

$$n_1 = 50, \qquad \sum n = 50, \qquad \text{Ac}_1 = \text{Re}_1 = 4$$

$$n_2 = 50, \qquad \sum n = 100, \qquad \text{Ac2} = 4, \qquad \text{Re}_2 = 5$$

and the multiple plan with n's of 20. Now consider the vertical scale: n, $\frac{3}{4}n$, and so on. Here $n = 80$ so that in sampled pieces these five points are 80, 60, 40, 20, and 0 for our example. Now for the horizontal scale: it is np_o, i.e., the expected number of nonconforming units in a sample. Since the n we are talking about is 80, these numbers represent $80p_o$. When $80p_o = 4$, for example, then dividing by 80 gives $p_o = 4/80 = 0.05$. Therefore, when in the horizontal scale we see 4, we could put a p_o value of 0.05 for a p_o scale. Similarly, when we see 2 in the scale shown, this is equivalent to $p_o = 0.025$, 6 to a $p_o = 0.075$, and so on. So we now can compare ASNs for the three plans. The single plan is uniformly at 80. The double plan starts off at $n = 50$ (just above $1/2n$) and reaches a maximum of about 75 when $np_o = 4.5$ or so, which corresponds to $p_o = 4.51/80 = 0.056$, then it starts to decrease back toward 50. Meanwhile the multiple plan starts off at 40, because it takes two samples, each of 20, before there is a zero acceptance number. Thereafter, the ASN increases, reaching a maximum of about 65 when $np_o = 3$ or $p_o = 3/80 = 0.038$.

Now what is the meaning of the arrow? It shows the relative position of the AQL in normal sampling. For $n = 80$, this plan with Ac = 3 has the arrow at about 1.3. Equating this to 80AQL = 1.3, we obtain in this example AQL = 1.3/80 = 0.016. It is listed for our example as 1.5%, which is close enough for the scale reading.

For the single normal plan for $n = 500$, Ac = 3, and AQL = 0.25, the vertical scale would read 500, 375, 250, 125, 0. Now when we have 4 on the horizontal scale this is $500p_o = 4$ or $p_o = 4/500 = 0.008$, and so on. Meanwhile, the AQL arrow is at 1.3/500 = 0.0026 = 0.26%, close to AQL = 0.25%.

Thus each of the various blocks show the relative ASNs for a whole collection of plans. Is not this a rather neat, compact summary form?

This completes our discussion of the ABC Standard with its aim of maintenance of outgoing quality at the AQL or better, but meanwhile giving a small risk of a lot being rejected when the producer is running at the AQL or better.

PARTIAL REPRODUCTION OF ANSI/ASQ Z1.4-2003

SAMPLING PROCEDURES AND TABLES FOR INSPECTION BY ATTRIBUTES

1. Scope

1.1. Purpose

This publication establishes sampling plans and procedures for inspection by attributes. When specified by the responsible authority, this publication shall be referenced in the specification, contract, inspection instructions, or other documents and the provisions set forth herein shall govern. The "responsible authority" shall be designated in one of the above documents, as agreed to by the purchaser and seller or producer and user.

1.2. Application

Sampling plans designated in this publication are applicable, but not limited, to inspection of the following:

a. End items.
b. Components and raw materials.

c. Operations.
d. Materials in process.
e. Supplies in storage.
f. Maintenance operations.
g. Data or records.
h. Administrative procedures.

These plans are intended primarily to be used for a continuing series of lots or batches. The plans may also be used for the inspection of isolated lots or batches, but, in this latter case, the user is cautioned to consult the operating character istic curves to find a plan which will yield the desired protection (see 11.6).

1.3. Inspection

Inspection is the process of measuring, examining, testing, or otherwise comparing the unit of product (see 1.5) with the requirements.

1.4. Inspection by Attributes

Inspection by attributes is inspection whereby either the unit of product is classified simply as conforming or nonconforming, or the number of nonconformities in the unit of products is counted, with respect to a given requirement or set of requirements.

1.5. Unit of Product

The unit of product is the unit inspected in order to determine its classification as conforming or nonconforming or to count the number of non-conformities. It may be a single article, a pair, a set, a length, an area, an operation, a volume, a component of an end product, or the end product itself. The unit of product may or may not be the same as the unit of purchase, supply, production, or shipment.

2. Definitions and Terminology

The definitions and terminology employed in this standard are in accord with ANSI/ASQ A3534-2-1993 (terms, symbols, and definitions for acceptance sampling). The following two definitions are particularly important in applying the standard.

> Defect: A departure of a quality characteristic from its intended level or state that occurs with a severity sufficient to cause an associated product or service not to satisfy intended normal, or foreseeable, usage requirements.
>
> Nonconformity: A departure of a quality characteristic from its intended level or state that occurs with severity sufficient to cause an associated product or service not to meet a specification requirement.

These acceptance sampling plans for attributes are given in terms of the percent or proportion of product in a lot or batch that depart from some requirement. The general terminology used within the document will be given in terms of percent of nonconforming units or number of nonconformities, since these terms are likely to constitute the most widely used criteria for acceptance sampling.

In the use of this standard, it is helpful to distinguish between:

a. An individual sampling plan—a specific plan that states the sample size or sizes to be used, and the associated acceptance criteria.
b. A sampling scheme—a combination of sampling plans with switching rules and possibly a provision for discontinuance of inspection. In this standard, the terms "sampling scheme" and "scheme performance" will be used in the restricted sense described in Section 11.1.
c. A sampling system—a collection of sampling schemes. This standard is a sampling system indexed by lot-size ranges, inspection levels, and AQLs.

3. Percent Nonconforming and Nonconformities per Hundred Units

3.1. Expression of Nonconformance:

The extend of nonconformance of product shall be expressed either in terms of percent nonconforming or in terms of nonconformities per hundred units.

3.2. Percent Nonconforming:

The percent nonconforming of any given quantity of units of product is one hundred times the number of nonconforming units divided by the total number of units of product, i.e.

$$\text{percent nonconforming} = \frac{\text{number nonconforming}}{\text{number of units inspected}} \times 100$$

3.3 Nonconformities per Hundred Units:

The number of nonconformities per hundred units of any given quantity of units of product is one hundred times the number of nonconformities contained therein (one or more nonconformities being possible in any unit of product) divided by the total number of units of product, i.e.

$$\text{nonconformities per hundred units} = \frac{\text{number of nonconformities}}{\text{number of units inspected}} \times 100$$

It is assumed that nonconformities occur randomly and with statistical independence within and between units.

4. Acceptance Quality Limit (AQL)

4.1. Use

The AQL together with the sample size code letter is used for indexing the sampling plans provided herein.

4.2. Definition

The AQL is the quality level that is the worst tolerable product average when a continuing series of lots is submitted for acceptance sampling.

Note: The use of the abbreviation AQL to mean acceptable quality level is no longer recommended.

4.3. Note on the Meaning of AQL

The concept of AQL only applies when an acceptance sampling scheme with rules for switching between normal, tightened, and reduced inspection and discontinuance of sampling inspection is used. These rules are designed to encourage suppliers to have process averages consistently better than the AQL. If suppliers fail to do so, there is a high probability of being switched from normal inspection to tightened inspection where lot acceptance becomes more difficult. Once on tightened inspection, unless corrective action is taken to improve product quality, it is very likely that the rule requiring discontinuance of sampling inspection will be invoked.

Although individual lots with quality as bad as the AQL can be accepted with fairly high probability, the designation of an AQL does not suggest that this is necessarily a desirable quality level. The AQL is a parameter of the sampling scheme and should not be confused with a process average which describes the operating level of a manufacturing process. It is expected that the product quality level will be less than the AQL to avoid excessive non-accepted lots.

The sampling plans in this standard are so arranged that the probability of lot acceptance at the designated AQL depends upon sample size, being generally higher for large samples than for small samples for a given AQL. To determine the specific protection to the consumer at a given AQL, it is necessary to refer to the operating characteristic curves (which are provided in this standard)of the corresponding scheme and its constituent plans.

The AQL alone does not describe the protection to the consumer for individual lots or batches, but more directly relates to what is expected from a series of lots or batches provided the provisions of this standard are satisfied.

4.4. Limitation

The designation of an AQL shall not imply that the supplier has the right to knowingly supply any nonconforming unit of product.

4.5. Specifying AQLs:

The AQL to be used will be designated in the contract or by the responsible authority. Different AQLs may be designated for groups of nonconformities considered collectively, or for individual nonconformities. For example, Group A may include nonconformities of a type felt to be of the highest concern for the product or service and therefore be assigned a small AQL value; Group B may include nonconformities of the next higher degree of concern and therefore be assigned a larger AQL value than for Group A and smaller than that of Group C, etc. The classification into groups should be appropriate to the quality requirements of the specific situation. An AQL for a group of nonconformities may be designated in addition to AQLs for individual nonconformities, or subgroups, within that group. AQL values of 10.0 or less may be expressed either in percent nonconforming or in nonconformities per hundred units; those over 10.0 shall be expressed in nonconformities per hundred units only.

4.6. Preferred AQLs

The values of AQLs given in these tables are known are preferred AQLs. If, for any product, an AQL be designated other than a preferred AQL, these tables are not applicable.

5. Submission of Product

5.1. Lot or Batch

The term lot or batch shall mean “inspection lot” or “inspection batch”, i.e., a collection of units of product from which a sample is to be drawn and inspected to determine conformance with the acceptability criteria, and may differ from a collection of units designated as a lot or batch for other purposes (e.g., production, shipment, etc.).

5.2. Formation of Lots or Batches

The product shall be assembled into identifiable lots, sublots, batches, or in such other manner as may be prescribed (see 5.4) . Each lot or batch shall, as far as is practicable, consists of units of product of a single type, grade, class, size, and composition, manufactured under essentially the same conditions, and at essentially the same time.

5.3. Lot or Batch Size

The lot or batch size is the number of units of product in a lot or batch.

5.4. Presentation of Lots or Batches

The formation of the lots or batches, lot or batch size, and the manner in which each lot or batch is to be presented and identified by the supplier shall be designated or approved by the responsible authority. As necessary, the supplier shall provide adequate and suitable storage space for each lot or batch, equipment needed for proper identification and presentation, and personnel for all handling of product required for drawing of samples.

6. Acceptance and Non-acceptance

6.1. Acceptability of Lots or Batches

Acceptability of a lot or batch will be determined by the use of a sampling plan or plans associated with the designated AQL or AQLs.

In the use of this standard' a statement that a lot is acceptable means simply that sample results satisfy the standard's acceptance criteria. The acceptance of a lot is not intended to provide information about lot quality. If a stream of lots from a given process is inspected under an acceptance sampling scheme such as provided in this standard, some lots will be accepted and others will not. If all incoming lots are assumed to be at the same process average and if the nonconforming items that are discovered and replaced by conforming items during sample inspection are ignored, it will be found that both the set of accepted lots and the set of non-accepted lots will have the same long run average quality as the original set of lots submitted for inspection. Inspection of incoming lots whose quality levels vary around a fixed long run average quality level will divide the lots into a set of accepted lots and a set of non-accepted lots, but it will be found that the long run average quality of the accepted lots is only slightly better than the long run average quality of the non-accepted lots. Replacement of the nonconforming items that are discovered during sample inspection does not alter this finding because the samples are a small fraction of the lots.

The purpose of this standard is, through the economic and psychological pressure of lot non-acceptance, to induce a supplier to maintain a process average at least as good as the specified AQL while at the same time providing an upper limit on the consideration of the consumer's risk of accepting occasional poor lots. The standard is not intended as a procedure for estimating lot quality or for segregating lots.

In acceptance sampling, when sample data do not meet the acceptance criteria, it is often stated that the lot is to be "rejected". In this connection, the words "to reject" generally are used. Rejection in an acceptance sampling sense means to decide that a batch, lot or quantity of product, material, or service has not been shown to satisfy the acceptance criteria based on the information obtained from the sample(s).

In acceptance sampling, the words "to reject" generally are used to mean "to not accept" without direct implication of product usability. Lots which are "rejected" may be scrapped, sorted (with or without nonconforming units being replaced), reworked, re-evaluated against more specific usability criteria, held for additional information, etc. Since the common language usage of "reject" often results in an inference of unsafe or unusable product, it is recommended that "not accept" be understood rather than "reject" in the use of this standard.

The word "non-acceptance" is used here for "rejection" when it refers to the result of following the procedure. Forms of the word "reject" are retained when they refer to actions the customer may take, as in "rejection number".

6.2. Nonconforming Units

The right is reserved to reject any unit of product found nonconforming during inspection whether that unit of product forms a part of a sample or not, and whether the lot of batch as a whole is accepted or rejected. Rejected units may be repaired or corrected and resubmitted for inspection with the approval of, and in the manner specified by, the responsible authority.

6.3. Special Reservation for Designated Nonconformities

Since most acceptance sampling involves evaluation of more than one quality characteristic, and since these may differ in importance in terms of quality and/or economic effects, it is often desirable to classify the types of nonconformity according to agreed upon groupings. Specific assignment of types of nonconformities to each class is a function of agreement on specific sampling applications. In general, the function of such classification is to permit the use of a set of sampling plans having a common sample size, but different acceptance numbers for each class having a different AQL, such as in Tables II, III.

The supplier may be required at the discretion of the responsible authority to inspect every unit of the lot or batch for designated classes of nonconformities. The right is reserved to inspect every unit submitted by the supplier for specified nonconformities, and to reject the lot or batch immediately, when a nonconformity of this class is found. The right is reserved also to sample, for specified classes of nonconformities, lots or batches submitted by the supplier and to reject any lot or batch if a sample drawn therefrom is found to contain one or more of these nonconformites.

6.4. Resubmitted Lots or Batches

Lots or batches found unacceptable shall be resubmitted for reinspection only after all units are re-examined or re-tested and all nonconforming units are removed or nonconformities corrected. The responsible authority shall determine whether normal or tightened inspection shall be used on reinspection and whether reinspection shall include all types or classes of nonconformities or only the particular types or classes of nonconformities which caused initial rejection.

7. Drawing of Samples

7.1. Sample

A sample consists of one or more units of product drawn from a lot or batch, the units of the sample being selected at random without regard to their quality. The number of units of product in the sample is the sample size.

7.2. Sampling

When appropriate, the number of units in the sample shall be selected in proportion to the size of sublots or subbatches, or parts of the lot or batch, identified by some rational criterion. In so doing, the units from each part of the lot or batch shall be selected at random, as defined in ANSI/ASQ A3534-2-1993.

7.3. Time of Sampling

Samples may be drawn after all the units comprising the lot or batch have been produced, or samples may be drawn during production of the lot or batch.

7.4. Double or Multiple Sampling

Where double or multiple sampling is to be used, each sample shall be selected over the entire lot or batch.

8. Normal, Tightened, and Reduced Inspection

8.1. Initiation of Inspection

Normal inspection will be used at the start of inspection unless otherwise directed by the responsible authority.

8.2. Continuation of Inspection

Normal, tightened, or reduced inspection shall continue unchanged on successive lots or batches except where the switching procedures given below require change.

8.3. Switching Procedures

8.3.1. Normal to Tightened

When normal inspection is in effect tightened inspection shall be instituted when two out of five consecutive lots or batches have been non-acceptable on original inspection (i.e., ignoring resubmitted lots or batches for this procedure).

8.3.2. Tightened to Normal

When tightened inspection is in effect, normal inspection shall be instituted when five consecutive lots or batches have been considered acceptable on original inspection.

8.3.3. Normal to Reduced

When normal inspection is in effect, reduced inspection shall be instituted providing that all of the following conditions are satisfied:

a. The preceding 10 lots or batches (or more, as indicated by the note to Table 10.8 have been on normal inspection and all have been accepted on original inspection.
b. The total number of nonconforming units (or nonconformities) in the samples from the preceding 10 lots or batches (or such other number as was used for condition "a" above) is equal to or less than the applicable number given in Table 10.8. If double or multiple sampling is in use, all samples inspected should be included, not "first" samples only.

c. Production is at a steady rate.
d. Reduced inspection is considered desirable by the responsible authority.

8.3.4. Reduced to Normal

When reduced inspection is in effect, normal inspection shall be instituted if any of the following occur on original inspection:

a. a lot or batch is rejected; or
b. a lot or batch is considered acceptable under the procedures for reduced inspection given in 10.1.4; or
c. production becomes irregular or delayed; or
d. other conditions warrant that normal inspection shall be instituted.

8.4. Discontinuation of Inspection

If the cumulative number of lots not accepted in a sequence of consecutive lots on tightened inspection reaches 5, the acceptance procedures of this standard shall be discontinued. Inspection under the provisions of this standard shall not be resumed until corrective action has been taken. Tightened inspection shall then be used as if 8.3.1 had been invoked.

8.5. Limit Numbers for Reduced Inspection

When agreed upon by responsible authority for both parties to the inspection, that is, the supplier and the end item customer, the requirements of 8.3.3(b) may be dropped. This action will have little effect on the operating properties of the scheme.

8.6. Switching Sequence

A schematic diagram describing the sequence of application of the switching rules is shown in Figure 10.1

9. Sampling Plans

9.1. Sampling Plan

A sampling plan indicates the number of units of product from each lot or batch which are to be inspected (sample size or series of sample sizes) and the criteria for determining the acceptability of the lot or batch (acceptance and rejection numbers).

9.2. Inspection Level

The inspection level determines the relationship between the lot or batch size and the sample size. The inspection level to be used for any particular requirement will be prescribed by the responsible authority. Three inspection levels; I, II, and III are given in Table 10.1 for general use. Unless otherwise specified, inspection level II will be used. However, inspection level I may be specified when less discrimination is needed, or Level III may be specified when less discrimination is needed, or Level III may be specified for greater discrimination. Four additional special levels: S-1, S-2, S-3, and S-4, are given in the same table and may be used where relatively small sample sizes are necessary and large sampling risks can or must be tolerated.

Note: In the designation of inspection levels S-1 to S-4, care must be exercised to avoid AQLs inconsistent with these inspection levels.

9.3. Code Letters

Sample sizes are designated by code letters. Table 10.1 shall be used to find the applicable code letter for the particular lot or batch size and the prescribed inspection level.

9.4. Obtaining Sampling Plan

The AQL and the code letter shall be used to obtain the sampling plan from Tables 10.2, 10.3, or 10.4. When no sampling plan is available for a given combination of AQL and code letter, the tables direct the user to a different letter. The sample size to be used is given by the new code letter, note by the original letter. If this procedure leads to different sample sizes for different classes of nonconformities, the code letter corresponding to the largest sample size derived may be used for all classes of nonconformities when designated or approved by the responsible authority. As an alternative to a single sampling plan with an acceptance number of 0, the plan with an acceptance number of 1 with its correspondingly larger sample size for a designated AQL (where available) may be used when designated or approved by the responsible authority.

9.5. Types of Sampling Plans

Three types of sampling plans: single, double and multiple, are given in Tables 10.2, 10.3 and 10.4, respectively. When several types of plans are available for a given AQL and code letter, any one may be used. A decision as to type of plan, either single, double, or multiple, when available for a given AQL and code letter, will usually be based upon the comparison between the administrative difficulty and the average sample sizes of the available plans. The average sample size of multiple plans is less than for double (except in the case corresponding to single acceptance number 1) and both of these are always less than a single sample size (see Figure 10.2). Usually the administrative difficulty for single sampling and the cost per unit of the sample are less than for double or multiple.

10. Determination of Acceptability

10.1 Percent Nonconforming Inspection

To determine acceptability of a lot or batch under percent non-conforming inspection, the applicable sampling plan shall be used in accordance with 10.1.1, 10.1.2, 10.1.3, and 10.1.4.

10.1.1. Single Sampling Plan

The number of sample units inspected shall be equal to the sample size given by the plan. If the number of nonconforming units found in the sample is equal to or less than the acceptance number, the lot or batch shall be considered acceptable. If the number of nonconforming units is equal to or greater than the rejection number, the lot or batch shall be considered not acceptable.

10.1.2. Double Sample Plan

The number of sample units first inspected shall be equal to the first sample size given by the plan. If the number of nonconforming units found in the first sample is equal to or less than the first acceptable number, the lot or batch shall be considered acceptable. If the number of nonconforming units found in the first sample is equal to or greater than the first rejection number, the lot or batch shall be considered not acceptable. If the number of nonconforming units found in the first sample is between the first acceptance and rejection numbers, a second sample of the size given by the plan shall be inspected. The number of nonconforming units found in the first and second samples shall be accumulated. If the cumulative number of nonconforming units is equal to or less than the second acceptance number, the lot or batch shall be considered acceptable. If the cumulative number of nonconforming units is equal to or greater than the second rejection number, the lot or batch shall be considered not acceptable.

10.1.3. Multiple Sample Plan

Under multiple sampling, the procedure shall be similar to that specified in 10.1.2, except that the number of successive samples required to reach a decision might be more than two.

10.1.4. Special Procedure for Reduced Inspection

Under reduced inspection, the sampling procedure may terminate without making a decision. In these circumstances, the lot or batch will be considered acceptable, but normal inspection will be reinstated starting with the next lot or batch (see 8.3.4(b)).

10.2 Non conformities Per Hundred Units Inspection

To determine the acceptability of a lot or batch under nonconformities per hundred units inspection, the procedure specified for percent nonconforming inspection above shall be used, except that the word "nonconformities" shall be substituted for "nonconforming units".

11. Supplementary Information

11.1 Operating Characteristic Curves

Operating characteristic curves and other measures of performance presented in this standard are of two types. Those for the individual plans that represent the elements of the schemes are presented in Tables 10.5, 10.6, 10.7, 10.9 and 10.10. Analogous curves and other measures of overall scheme performance when the switching rules are used are given in Tables 10.11–10.15. Scheme performance is defined as the composite proportion of lots accepted at a stated percent nonconforming when the switching rules are applied. The term scheme performance is used here in a very restrictive sense. It refers to how the ANSI Z1.4 scheme of switching rules would operate at a given process level under the assumption that the process stays at that level even after switching to tightened inspection or discontinuation of inspection. This gives a conservative "worst case" description of the performance of the scheme for use as a baseline in the sense that if the psychological and economic pressures associated with the switching rules are considered, the protection of the scheme may be somewhat better than that shown.

Operating characteristic curves are given in Table 10.10 for individual sampling plans for normal and tightened inspection. The operating characteristic curve for unqualified acceptance under reduced inspection can be found by using the AQL index of the normal plan with the sample size(s) and acceptance number(s) of the reduced plan. The curves shown are for single sampling; curves for double and multiple sampling are matched as closely as practicable. The OC curves shown for AQLs greater than 10.0 are based on the Poisson distribution and apply for nonconformities per hundred units inspection; those for AQLs of 10.0 or less and sample sizes of 80 or less are based on the binomial distribution and apply for percent nonconforming inspection; those for AQLs of 10.0 or less and sample sizes larger than 80 are based on the Poisson distribution and apply either for nonconformities per hundred units inspection, or for percent nonconforming inspection (the Poisson distribution being an adequate approximation to the binomial distribution under these conditions). Tabulated values corresponding to selected values of probabilities of acceptance (P_a in percent) are given for each of the curves shown, and, in addition, are indexed for tightened inspection, and also show values for nonconformities per hundred units for AQLs of 10.0 or less and sample sizes of 80 or less. The operating characteristic curves for scheme performance shown in Table 10.15 indicate the percentage of lots or batches which may be expected to be accepted under use of the switching rules with the various sampling plans for a given process quality subject to the restrictions stated above.

The operating characteristic curves of scheme performance are based on the use of limit numbers in switching to reduced inspection and are approximately correct when the limit numbers for reduced inspection are not used under option 8.5. The curves also assume a return to tightened inspection when inspection is resumed after discontinuation has been imposed. This is also true of average outgoing quality limit and average sample size for ANSI Z1.4 scheme performance.

Note that the operating characteristic curve for scheme performance is approximately that of the normal plan for low levels of percent nonconforming and that the tightened plan for high levels of percent nonconforming. Use of the reduced plan increases scheme probability of acceptance only for extremely low levels of percent nonconforming.

11.2 Process Average

The process average is the average percent nonconforming or average number of non-conformities per hundred units (whichever is applicable) of product submitted by the supplier for original inspection. Original inspection is the first inspection of a particular quantity of product as distinguished from the inspection of product which has been resubmitted after prior rejection. When double or multiple sampling is used, only first sample results shall be included in the process average calculation.

11.3 Average Outgoing Quality (AOQ)

The AOQ is the average quality of outgoing product including all accepted lots or batches, plus all lots or batches which are not accepted after such lots or batches have been effectively 100% inspected and all nonconforming units replaced by conforming units.

11.4 Average Outgoing Quality Limit (AOQL)

The AOQL is the maximum of the AOQs for all possible incoming qualities for a given acceptance sampling plan. AOQL values are given in Table 10.5(A) for each of the single sampling plans for normal inspection and in Table 10.5(B) for each of the single sampling plans for tightened inspections. AOQL values for ANSI Z1.4 scheme performance are given in Table 10.11 subject to the restrictions of 11.1. They show the average outgoing quality limits for scheme performance when using single sampling. AOQL will be slightly higher when the limit numbers for reduced inspection are not used under option 8.5.

11.5 Average Sample Size Curves

Average sample size curves for double and multiple sampling as compared to the single sampling plan for each acceptance number are in Figure 10.2. These show the average sample sizes which may be expected to occur under the various sampling plans for a given process quality level. The curves assume no curtailment of inspection and are approximate to the extent that they are based upon the Poisson distribution, and that the sample sizes at each stage for double and multiple sampling are assumed to be $0.631n$ and $0.25n$, respectively, where n is the equivalent single sample size. Average sample size tables for ANSI Z1.4 scheme performance are given in Table 10.14. They show the average sample size for scheme performance when using single sampling.

11.6 Limiting Quality Protection

11.6.1. Use of Individual Plans

This standard is intended to be used as a system employing tightened, normal, and reduced inspection on a continuing series of lots to achieve consumer protection while assuring the producer that acceptance will occur most of the time if quality is better than the AQL.

11.6.2. Importance of Switching Rules

Occasionally specific individual plans are selected from the standard and used without the switching rules. This is not the intended application of the ANSI Z1.4 system and its use in this way should not be referred to as inspection under ANSI Z1.4. When employed in this way, this document simply represents a repository for a collection of individual plans indexed by AQL. The operating characteristics and other measures of a plan so chosen must be assessed individually for that plan from the tables provided.

11.6.3. Limiting Quality Tables

If the lot or batch is of an isolated nature, it is desirable to limit the selection of sampling plans to those, associated with a designated AQL value, that provide not less than a specified limiting quality protection. Sampling plans for this purpose can be selected by choosing a limiting quality (LQ) and a consumer's risk to be associated with it. Limiting quality is the percentage of nonconforming units (or nonconformities) in a batch or lot for which for purposes of acceptance sampling, the consumer wishes the probability of acceptance to be restricted to a specified low value.

Tables VI and VII give process levels for which the probabilities of lot acceptance under various sampling plans are 10% and 5%, respectively. If a different value of consumer's risk is required, the OC curves and their tabulated values may be used. For individual lots with percents nonconforming or nonconformities per 100 units equal to the specified limiting quality (LQ) values, the probabilities of lot acceptance are less than 10% in the case of plans listed in Table 10.6 and less than 5% in the case of plans listed in Table 10.7

Table I Sample Size Code Letters

(See 9.2 and 9.3)

Lot or batch size			Special inspection levels				General inspection levels		
			S-1	S-2	S-3	S-4	I	II	III
2	to	8	A	A	A	A	A	A	B
9	to	15	A	A	A	A	A	B	C
16	to	25	A	A	B	B	B	C	D
26	to	50	A	B	B	C	C	D	E
51	to	90	B	B	C	C	C	E	F
91	to	150	B	B	C	D	D	F	G
151	to	280	B	C	D	E	E	G	H
281	to	500	B	C	D	E	F	H	J
501	to	1200	C	C	E	F	G	J	K
1201	to	3200	C	D	E	G	H	K	L
3201	to	10000	C	D	F	G	J	L	M
10001	to	35000	C	D	F	H	K	M	N
35001	to	150000	D	E	G	J	L	N	P
150001	to	500000	D	E	G	J	M	P	Q
500001	and	over	D	E	H	K	N	Q	R

Table II(A) Single Sampling Plans for Normal Inspection (Master Table)

SINGLE NORMAL PLANS

(See 9.4 and 9.5)

Sample size code letter	Sample size	Acceptance Quality Limits, *AQLs*, in Percent Nonconforming Items and Nonconformities per 100 Items (Normal Inspection)																									
		0.010	0.015	0.025	0.040	0.065	0.10	0.15	0.25	0.40	0.65	1.0	1.5	2.5	4.0	6.5	10	15	25	40	65	100	150	250	400	650	1000
		Ac Re	Ac Re	Ac Re	Ac Re	Ac Re	Ac Re	Ac Re	Ac Re	Ac Re	Ac Re	Ac Re	Ac Re	Ac Re	Ac Re	Ac Re	Ac Re	Ac Re	Ac Re	Ac Re	Ac Re	Ac Re	Ac Re	Ac Re	Ac Re	Ac Re	Ac Re
A	2	↓	↓	↓	↓	↓	↓	↓	↓	↓	↓	↓	↓	↓	↓	0 1	↓	↓	1 2	2 3	3 4	5 6	7 8	10 11	14 15	21 22	30 31
B	3	↓	↓	↓	↓	↓	↓	↓	↓	↓	↓	↓	↓	↓	0 1	↑	↓	1 2	2 3	3 4	5 6	7 8	10 11	14 15	21 22	30 31	44 45
C	5	↓	↓	↓	↓	↓	↓	↓	↓	↓	↓	↓	↓	0 1	↑	↓	1 2	2 3	3 4	5 6	7 8	10 11	14 15	21 22	30 31	44 45	↑
D	8	↓	↓	↓	↓	↓	↓	↓	↓	↓	↓	↓	0 1	↑	↓	1 2	2 3	3 4	5 6	7 8	10 11	14 15	21 22	30 31	44 45	↑	↑
E	13	↓	↓	↓	↓	↓	↓	↓	↓	↓	↓	0 1	↑	↓	1 2	2 3	3 4	5 6	7 8	10 11	14 15	21 22	30 31	44 45	↑	↑	↑
F	20	↓	↓	↓	↓	↓	↓	↓	↓	↓	0 1	↑	↓	1 2	2 3	3 4	5 6	7 8	10 11	14 15	21 22	↑	↑	↑	↑	↑	↑
G	32	↓	↓	↓	↓	↓	↓	↓	↓	0 1	↑	↓	1 2	2 3	3 4	5 6	7 8	10 11	14 15	21 22	↑	↑	↑	↑	↑	↑	↑
H	50	↓	↓	↓	↓	↓	↓	↓	0 1	↑	↓	1 2	2 3	3 4	5 6	7 8	10 11	14 15	21 22	↑	↑	↑	↑	↑	↑	↑	↑
J	80	↓	↓	↓	↓	↓	↓	0 1	↑	↓	1 2	2 3	3 4	5 6	7 8	10 11	14 15	21 22	↑	↑	↑	↑	↑	↑	↑	↑	↑
K	125	↓	↓	↓	↓	↓	0 1	↑	↓	1 2	2 3	3 4	5 6	7 8	10 11	14 15	21 22	↑	↑	↑	↑	↑	↑	↑	↑	↑	↑
L	200	↓	↓	↓	↓	0 1	↑	↓	1 2	2 3	3 4	5 6	7 8	10 11	14 15	21 22	↑	↑	↑	↑	↑	↑	↑	↑	↑	↑	↑
M	315	↓	↓	↓	0 1	↑	↓	1 2	2 3	3 4	5 6	7 8	10 11	14 15	21 22	↑	↑	↑	↑	↑	↑	↑	↑	↑	↑	↑	↑
N	500	↓	↓	0 1	↑	↓	1 2	2 3	3 4	5 6	7 8	10 11	14 15	21 22	↑	↑	↑	↑	↑	↑	↑	↑	↑	↑	↑	↑	↑
P	800	↓	0 1	↑	↓	1 2	2 3	3 4	5 6	7 8	10 11	14 15	21 22	↑	↑	↑	↑	↑	↑	↑	↑	↑	↑	↑	↑	↑	↑
Q	1250	0 1	↑	↓	1 2	2 3	3 4	5 6	7 8	10 11	14 15	21 22	↑	↑	↑	↑	↑	↑	↑	↑	↑	↑	↑	↑	↑	↑	↑
R	2000	↑	↑	1 2	2 3	3 4	5 6	7 8	10 11	14 15	21 22	↑	↑	↑	↑	↑	↑	↑	↑	↑	↑	↑	↑	↑	↑	↑	↑

↓ = Use the first sampling plan below the arrow. If sample size equals, or exceeds, lot size, carry out 100 percent inspection.

↑ = Use the first sampling plan above the arrow.

Ac = Acceptance number.

Re = Rejection number.

Table II(B) Single Sampling Plans for Normal Inspection (Master Table)
SINGLE TIGHTENED PLANS

(See 9.4 and 9.5)

Sample size code letter	Sample size	Acceptance Quality Limits (tightened inspection)																									
		0.010	0.015	0.025	0.040	0.065	0.10	0.15	0.25	0.40	0.65	1.0	1.5	2.5	4.0	6.5	10	15	25	40	65	100	150	250	400	650	1000
		Ac Re	Ac Re	Ac Re	Ac Re	Ac Re	Ac Re	Ac Re	Ac Re	Ac Re	Ac Re	Ac Re	Ac Re	Ac Re	Ac Re	Ac Re	Ac Re	Ac Re	Ac Re	Ac Re	Ac Re	Ac Re	Ac Re	Ac Re	Ac Re	Ac Re	Ac Re
A	2	↓	↓	↓	↓	↓	↓	↓	↓	↓	↓	↓	↓	↓	↓	↓	↓	↓	↓	1 2	2 3	3 4	5 6	8 9	12 13	18 19	27 28
B	3	↓	↓	↓	↓	↓	↓	↓	↓	↓	↓	↓	↓	↓	↓	0 1	↓	↓	1 2	2 3	3 4	5 6	8 9	12 13	18 19	27 28	41 42
C	5	↓	↓	↓	↓	↓	↓	↓	↓	↓	↓	↓	↓	↓	0 1	↓	↓	1 2	2 3	3 4	5 6	8 9	12 13	18 19	27 28	41 42	↑
D	8	↓	↓	↓	↓	↓	↓	↓	↓	↓	↓	↓	↓	0 1	↓	↓	1 2	2 3	3 4	5 6	8 9	12 13	18 19	27 28	41 42	↑	↑
E	13	↓	↓	↓	↓	↓	↓	↓	↓	↓	↓	↓	0 1	↓	↓	1 2	2 3	3 4	5 6	8 9	12 13	18 19	27 28	41 42	↑	↑	↑
F	20	↓	↓	↓	↓	↓	↓	↓	↓	↓	↓	0 1	↓	↓	1 2	2 3	3 4	5 6	8 9	12 13	18 19	↑	↑	↑	↑	↑	↑
G	32	↓	↓	↓	↓	↓	↓	↓	↓	↓	0 1	↓	↓	1 2	2 3	3 4	5 6	8 9	12 13	18 19	↑	↑	↑	↑	↑	↑	↑
H	50	↓	↓	↓	↓	↓	↓	↓	↓	0 1	↓	↓	1 2	2 3	3 4	5 6	8 9	12 13	18 19	↑	↑	↑	↑	↑	↑	↑	↑
J	80	↓	↓	↓	↓	↓	↓	↓	0 1	↓	↓	1 2	2 3	3 4	5 6	8 9	12 13	18 19	↑	↑	↑	↑	↑	↑	↑	↑	↑
K	125	↓	↓	↓	↓	↓	↓	0 1	↓	↓	1 2	2 3	3 4	5 6	8 9	12 13	18 19	↑	↑	↑	↑	↑	↑	↑	↑	↑	↑
L	200	↓	↓	↓	↓	↓	0 1	↓	↓	1 2	2 3	3 4	5 6	8 9	12 13	18 19	↑	↑	↑	↑	↑	↑	↑	↑	↑	↑	↑
M	315	↓	↓	↓	↓	0 1	↓	↓	1 2	2 3	3 4	5 6	8 9	12 13	18 19	↑	↑	↑	↑	↑	↑	↑	↑	↑	↑	↑	↑
N	500	↓	↓	↓	0 1	↓	↓	1 2	2 3	3 4	5 6	8 9	12 13	18 19	↑	↑	↑	↑	↑	↑	↑	↑	↑	↑	↑	↑	↑
P	800	↓	↓	0 1	↓	↓	1 2	2 3	3 4	5 6	8 9	12 13	18 19	↑	↑	↑	↑	↑	↑	↑	↑	↑	↑	↑	↑	↑	↑
Q	1250	↓	0 1	↓	↓	1 2	2 3	3 4	5 6	8 9	12 13	18 19	↑	↑	↑	↑	↑	↑	↑	↑	↑	↑	↑	↑	↑	↑	↑
R	2000	0 1	↑	↓	1 2	2 3	3 4	5 6	8 9	12 13	18 19	↑	↑	↑	↑	↑	↑	↑	↑	↑	↑	↑	↑	↑	↑	↑	↑
S	3150		↑	1 2								↑	↑	↑	↑	↑	↑	↑	↑	↑	↑	↑	↑	↑	↑	↑	↑

↓ = Use first sampling plan below arrow. If sample size equals or exceeds lot or batch size, do 100 percent inspection.

↑ = Use first sampling plan above arrow.

Ac = Acceptance number.

Re = Rejection number.

Table II(C) Single Sampling Plans for Normal Inspection (Master Table)
SINGLE REDUCED PLANS

(See 9.4 and 9.5)

Acceptance Quality Limits (reduced inspection)†

Sample size code letter	Sample size	0.010 Ac Re	0.015 Ac Re	0.025 Ac Re	0.040 Ac Re	0.065 Ac Re	0.10 Ac Re	0.15 Ac Re	0.25 Ac Re	0.40 Ac Re	0.65 Ac Re	1.0 Ac Re	1.5 Ac Re	2.5 Ac Re	4.0 Ac Re	6.5 Ac Re	10 Ac Re	15 Ac Re	25 Ac Re	40 Ac Re	65 Ac Re	100 Ac Re	150 Ac Re	250 Ac Re	400 Ac Re	650 Ac Re	1000 Ac Re
A	2	↓	↓	↓	↓	↓	↓	↓	↓	↓	↓	↓	↓	↓	↓	0 1	↓	↓	1 2	2 3	3 4	5 6	7 8	10 11	14 15	21 22	30 31
B	2	↓	↓	↓	↓	↓	↓	↓	↓	↓	↓	↓	↓	↓	0 1	↑	↓	0 2	1 3	2 4	3 5	5 6	7 8	10 11	14 15	21 22	30 31
C	2	↓	↓	↓	↓	↓	↓	↓	↓	↓	↓	↓	↓	0 1	↑	↓	0 2	1 3	1 4	2 5	3 6	5 8	7 10	10 13	14 17	21 24	↑
D	3	↓	↓	↓	↓	↓	↓	↓	↓	↓	↓	↓	0 1	↑	↓	0 2	1 3	1 4	2 5	3 6	5 8	7 10	10 13	14 17	21 24	↑	↑
E	5	↓	↓	↓	↓	↓	↓	↓	↓	↓	↓	0 1	↑	↓	0 2	1 3	1 4	2 5	3 6	5 8	7 10	10 13	14 17	21 24	↑	↑	↑
F	8	↓	↓	↓	↓	↓	↓	↓	↓	↓	0 1	↑	↓	0 2	1 3	1 4	2 5	3 6	5 8	7 10	10 13	↑	↑	↑	↑	↑	↑
G	13	↓	↓	↓	↓	↓	↓	↓	↓	0 1	↑	↓	0 2	1 3	1 4	2 5	3 6	5 8	7 10	10 13	↑	↑	↑	↑	↑	↑	↑
H	20	↓	↓	↓	↓	↓	↓	↓	0 1	↑	↓	0 2	1 3	1 4	2 5	3 6	5 8	7 10	10 13	↑	↑	↑	↑	↑	↑	↑	↑
J	32	↓	↓	↓	↓	↓	↓	0 1	↑	↓	0 2	1 3	1 4	2 5	3 6	5 8	7 10	10 13	↑	↑	↑	↑	↑	↑	↑	↑	↑
K	50	↓	↓	↓	↓	↓	0 1	↑	↓	0 2	1 3	1 4	2 5	3 6	5 8	7 10	10 13	↑	↑	↑	↑	↑	↑	↑	↑	↑	↑
L	80	↓	↓	↓	↓	0 1	↑	↓	0 2	1 3	1 4	2 5	3 6	5 8	7 10	10 13	↑	↑	↑	↑	↑	↑	↑	↑	↑	↑	↑
M	125	↓	↓	↓	0 1	↑	↓	0 2	1 3	1 4	2 5	3 6	5 8	7 10	10 13	↑	↑	↑	↑	↑	↑	↑	↑	↑	↑	↑	↑
N	200	↓	↓	0 1	↑	↓	0 2	1 3	1 4	2 5	3 6	5 8	7 10	10 13	↑	↑	↑	↑	↑	↑	↑	↑	↑	↑	↑	↑	↑
P	315	↓	0 1	↑	↓	0 2	1 3	1 4	2 5	3 6	5 8	7 10	10 13	↑	↑	↑	↑	↑	↑	↑	↑	↑	↑	↑	↑	↑	↑
Q	500	0 1	↑	↓	0 2	1 3	1 4	2 5	3 6	5 8	7 10	10 13	↑	↑	↑	↑	↑	↑	↑	↑	↑	↑	↑	↑	↑	↑	↑
R	800	↑	↑	0 2	1 3	1 4	2 5	3 6	5 8	7 10	10 13	↑	↑	↑	↑	↑	↑	↑	↑	↑	↑	↑	↑	↑	↑	↑	↑

↓ = Use first sampling plan below arrow. If sample size equals or exceeds lot or batch size, do 100 percent inspection.

↑ = Use first sampling plan above arrow.

Ac = Acceptance number.

Re = Rejection number.

† = If the acceptance number has been exceeded, but the rejection number has not been reached, accept the lot, but reinstate normal inspection (see 10.1.4).

Table III(A) Double Sampling Plans for Normal Inspection (Master Table)

DOUBLE NORMAL PLANS

(See 9.4 and 9.5)

Sample size code letter	Sample	Sample size	Cumulative sample size	Acceptance Quality Limits (reduced inspection)																									
				0.010	0.015	0.025	0.040	0.065	0.10	0.15	0.25	0.40	0.65	1.0	1.5	2.5	4.0	6.5	10	15	25	40	65	100	150	250	400	650	1000
				Ac Re	Ac Re	Ac Re	Ac Re	Ac Re	Ac Re	Ac Re	Ac Re	Ac Re	Ac Re	Ac Re	Ac Re	Ac Re	Ac Re	Ac Re	Ac Re	Ac Re	Ac Re	Ac Re	Ac Re	Ac Re	Ac Re	Ac Re	Ac Re	Ac Re	Ac Re
A				↓	↓	↓	↓	↓	↓	↓	↓	↓	↓	↓	↓	↓	↓	*	↓	↓	*	*	*	*	*	*	*	*	*
B	First	2	2	↓	↓	↓	↓	↓	↓	↓	↓	↓	↓	↓	↓	↓	*	↑	↓	0 2	0 3	1 4	2 5	3 7	5 9	7 11	11 16	17 22	25 31
	Second	2	4																	1 2	3 4	4 5	6 7	8 9	12 13	18 19	26 27	37 38	56 57
C	First	3	3	↓	↓	↓	↓	↓	↓	↓	↓	↓	↓	↓	↓	*	↑	↓	0 2	0 3	1 4	2 5	3 7	5 9	7 11	11 16	17 22	25 31	↑
	Second	3	6																1 2	3 4	4 5	6 7	8 9	12 13	18 19	26 27	37 38	56 57	
D	First	5	5	↓	↓	↓	↓	↓	↓	↓	↓	↓	↓	↓	*	↑	↓	0 2	0 3	1 4	2 5	3 7	5 9	7 11	11 16	17 22	25 31	↑	↑
	Second	5	10															1 2	3 4	4 5	6 7	8 9	12 13	18 19	26 27	37 38	56 57		
E	First	8	8	↓	↓	↓	↓	↓	↓	↓	↓	↓	↓	*	↑	↓	0 2	0 3	1 4	2 5	3 7	5 9	7 11	11 16	17 22	25 31	↑	↑	↑
	Second	8	16														1 2	3 4	4 5	6 7	8 9	12 13	18 19	26 27	37 38	56 57			
F	First	13	13	↓	↓	↓	↓	↓	↓	↓	↓	↓	*	↑	↓	0 2	0 3	1 4	2 5	3 7	5 9	7 11	11 16	↑	↑	↑	↑	↑	↑
	Second	13	26													1 2	3 4	4 5	6 7	8 9	12 13	18 19	26 27						
G	First	20	20	↓	↓	↓	↓	↓	↓	↓	↓	*	↑	↓	0 2	0 3	1 4	2 5	3 7	5 9	7 11	11 16	↑	↑	↑	↑	↑	↑	↑
	Second	20	40											1 2	3 4	4 5	6 7	8 9	12 13	18 19	26 27								
H	First	32	32	↓	↓	↓	↓	↓	↓	↓	*	↑	↓	0 2	0 3	1 4	2 5	3 7	5 9	7 11	11 16	↑	↑	↑	↑	↑	↑	↑	↑
	Second	32	64											1 2	3 4	4 5	6 7	8 9	12 13	18 19	26 27								
J	First	50	50	↓	↓	↓	↓	↓	↓	*	↑	↓	0 2	0 3	1 4	2 5	3 7	5 9	7 11	11 16	↑	↑	↑	↑	↑	↑	↑	↑	↑
	Second	50	100									1 2	3 4	4 5	6 7	8 9	12 13	18 19	26 27										
K	First	80	80	↓	↓	↓	↓	↓	*	↑	↓	0 2	0 3	1 4	2 5	3 7	5 9	7 11	11 16	↑	↑	↑	↑	↑	↑	↑	↑	↑	↑
	Second	80	160									1 2	3 4	4 5	6 7	8 9	12 13	18 19	26 27										
L	First	125	125	↓	↓	↓	↓	*	↑	↓	0 2	0 3	1 4	2 5	3 7	5 9	7 11	11 16	↑	↑	↑	↑	↑	↑	↑	↑	↑	↑	↑
	Second	125	250								1 2	3 4	4 5	6 7	8 9	12 13	18 19	26 27											
M	First	200	200	↓	↓	↓	*	↑	↓	0 2	0 3	1 4	2 5	3 7	5 9	7 11	11 16	↑	↑	↑	↑	↑	↑	↑	↑	↑	↑	↑	↑
	Second	200	400							1 2	3 4	4 5	6 7	8 9	12 13	18 19	26 27												
N	First	315	315	↓	↓	*	↑	↓	0 2	0 3	1 4	2 5	3 7	5 9	7 11	11 16	↑	↑	↑	↑	↑	↑	↑	↑	↑	↑	↑	↑	↑
	Second	315	630						1 2	3 4	4 5	6 7	8 9	12 13	18 19	26 27													
P	First	500	500	↓	*	↑	↓	0 2	0 3	1 4	2 5	3 7	5 9	7 11	11 16	↑	↑	↑	↑	↑	↑	↑	↑	↑	↑	↑	↑	↑	↑
	Second	500	1000					1 2	3 4	4 5	6 7	8 9	12 13	18 19	26 27														
Q	First	800	800	*	↑	↓	0 2	0 3	1 4	2 5	3 7	5 9	5 9	11 16	↑	↑	↑	↑	↑	↑	↑	↑	↑	↑	↑	↑	↑	↑	↑
	Second	800	1600				1 2	3 4	4 5	6 7	8 9	12 13	18 19	26 27															
R	First	1250	1250	↑	↑	0 2	0 3	1 4	2 5	3 7	5 9	7 11	11 16	↑	↑	↑	↑	↑	↑	↑	↑	↑	↑	↑	↑	↑	↑	↑	↑
	Second	1250	2500			1 2	3 4	4 5	6 7	8 9	12 13	18 19	26 27																

↓ = Use first sampling plan below arrow. If sample size equals or exceeds lot or batch size, do 100 percent inspection.

↑ = Use first sampling plan above arrow.

Ac = Acceptance number.

Re = Rejection number.

* = Use corresponding single sampling plan (or alternatively, use double sampling plan below, where available).

Table III(B) Double Sampling Plans for Normal Inspection (Master Table)

DOUBLE TIGHTENED PLANS

(See 9.4 and 9.5)

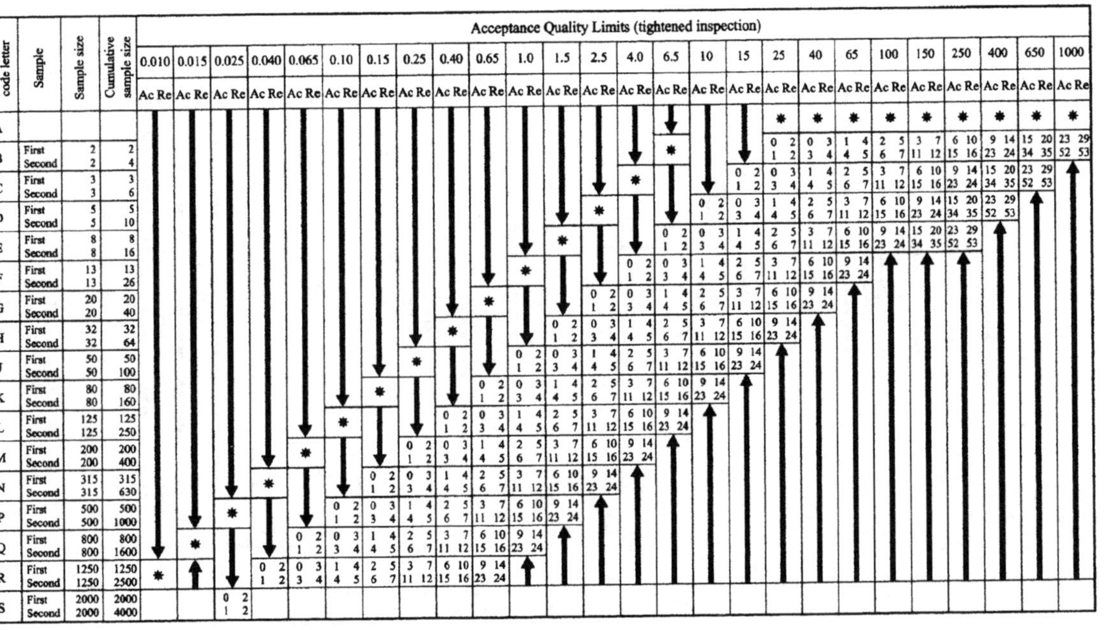

Sample size code letter	Sample	Sample size	Cumulative sample size	Acceptance Quality Limits (tightened inspection)																									
				0.010	0.015	0.025	0.040	0.065	0.10	0.15	0.25	0.40	0.65	1.0	1.5	2.5	4.0	6.5	10	15	25	40	65	100	150	250	400	650	1000
				Ac Re	Ac Re	Ac Re	Ac Re	Ac Re	Ac Re	Ac Re	Ac Re	Ac Re	Ac Re	Ac Re	Ac Re	Ac Re	Ac Re	Ac Re	Ac Re	Ac Re	Ac Re	Ac Re	Ac Re	Ac Re	Ac Re	Ac Re	Ac Re	Ac Re	Ac Re
A				↓	↓	↓	↓	↓	↓	↓	↓	↓	↓	↓	↓	↓	↓	↓	↓	↓	*	*	*	*	*	*	*	*	*
B	First	2	2	↓	↓	↓	↓	↓	↓	↓	↓	↓	↓	↓	↓	↓	↓	*	↓	↓	0 2	0 3	1 4	2 5	3 7	6 10	9 14	15 20	23 29
	Second	2	4																		1 2	3 4	4 5	6 7	11 12	15 16	23 24	34 35	52 53
C	First	3	3	↓	↓	↓	↓	↓	↓	↓	↓	↓	↓	↓	↓	↓	*	↓	↓	0 2	0 3	1 4	2 5	3 7	6 10	9 14	15 20	23 29	↑
	Second	3	6																	1 2	3 4	4 5	6 7	11 12	15 16	23 24	34 35	52 53	
D	First	5	5	↓	↓	↓	↓	↓	↓	↓	↓	↓	↓	↓	↓	*	↓	↓	0 2	0 3	1 4	2 5	3 7	6 10	9 14	15 20	23 29	↑	↑
	Second	5	10															1 2	3 4	4 5	6 7	11 12	15 16	23 24	34 35	52 53			
E	First	8	8	↓	↓	↓	↓	↓	↓	↓	↓	↓	↓	↓	*	↓	↓	0 2	0 3	1 4	2 5	3 7	6 10	9 14	15 20	23 29	↑	↑	↑
	Second	8	16														1 2	3 4	4 5	6 7	11 12	15 16	23 24	34 35	52 53				
F	First	13	13	↓	↓	↓	↓	↓	↓	↓	↓	↓	↓	*	↓	↓	0 2	0 3	1 4	2 5	3 7	6 10	9 14	↑	↑	↑	↑	↑	↑
	Second	13	26													1 2	3 4	4 5	6 7	11 12	15 16	23 24							
G	First	20	20	↓	↓	↓	↓	↓	↓	↓	↓	↓	*	↓	↓	0 2	0 3	1 4	2 5	3 7	6 10	9 14	↑	↑	↑	↑	↑	↑	↑
	Second	20	40												1 2	3 4	4 5	6 7	11 12	15 16	23 24								
H	First	32	32	↓	↓	↓	↓	↓	↓	↓	↓	*	↓	↓	0 2	0 3	1 4	2 5	3 7	6 10	9 14	↑	↑	↑	↑	↑	↑	↑	↑
	Second	32	64											1 2	3 4	4 5	6 7	11 12	15 16	23 24									
J	First	50	50	↓	↓	↓	↓	↓	↓	↓	*	↓	↓	0 2	0 3	1 4	2 5	3 7	6 10	9 14	↑	↑	↑	↑	↑	↑	↑	↑	↑
	Second	50	100										1 2	3 4	4 5	6 7	11 12	15 16	23 24										
K	First	80	80	↓	↓	↓	↓	↓	↓	*	↓	↓	0 2	0 3	1 4	2 5	3 7	6 10	9 14	↑	↑	↑	↑	↑	↑	↑	↑	↑	↑
	Second	80	160									1 2	3 4	4 5	6 7	11 12	15 16	23 24											
L	First	125	125	↓	↓	↓	↓	↓	*	↓	↓	0 2	0 3	1 4	2 5	3 7	6 10	9 14	↑	↑	↑	↑	↑	↑	↑	↑	↑	↑	↑
	Second	125	250								1 2	3 4	4 5	6 7	11 12	15 16	23 24												
M	First	200	200	↓	↓	↓	↓	*	↓	↓	0 2	0 3	1 4	2 5	3 7	6 10	9 14	↑	↑	↑	↑	↑	↑	↑	↑	↑	↑	↑	↑
	Second	200	400							1 2	3 4	4 5	6 7	11 12	15 16	23 24													
N	First	315	315	↓	↓	↓	*	↓	↓	0 2	0 3	1 4	2 5	3 7	6 10	9 14	↑	↑	↑	↑	↑	↑	↑	↑	↑	↑	↑	↑	↑
	Second	315	630							1 2	3 4	4 5	6 7	11 12	15 16	23 24													
P	First	500	500	↓	↓	*	↓	↓	0 2	0 3	1 4	2 5	3 7	6 10	9 14	↑	↑	↑	↑	↑	↑	↑	↑	↑	↑	↑	↑	↑	↑
	Second	500	1000						1 2	3 4	4 5	6 7	11 12	15 16	23 24														
Q	First	800	800	↓	*	↓	↓	0 2	0 3	1 4	2 5	3 7	6 10	9 14	↑	↑	↑	↑	↑	↑	↑	↑	↑	↑	↑	↑	↑	↑	↑
	Second	800	1600					1 2	3 4	4 5	6 7	11 12	15 16	23 24															
R	First	1250	1250	*	↑	↓	0 2	0 3	1 4	2 5	3 7	6 10	9 14	↑	↑	↑	↑	↑	↑	↑	↑	↑	↑	↑	↑	↑	↑	↑	↑
	Second	1250	2500				1 2	3 4	4 5	6 7	11 12	15 16	23 24																
S	First	2000	2000			0 2																							
	Second	2000	4000			1 2																							

↓ = Use first sampling plan below arrow. If sample size equals or exceeds lot or batch size, do 100 percent inspection.

↑ = Use first sampling plan above arrow.

Ac = Acceptance number.

Re = Rejection number.

* = Use corresponding single sampling plan (or alternatively, use double sampling plan below, where available).

Table III(C) Double Sampling Plans for Normal Inspection (Master Table)
DOUBLE REDUCED PLANS

(See 9.4 and 9.5)

Sample size code letter	Sample	Sample size	Cumulative sample size	Acceptance Quality Limits (reduced inspection)†																									
				0.010	0.015	0.025	0.040	0.065	0.10	0.15	0.25	0.40	0.65	1.0	1.5	2.5	4.0	6.5	10	15	25	40	65	100	150	250	400	650	1000
				Ac Re	Ac Re	Ac Re	Ac Re	Ac Re	Ac Re	Ac Re	Ac Re	Ac Re	Ac Re	Ac Re	Ac Re	Ac Re	Ac Re	Ac Re	Ac Re	Ac Re	Ac Re	Ac Re	Ac Re	Ac Re	Ac Re	Ac Re	Ac Re	Ac Re	Ac Re
A				↓	↓	↓	↓	↓	↓	↓	↓	↓	↓	↓	↓	↓	↓	*	↓	↓	*	*	*	*	*	*	*	*	*
B				↓	↓	↓	↓	↓	↓	↓	↓	↓	↓	↓	↓	↓	*	↑	↓	*	*	*	*	*	*	*	*	*	*
C				↓	↓	↓	↓	↓	↓	↓	↓	↓	↓	↓	↓	*	↑	↓	*	*	*	*	*	*	*	*	*	*	↑
D	First	2	2	↓	↓	↓	↓	↓	↓	↓	↓	↓	↓	↓	*	↑	↓	0 2	0 3	0 4	0 4	1 5	2 7	3 8	5 10	7 12	11 17	↑	↑
	Second	2	4															0 2	0 4	1 5	3 6	4 7	6 9	8 12	12 16	18 22	26 30		
E	First	3	3	↓	↓	↓	↓	↓	↓	↓	↓	↓	↓	*	↑	↓	0 2	0 3	0 4	0 4	1 5	2 7	3 8	5 10	7 12	11 17	↑	↑	↑
	Second	3	6														0 2	0 4	1 5	3 6	4 7	6 9	8 12	12 16	18 22	26 30			
F	First	5	5	↓	↓	↓	↓	↓	↓	↓	↓	↓	*	↑	↓	0 2	0 3	0 4	0 4	1 5	2 7	3 8	5 10	↑	↑	↑	↑	↑	↑
	Second	5	10													0 2	0 4	1 5	3 6	4 7	6 9	8 12	12 16						
G	First	8	8	↓	↓	↓	↓	↓	↓	↓	↓	*	↑	↓	0 2	0 3	0 4	0 4	1 5	2 7	3 8	5 10	↑	↑	↑	↑	↑	↑	↑
	Second	8	16												0 2	0 4	1 5	3 6	4 7	6 9	8 12	12 16							
H	First	13	13	↓	↓	↓	↓	↓	↓	↓	*	↑	↓	0 2	0 3	0 4	0 4	1 5	2 7	3 8	5 10	↑	↑	↑	↑	↑	↑	↑	↑
	Second	13	26											0 2	0 4	1 5	3 6	4 7	6 9	8 12	12 16								
J	First	20	20	↓	↓	↓	↓	↓	↓	*	↑	↓	0 2	0 3	0 4	0 4	1 5	2 7	3 8	5 10	↑	↑	↑	↑	↑	↑	↑	↑	↑
	Second	20	40										0 2	0 4	1 5	3 6	4 7	6 9	8 12	12 16									
K	First	32	32	↓	↓	↓	↓	↓	*	↑	↓	0 2	0 3	0 4	0 4	1 5	2 7	3 8	5 10	↑	↑	↑	↑	↑	↑	↑	↑	↑	↑
	Second	32	64									0 2	0 4	1 5	3 6	4 7	6 9	8 12	12 16										
L	First	50	50	↓	↓	↓	↓	*	↑	↓	0 2	0 3	0 4	0 4	1 5	2 7	3 8	5 10	↑	↑	↑	↑	↑	↑	↑	↑	↑	↑	↑
	Second	50	100								0 2	0 4	1 5	3 6	4 7	6 9	8 12	12 16											
M	First	80	80	↓	↓	↓	*	↑	↓	0 2	0 3	0 4	0 4	1 5	2 7	3 8	5 10	↑	↑	↑	↑	↑	↑	↑	↑	↑	↑	↑	↑
	Second	80	160							0 2	0 4	1 5	3 6	4 7	6 9	8 12	12 16												
N	First	125	125	↓	↓	*	↑	↓	0 2	0 3	0 4	0 4	1 5	2 7	3 8	5 10	↑	↑	↑	↑	↑	↑	↑	↑	↑	↑	↑	↑	↑
	Second	125	250						0 2	0 4	1 5	3 6	4 7	6 9	8 12	12 16													
P	First	200	200	↓	*	↑	↓	0 2	0 3	0 4	0 4	1 5	2 7	3 8	5 10	↑	↑	↑	↑	↑	↑	↑	↑	↑	↑	↑	↑	↑	↑
	Second	200	400					0 2	0 4	1 5	3 6	4 7	6 9	8 12	12 16														
Q	First	315	315	*	↑	↓	0 2	0 3	0 4	0 4	1 5	2 7	3 8	5 10	↑	↑	↑	↑	↑	↑	↑	↑	↑	↑	↑	↑	↑	↑	↑
	Second	315	630				0 2	0 4	1 5	3 6	4 7	6 9	8 12	12 16															
R	First	500	500	↑	↑	0 2	0 3	0 4	0 4	1 5	2 7	3 8	5 10	↑	↑	↑	↑	↑	↑	↑	↑	↑	↑	↑	↑	↑	↑	↑	↑
	Second	500	1000			0 2	0 4	1 5	3 6	4 7	6 9	8 12	12 16																

↓ = Use first sampling plan below arrow. If sample size equals or exceeds lot or batch size, do 100 percent inspection.

↑ = Use first sampling plan above arrow.

Ac = Acceptance number.

Re = Rejection number.

* = Use corresponding single sampling plan (or alternatively, use double sampling plan below, where available).

† = If, after the second sample, the acceptance number has been exceeded, but the rejection number has not been reached, accept the lot, but reinstate normal inspection (see 10.1.4).

Table IV(A) Multiple Sampling Plans for Normal Inspection (Master Table)
MULTIPLE NORMAL PLANS

(See 9.4 and 9.5)

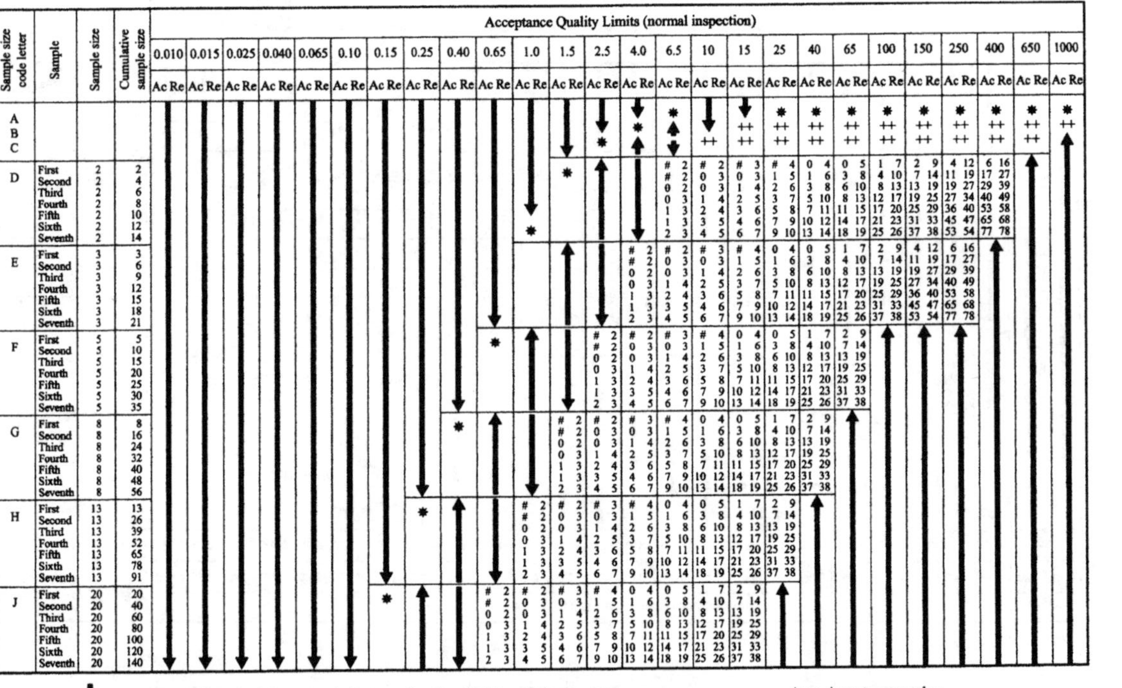

Sample size code letter	Sample	Sample size	Cumulative sample size	Acceptance Quality Limits (normal inspection)																									
				0.010	0.015	0.025	0.040	0.065	0.10	0.15	0.25	0.40	0.65	1.0	1.5	2.5	4.0	6.5	10	15	25	40	65	100	150	250	400	650	1000
				Ac Re	Ac Re	Ac Re	Ac Re	Ac Re	Ac Re	Ac Re	Ac Re	Ac Re	Ac Re	Ac Re	Ac Re	Ac Re	Ac Re	Ac Re	Ac Re	Ac Re	Ac Re	Ac Re	Ac Re	Ac Re	Ac Re	Ac Re	Ac Re	Ac Re	Ac Re
A				↓	↓	↓	↓	↓	↓	↓	↓	↓	↓	↓	↓	↓	↓	*	↓	↓	*	*	*	*	*	*	*	*	*
B																	*	↑		++	++	++	++	++	++	++	++	++	++
C																*	↑	↓	++	++	++	++	++	++	++	++	++	++	↑
D	First	2	2	↓	↓	↓	↓	↓	↓	↓	↓	↓	↓	↓	*	↑	↓	# 2	# 2	# 3	# 4	0 4	0 5	1 7	2 9	4 12	6 16	↑	↑
	Second	2	4															# 2	0 3	0 3	1 5	1 6	3 8	4 10	7 14	11 19	17 27		
	Third	2	6															0 2	0 3	1 4	2 6	3 8	6 10	8 13	13 19	19 27	29 39		
	Fourth	2	8															0 3	1 4	2 5	3 7	5 10	8 13	12 17	19 25	27 34	40 49		
	Fifth	2	10															1 3	2 4	3 6	5 8	7 11	11 15	17 20	25 29	36 40	53 58		
	Sixth	2	12															1 3	3 5	4 6	7 9	10 12	14 17	21 23	31 33	45 47	65 68		
	Seventh	2	14											*				2 3	4 5	6 7	9 10	13 14	18 19	25 26	37 38	53 54	77 78		
E	First	3	3	↓	↓	↓	↓	↓	↓	↓	↓	↓	↓		↑	↓	# 2	# 2	# 3	# 4	0 4	0 5	1 7	2 9	4 12	6 16	↑	↑	↑
	Second	3	6														# 2	0 3	0 3	1 5	1 6	3 8	4 10	7 14	11 19	17 27			
	Third	3	9														0 2	0 3	1 4	2 6	3 8	6 10	8 13	13 19	19 27	29 39			
	Fourth	3	12														0 3	1 4	2 5	3 7	5 10	8 13	12 17	19 25	27 34	40 49			
	Fifth	3	15														1 3	2 4	3 6	5 8	7 11	11 15	17 20	25 29	36 40	53 58			
	Sixth	3	18														1 3	3 5	4 6	7 9	10 12	14 17	21 23	31 33	45 47	65 68			
	Seventh	3	21														2 3	4 5	6 7	9 10	13 14	18 19	25 26	37 38	53 54	77 78			
F	First	5	5	↓	↓	↓	↓	↓	↓	↓	↓	↓	*	↑	↓	# 2	# 2	# 3	# 4	0 4	0 5	1 7	2 9	↑	↑	↑	↑	↑	↑
	Second	5	10													# 2	0 3	0 3	1 5	1 6	3 8	4 10	7 14						
	Third	5	15													0 2	0 3	1 4	2 6	3 8	6 10	8 13	13 19						
	Fourth	5	20													0 3	1 4	2 5	3 7	5 10	8 13	12 17	19 25						
	Fifth	5	25													1 3	2 4	3 6	5 8	7 11	11 15	17 20	25 29						
	Sixth	5	30													1 3	3 5	4 6	7 9	10 12	14 17	21 23	31 33						
	Seventh	5	35													2 3	4 5	6 7	9 10	13 14	18 19	25 26	37 38						
G	First	8	8	↓	↓	↓	↓	↓	↓	↓	↓	*	↑	↓	# 2	# 2	# 3	# 4	0 4	0 5	1 7	2 9	↑	↑	↑	↑	↑	↑	↑
	Second	8	16												# 2	0 3	0 3	1 5	1 6	3 8	4 10	7 14							
	Third	8	24												0 2	0 3	1 4	2 6	3 8	6 10	8 13	13 19							
	Fourth	8	32												0 3	1 4	2 5	3 7	5 10	8 13	12 17	19 25							
	Fifth	8	40												1 3	2 4	3 6	5 8	7 11	11 15	17 20	25 29							
	Sixth	8	48												1 3	3 5	4 6	7 9	10 12	14 17	21 23	31 33							
	Seventh	8	56												2 3	4 5	6 7	9 10	13 14	18 19	25 26	37 38							
H	First	13	13	↓	↓	↓	↓	↓	↓	↓	*	↑	↓	# 2	# 2	# 3	# 4	0 4	0 5	1 7	2 9	↑	↑	↑	↑	↑	↑	↑	↑
	Second	13	26											# 2	0 3	0 3	1 5	1 6	3 8	4 10	7 14								
	Third	13	39											0 2	0 3	1 4	2 6	3 8	6 10	8 13	13 19								
	Fourth	13	52											0 3	1 4	2 5	3 7	5 10	8 13	12 17	19 25								
	Fifth	13	65											1 3	2 4	3 6	5 8	7 11	11 15	17 20	25 29								
	Sixth	13	78											1 3	3 5	4 6	7 9	10 12	14 17	21 23	31 33								
	Seventh	13	91											2 3	4 5	6 7	9 10	13 14	18 19	25 26	37 38								
J	First	20	20	↓	↓	↓	↓	↓	↓	*	↑	↓	# 2	# 2	# 3	# 4	0 4	0 5	1 7	2 9	↑	↑	↑	↑	↑	↑	↑	↑	↑
	Second	20	40										# 2	0 3	0 3	1 5	1 6	3 8	4 10	7 14									
	Third	20	60										0 2	0 3	1 4	2 6	3 8	6 10	8 13	13 19									
	Fourth	20	80										0 3	1 4	2 5	3 7	5 10	8 13	12 17	19 25									
	Fifth	20	100										1 3	2 4	3 6	5 8	7 11	11 15	17 20	25 29									
	Sixth	20	120										1 3	3 5	4 6	7 9	10 12	14 17	21 23	31 33									
	Seventh	20	140										2 3	4 5	6 7	9 10	13 14	18 19	25 26	37 38									

↓ = Use first sampling plan below arrow (refer to continuation of table on following page, when necessary). If sample size equals or exceeds lot or batch size, do 100 percent inspection.

↑ = Use first sampling plan above arrow.

* = Use corresponding single sampling plan (or alternatively, use double sampling plan below, where available).

++ = Use corresponding double sampling plan (or alternatively, use multiple sampling plan below, where available).

Ac = Acceptance number.

Re = Rejection number.

= Acceptance not permitted at this sample size.

(See 9.4 and 9.5)

Sample size code letter	Sample	Sample size	Cumulative sample size	Acceptance Quality Limits (normal inspection)																									
				0.010	0.015	0.025	0.040	0.065	0.10	0.15	0.25	0.40	0.65	1.0	1.5	2.5	4.0	6.5	10	15	25	40	65	100	150	250	400	650	1000
				Ac Re	Ac Re	Ac Re	Ac Re	Ac Re	Ac Re	Ac Re	Ac Re	Ac Re	Ac Re	Ac Re	Ac Re	Ac Re	Ac Re	Ac Re	Ac Re	Ac Re	Ac Re	Ac Re	Ac Re	Ac Re	Ac Re	Ac Re	Ac Re	Ac Re	Ac Re
K	First	32	32	↓	↓	↓	↓	↓	*	↑	↓	# 2	# 2	# 3	# 4	0 4	0 5	1 7	2 9	↑	↑	↑	↑	↑	↑	↑	↑	↑	↑
	Second	32	64									# 2	0 3	0 3	1 5	1 6	3 8	4 10	7 14										
	Third	32	96									0 2	0 3	1 4	2 6	3 8	6 10	8 13	13 19										
	Fourth	32	128									0 3	1 4	2 5	3 7	5 10	8 13	12 17	19 25										
	Fifth	32	160									1 3	2 4	3 6	5 8	7 11	11 15	17 20	25 29										
	Sixth	32	192									1 3	3 5	4 6	7 9	10 12	14 17	21 23	31 33										
	Seventh	32	224									2 3	4 5	6 7	9 10	13 14	18 19	25 26	37 38										
L	First	50	50	↓	↓	↓	↓	*	↑	↓	# 2	# 2	# 3	# 4	0 4	0 5	1 7	2 9	↑	↑	↑	↑	↑	↑	↑	↑	↑	↑	↑
	Second	50	100								# 2	0 3	0 3	1 5	1 6	3 8	4 10	7 14											
	Third	50	150								0 2	0 3	1 4	2 6	3 8	6 10	8 13	13 19											
	Fourth	50	200								0 3	1 4	2 5	3 7	5 10	8 13	12 17	19 25											
	Fifth	50	250								1 3	2 4	3 6	5 8	7 11	11 15	17 20	25 29											
	Sixth	50	300								1 3	3 5	4 6	7 9	10 12	14 17	21 23	31 33											
	Seventh	50	350								2 3	4 5	6 7	9 10	13 14	18 19	25 26	37 38											
M	First	80	80	↓	↓	↓	*	↑	↓	# 2	# 2	# 3	# 4	0 4	0 5	1 7	2 9	↑	↑	↑	↑	↑	↑	↑	↑	↑	↑	↑	↑
	Second	80	160							# 2	0 3	0 3	1 5	1 6	3 8	4 10	7 14												
	Third	80	240							0 2	0 3	1 4	2 6	3 8	6 10	8 13	13 19												
	Fourth	80	320							0 3	1 4	2 5	3 7	5 10	8 13	12 17	19 25												
	Fifth	80	400							1 3	2 4	3 6	5 8	7 11	11 15	17 20	25 29												
	Sixth	80	480							1 3	3 5	4 6	7 9	10 12	14 17	21 23	31 33												
	Seventh	80	560							2 3	4 5	6 7	9 10	13 14	18 19	25 26	37 38												
N	First	125	125	↓	↓	*	↑	↓	# 2	# 2	# 3	# 4	0 4	0 5	1 7	2 9	↑	↑	↑	↑	↑	↑	↑	↑	↑	↑	↑	↑	↑
	Second	125	250						# 2	0 3	0 3	1 5	1 6	3 8	4 10	7 14													
	Third	125	375						0 2	0 3	1 4	2 6	3 8	6 10	8 13	13 19													
	Fourth	125	500						0 3	1 4	2 5	3 7	5 10	8 13	12 17	19 25													
	Fifth	125	625						1 3	2 4	3 6	5 8	7 11	11 15	17 20	25 29													
	Sixth	125	750						1 3	3 5	4 6	7 9	10 12	14 17	21 23	31 33													
	Seventh	125	875						2 3	4 5	6 7	9 10	13 14	18 19	25 26	37 38													
P	First	200	200	↓	*	↑	↓	# 2	# 2	# 3	# 4	0 4	0 5	1 7	2 9	↑	↑	↑	↑	↑	↑	↑	↑	↑	↑	↑	↑	↑	↑
	Second	200	400					# 2	0 3	0 3	1 5	1 6	3 8	4 10	7 14														
	Third	200	600					0 2	0 3	1 4	2 6	3 8	6 10	8 13	13 19														
	Fourth	200	800					0 3	1 4	2 5	3 7	5 10	8 13	12 17	19 25														
	Fifth	200	1000					1 3	2 4	3 6	5 8	7 11	11 15	17 20	25 29														
	Sixth	200	1200					1 3	3 5	4 6	7 9	10 12	14 17	21 23	31 33														
	Seventh	200	1400					2 3	4 5	6 7	9 10	13 14	18 19	25 26	37 38														
Q	First	315	315	*	↑	↓	# 2	# 2	# 3	# 4	0 4	0 5	1 7	2 9	↑	↑	↑	↑	↑	↑	↑	↑	↑	↑	↑	↑	↑	↑	↑
	Second	315	630				# 2	0 3	0 3	1 5	1 6	3 8	4 10	7 14															
	Third	315	945				0 2	0 3	1 4	2 6	3 8	6 10	8 13	13 19															
	Fourth	315	1260				0 3	1 4	2 5	3 7	5 10	8 13	12 17	19 25															
	Fifth	315	1575				1 3	2 4	3 6	5 8	7 11	11 15	17 20	25 29															
	Sixth	315	1890				1 3	3 5	4 6	7 9	10 12	14 17	21 23	31 33															
	Seventh	315	2205				2 3	4 5	6 7	9 10	13 14	18 19	25 26	37 38															
R	First	500	500	↑	↓	# 2	# 2	# 3	# 4	0 4	0 5	1 7	2 9	↑	↑	↑	↑	↑	↑	↑	↑	↑	↑	↑	↑	↑	↑	↑	↑
	Second	500	1000			# 2	0 3	0 3	1 5	1 6	3 8	4 10	7 14																
	Third	500	1500			0 2	0 3	1 4	2 6	3 8	6 10	8 13	13 19																
	Fourth	500	2000			0 3	1 4	2 5	3 7	5 10	8 13	12 17	19 25																
	Fifth	500	2500			1 3	2 4	3 6	5 8	7 11	11 15	17 20	25 29																
	Sixth	500	3000			1 3	3 5	4 6	7 9	10 12	14 17	21 23	31 33																
	Seventh	500	3500			2 3	4 5	6 7	9 10	13 14	18 19	25 26	37 38																

↓ = Use first sampling plan below arrow. If sample size equals or exceeds lot or batch size, do 100 percent inspection.

↑ = Use first sampling plan above arrow (refer to preceding page, when necessary).

* = Use corresponding single sampling plan (or alternatively, use double sampling plan below, where available).

\# = Acceptance not permitted at this sample size.

Ac = Acceptance number.

Re = Rejection number.

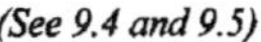

Table IV(B) Multiple Sampling Plans for Normal Inspection (Master Table)

MULTIPLE TIGHTENED PLANS

(See 9.4 and 9.5)

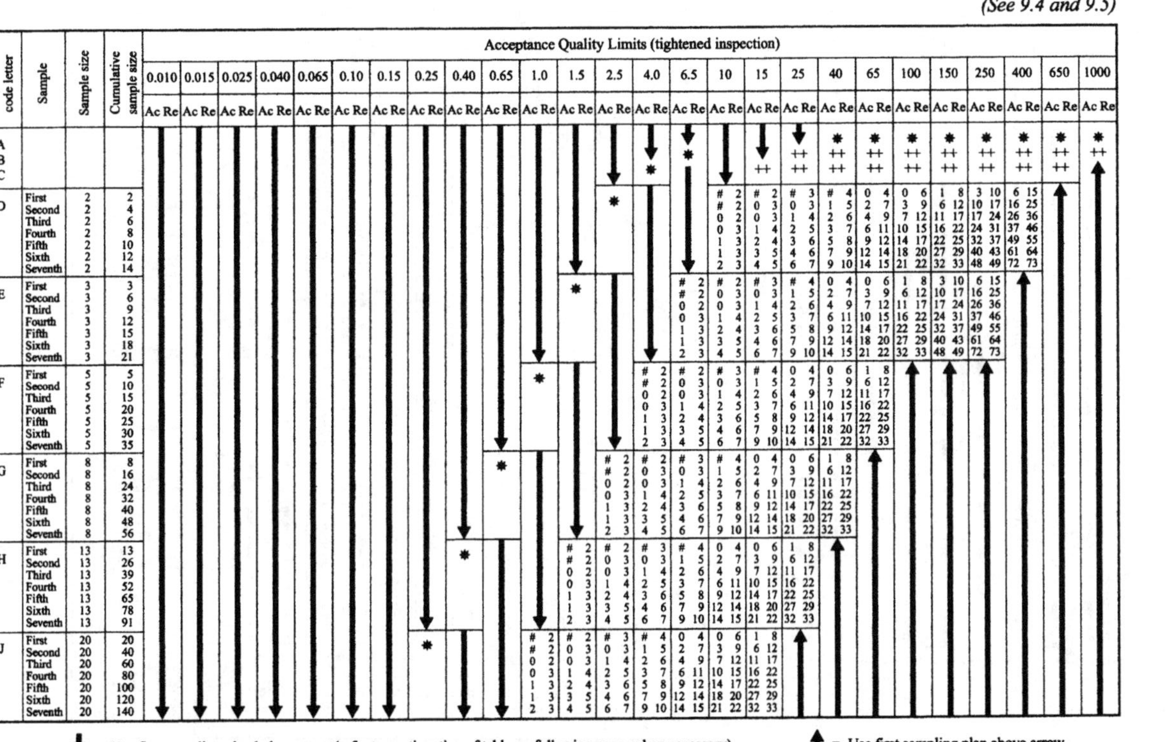

Sample size code letter	Sample	Sample size	Cumulative sample size	Acceptance Quality Limits (tightened inspection)																									
				0.010	0.015	0.025	0.040	0.065	0.10	0.15	0.25	0.40	0.65	1.0	1.5	2.5	4.0	6.5	10	15	25	40	65	100	150	250	400	650	1000
				Ac Re	Ac Re	Ac Re	Ac Re	Ac Re	Ac Re	Ac Re	Ac Re	Ac Re	Ac Re	Ac Re	Ac Re	Ac Re	Ac Re	Ac Re	Ac Re	Ac Re	Ac Re	Ac Re	Ac Re	Ac Re	Ac Re	Ac Re	Ac Re	Ac Re	Ac Re
A				↓	↓	↓	↓	↓	↓	↓	↓	↓	↓	↓	↓	↓	↓	↓	↓	↓	↓	*	*	*	*	*	*	*	*
B																		*			++	++	++	++	++	++	++	++	++
C																	*			++	++	++	++	++	++	++	++	++	↑
D	First	2	2	↓	↓	↓	↓	↓	↓	↓	↓	↓	↓	↓	↓	*	↓	↓	# 2	# 2	# 3	# 4	0 4	0 6	1 8	3 10	6 15	↑	↑
	Second	2	4																# 2	0 3	0 3	1 5	2 7	3 9	6 12	10 17	16 25		
	Third	2	6																0 2	0 3	1 4	2 6	4 9	7 12	11 17	17 24	26 36		
	Fourth	2	8																0 3	1 4	2 5	3 7	6 11	10 15	16 22	24 31	37 46		
	Fifth	2	10																1 3	2 4	3 6	5 8	9 12	14 17	22 25	32 37	49 55		
	Sixth	2	12																1 3	3 5	4 6	7 9	12 14	18 20	27 29	40 43	61 64		
	Seventh	2	14																2 3	4 5	6 7	9 10	14 15	21 22	32 33	48 49	72 73		
E	First	3	3	↓	↓	↓	↓	↓	↓	↓	↓	↓	↓	↓	*	↓	↓	# 2	# 2	# 3	# 4	0 4	0 6	1 8	3 10	6 15	↑	↑	↑
	Second	3	6															# 2	0 3	0 3	1 5	2 7	3 9	6 12	10 17	16 25			
	Third	3	9															0 2	0 3	1 4	2 6	4 9	7 12	11 17	17 24	26 36			
	Fourth	3	12															0 3	1 4	2 5	3 7	6 11	10 15	16 22	24 31	37 46			
	Fifth	3	15															1 3	2 4	3 6	5 8	9 12	14 17	22 25	32 37	49 55			
	Sixth	3	18															1 3	3 5	4 6	7 9	12 14	18 20	27 29	40 43	61 64			
	Seventh	3	21															2 3	4 5	6 7	9 10	14 15	21 22	32 33	48 49	72 73			
F	First	5	5	↓	↓	↓	↓	↓	↓	↓	↓	↓	↓	*	↓	↓	# 2	# 2	# 3	# 4	0 4	0 6	1 8	↑	↑	↑	↑	↑	↑
	Second	5	10														# 2	0 3	0 3	1 5	2 7	3 9	6 12						
	Third	5	15														0 2	0 3	1 4	2 6	4 9	7 12	11 17						
	Fourth	5	20														0 3	1 4	2 5	3 7	6 11	10 15	16 22						
	Fifth	5	25														1 3	2 4	3 6	5 8	9 12	14 17	22 25						
	Sixth	5	30														1 3	3 5	4 6	7 9	12 14	18 20	27 29						
	Seventh	5	35														2 3	4 5	6 7	9 10	14 15	21 22	32 33						
G	First	8	8	↓	↓	↓	↓	↓	↓	↓	↓	↓	*	↓	↓	# 2	# 2	# 3	# 4	0 4	0 6	1 8	↑	↑	↑	↑	↑	↑	↑
	Second	8	16													# 2	0 3	0 3	1 5	2 7	3 9	6 12							
	Third	8	24													0 2	0 3	1 4	2 6	4 9	7 12	11 17							
	Fourth	8	32													0 3	1 4	2 5	3 7	6 11	10 15	16 22							
	Fifth	8	40													1 3	2 4	3 6	5 8	9 12	14 17	22 25							
	Sixth	8	48													1 3	3 5	4 6	7 9	12 14	18 20	27 29							
	Seventh	8	56													2 3	4 5	6 7	9 10	14 15	21 22	32 33							
H	First	13	13	↓	↓	↓	↓	↓	↓	↓	↓	*	↓	↓	# 2	# 2	# 3	# 4	0 4	0 6	1 8	↑	↑	↑	↑	↑	↑	↑	↑
	Second	13	26												# 2	0 3	0 3	1 5	2 7	3 9	6 12								
	Third	13	39												0 2	0 3	1 4	2 6	4 9	7 12	11 17								
	Fourth	13	52												0 3	1 4	2 5	3 7	6 11	10 15	16 22								
	Fifth	13	65												1 3	2 4	3 6	5 8	9 12	14 17	22 25								
	Sixth	13	78												1 3	3 5	4 6	7 9	12 14	18 20	27 29								
	Seventh	13	91												2 3	4 5	6 7	9 10	14 15	21 22	32 33								
J	First	20	20	↓	↓	↓	↓	↓	↓	↓	*	↓	↓	# 2	# 2	# 3	# 4	0 4	0 6	1 8	↑	↑	↑	↑	↑	↑	↑	↑	↑
	Second	20	40											# 2	0 3	0 3	1 5	2 7	3 9	6 12									
	Third	20	60											0 2	0 3	1 4	2 6	4 9	7 12	11 17									
	Fourth	20	80											0 3	1 4	2 5	3 7	6 11	10 15	16 22									
	Fifth	20	100											1 3	2 4	3 6	5 8	9 12	14 17	22 25									
	Sixth	20	120											1 3	3 5	4 6	7 9	12 14	18 20	27 29									
	Seventh	20	140											2 3	4 5	6 7	9 10	14 15	21 22	32 33									

↓ = Use first sampling plan below arrow (refer to continuation of table on following page, when necessary). If sample size equals or exceeds lot or batch size, do 100 percent inspection.

* = Use corresponding single sampling plan (or alternatively, use double sampling plan below, where available).

↑ = Use first sampling plan above arrow.

= Acceptance not permitted at this sample size.

Ac = Acceptance number.

Re = Rejection number.

(See 9.4 and 9.5)

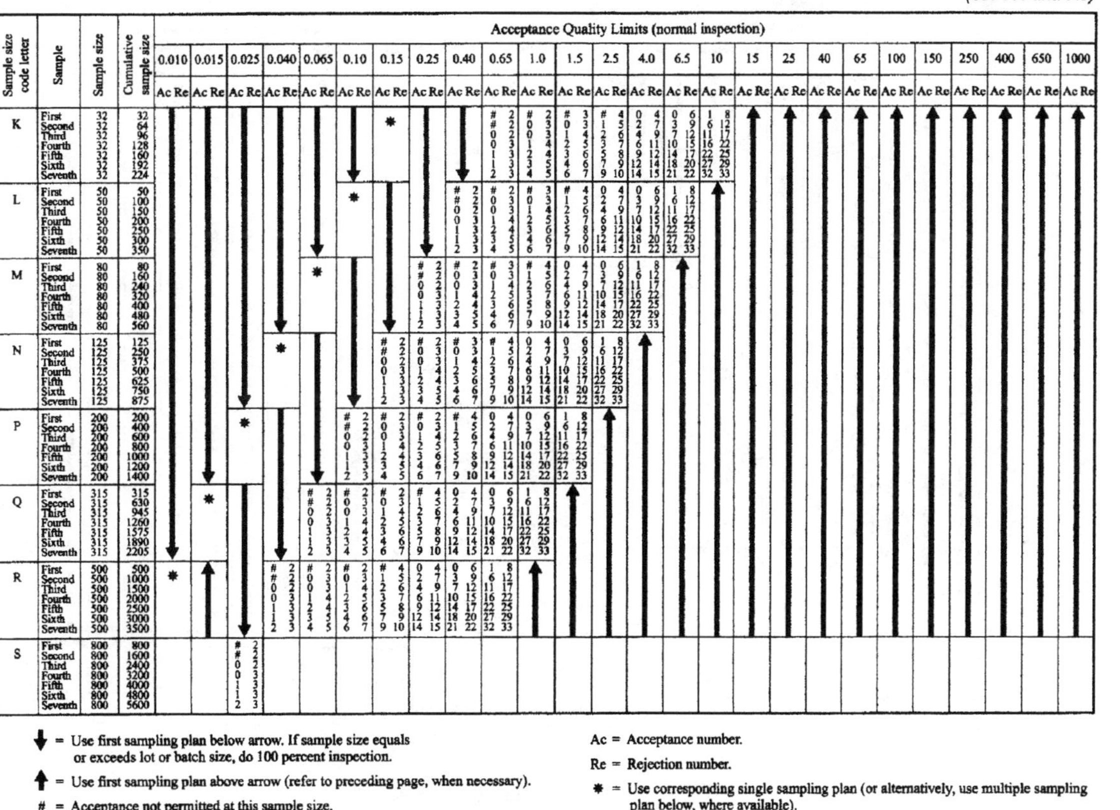

Sample size code letter	Sample	Sample size	Cumulative sample size	Acceptance Quality Limits (normal inspection)																									
				0.010	0.015	0.025	0.040	0.065	0.10	0.15	0.25	0.40	0.65	1.0	1.5	2.5	4.0	6.5	10	15	25	40	65	100	150	250	400	650	1000
				Ac Re	Ac Re	Ac Re	Ac Re	Ac Re	Ac Re	Ac Re	Ac Re	Ac Re	Ac Re	Ac Re	Ac Re	Ac Re	Ac Re	Ac Re	Ac Re	Ac Re	Ac Re	Ac Re	Ac Re	Ac Re	Ac Re	Ac Re	Ac Re	Ac Re	Ac Re
K	First	32	32	↓	↓	↓	↓	↓	↓	*	↓	↓	# 2	# 2	# 3	# 4	0 4	0 6	1 8	↑	↑	↑	↑	↑	↑	↑	↑	↑	↑
	Second	32	64										# 2	0 3	0 3	1 5	2 7	3 9	6 12										
	Third	32	96										0 2	0 3	1 4	2 6	4 9	7 12	11 17										
	Fourth	32	128										0 3	1 4	2 5	3 7	6 11	10 15	16 22										
	Fifth	32	160										1 3	2 4	3 6	5 8	9 12	14 17	22 25										
	Sixth	32	192										1 3	3 5	4 6	7 9	12 14	18 20	27 29										
	Seventh	32	224										2 3	4 5	6 7	9 10	14 15	21 22	32 33										
L	First	50	50	↓	↓	↓	↓	↓	*	↓	↓	# 2	# 2	# 3	# 4	0 4	0 6	1 8	↑	↑	↑	↑	↑	↑	↑	↑	↑	↑	↑
	Second	50	100									# 2	0 3	0 3	1 5	2 7	3 9	6 12											
	Third	50	150									0 2	0 3	1 4	2 6	4 9	7 12	11 17											
	Fourth	50	200									0 3	1 4	2 5	3 7	6 11	10 15	16 22											
	Fifth	50	250									1 3	2 4	3 6	5 8	9 12	14 17	22 25											
	Sixth	50	300									1 3	3 5	4 6	7 9	12 14	18 20	27 29											
	Seventh	50	350									2 3	4 5	6 7	9 10	14 15	21 22	32 33											
M	First	80	80	↓	↓	↓	↓	*	↓	↓	# 2	# 2	# 3	# 4	0 4	0 6	1 8	↑	↑	↑	↑	↑	↑	↑	↑	↑	↑	↑	↑
	Second	80	160								# 2	0 3	0 3	1 5	2 7	3 9	6 12												
	Third	80	240								0 2	0 3	1 4	2 6	4 9	7 12	11 17												
	Fourth	80	320								0 3	1 4	2 5	3 7	6 11	10 15	16 22												
	Fifth	80	400								1 3	2 4	3 6	5 8	9 12	14 17	22 25												
	Sixth	80	480								1 3	3 5	4 6	7 9	12 14	18 20	27 29												
	Seventh	80	560								2 3	4 5	6 7	9 10	14 15	21 22	32 33												
N	First	125	125	↓	↓	↓	*	↓	↓	# 2	# 2	# 3	# 4	0 4	0 6	1 8	↑	↑	↑	↑	↑	↑	↑	↑	↑	↑	↑	↑	↑
	Second	125	250							# 2	0 3	0 3	1 5	2 7	3 9	6 12													
	Third	125	375							0 2	0 3	1 4	2 6	4 9	7 12	11 17													
	Fourth	125	500							0 3	1 4	2 5	3 7	6 11	10 15	16 22													
	Fifth	125	625							1 3	2 4	3 6	5 8	9 12	14 17	22 25													
	Sixth	125	750							1 3	3 5	4 6	7 9	12 14	18 20	27 29													
	Seventh	125	875							2 3	4 5	6 7	9 10	14 15	21 22	32 33													
P	First	200	200	↓	↓	*	↓	↓	# 2	# 2	# 3	# 4	0 4	0 6	1 8	↑	↑	↑	↑	↑	↑	↑	↑	↑	↑	↑	↑	↑	↑
	Second	200	400						# 2	0 3	0 3	1 5	2 7	3 9	6 12														
	Third	200	600						0 2	0 3	1 4	2 6	4 9	7 12	11 17														
	Fourth	200	800						0 3	1 4	2 5	3 7	6 11	10 15	16 22														
	Fifth	200	1000						1 3	2 4	3 6	5 8	9 12	14 17	22 25														
	Sixth	200	1200						1 3	3 5	4 6	7 9	12 14	18 20	27 29														
	Seventh	200	1400						2 3	4 5	6 7	9 10	14 15	21 22	32 33														
Q	First	315	315	↓	*	↓	↓	# 2	# 2	# 3	# 4	0 4	0 6	1 8	↑	↑	↑	↑	↑	↑	↑	↑	↑	↑	↑	↑	↑	↑	↑
	Second	315	630					# 2	0 3	0 3	1 5	2 7	3 9	6 12															
	Third	315	945					0 2	0 3	1 4	2 6	4 9	7 12	11 17															
	Fourth	315	1260					0 3	1 4	2 5	3 7	6 11	10 15	16 22															
	Fifth	315	1575					1 3	2 4	3 6	5 8	9 12	14 17	22 25															
	Sixth	315	1890					1 3	3 5	4 6	7 9	12 14	18 20	27 29															
	Seventh	315	2205					2 3	4 5	6 7	9 10	14 15	21 22	32 33															
R	First	500	500	*	↑	↓	# 2	# 2	# 3	# 4	0 4	0 6	1 8	↑	↑	↑	↑	↑	↑	↑	↑	↑	↑	↑	↑	↑	↑	↑	↑
	Second	500	1000				# 2	0 3	0 3	1 5	2 7	3 9	6 12																
	Third	500	1500				0 2	0 3	1 4	2 6	4 9	7 12	11 17																
	Fourth	500	2000				0 3	1 4	2 5	3 7	6 11	10 15	16 22																
	Fifth	500	2500				1 3	2 4	3 6	5 8	9 12	14 17	22 25																
	Sixth	500	3000				1 3	3 5	4 6	7 9	12 14	18 20	27 29																
	Seventh	500	3500				2 3	4 5	6 7	9 10	14 15	21 22	32 33																
S	First	800	800			# 2																							
	Second	800	1600			# 2																							
	Third	800	2400			0 2																							
	Fourth	800	3200			0 3																							
	Fifth	800	4000			1 3																							
	Sixth	800	4800			1 3																							
	Seventh	800	5600			2 3																							

↓ = Use first sampling plan below arrow. If sample size equals or exceeds lot or batch size, do 100 percent inspection.

↑ = Use first sampling plan above arrow (refer to preceding page, when necessary).

= Acceptance not permitted at this sample size.

Ac = Acceptance number.

Re = Rejection number.

* = Use corresponding single sampling plan (or alternatively, use multiple sampling plan below, where available).

Table IV(C) Multiple Sampling Plans for Normal Inspection (Master Table)

MULTIPLE REDUCED PLANS

(See 9.4 and 9.5)

Sample size code letter	Sample	Sample size	Cumulative sample size	Acceptance Quality Limits (reduced inspection)†																									
				0.010	0.015	0.025	0.040	0.065	0.10	0.15	0.25	0.40	0.65	1.0	1.5	2.5	4.0	6.5	10	15	25	40	65	100	150	250	400	650	1000
				Ac Re	Ac Re	Ac Re	Ac Re	Ac Re	Ac Re	Ac Re	Ac Re	Ac Re	Ac Re	Ac Re	Ac Re	Ac Re	Ac Re	Ac Re	Ac Re	Ac Re	Ac Re	Ac Re	Ac Re	Ac Re	Ac Re	Ac Re	Ac Re	Ac Re	Ac Re
A				↓	↓	↓	↓	↓	↓	↓	↓	↓	↓	↓	↓	↓	↓	*	↓	↓	*	*	*	*	*	*	*	*	*
B				↓	↓	↓	↓	↓	↓	↓	↓	↓	↓	↓	↓	↓	*	↑	↓	*	*	*	*	*	*	*	*	*	*
C				↓	↓	↓	↓	↓	↓	↓	↓	↓	↓	↓	↓	*	↑	↓	*	*	*	*	*	*	*	*	*	*	↑
D				↓	↓	↓	↓	↓	↓	↓	↓	↓	↓	↓	*	↑	↓	++	++	++	++	++	++	++	++	++	++	↑	↑
E				↓	↓	↓	↓	↓	↓	↓	↓	↓	↓	*	↑	↓	++	++	++	++	++	++	++	++	++	++	↑	↑	↑
F	First			↓	↓	↓	↓	↓	↓	↓	↓	↓	*	↑	↓	# 2	# 2	# 3	# 3	# 4	# 4	0 5	0 6	↑	↑	↑	↑	↑	↑
	Second															# 2	# 3	# 3	0 4	0 5	1 6	1 7	3 9						
	Third															0 2	0 3	0 4	0 5	1 6	2 8	3 9	6 12						
	Fourth															0 3	0 4	0 5	1 6	2 7	3 10	5 12	8 15						
	Fifth															0 3	0 4	1 6	2 7	3 8	5 11	7 13	11 17						
	Sixth															0 3	1 5	1 6	3 7	4 9	7 12	10 15	14 20						
	Seventh															1 3	1 5	2 7	4 8	6 10	9 14	13 17	18 22						
G	First	3	3	↓	↓	↓	↓	↓	↓	↓	↓	*	↑	↓	# 2	# 2	# 3	# 3	# 4	# 4	0 5	0 6	↑	↑	↑	↑	↑	↑	↑
	Second	3	6												# 2	# 3	# 3	0 4	0 5	1 6	1 7	3 9							
	Third	3	9												0 2	0 3	0 4	0 5	1 6	2 8	3 9	6 12							
	Fourth	3	12												0 3	0 4	0 5	1 6	2 7	3 10	5 12	8 15							
	Fifth	3	15												0 3	0 4	1 6	2 7	3 8	5 11	7 13	11 17							
	Sixth	3	18												0 3	1 5	1 6	3 7	4 9	7 12	10 15	14 20							
	Seventh	3	21												1 3	1 5	2 7	4 8	6 10	9 14	13 17	18 22							
H	First	5	5	↓	↓	↓	↓	↓	↓	↓	*	↑	↓	# 2	# 2	# 3	# 3	# 4	# 4	0 5	0 6	↑	↑	↑	↑	↑	↑	↑	↑
	Second	5	10											# 2	# 3	# 3	0 4	0 5	1 6	1 7	3 9								
	Third	5	15											0 2	0 3	0 4	0 5	1 6	2 8	3 9	6 12								
	Fourth	5	20											0 3	0 4	0 5	1 6	2 7	3 10	5 12	8 15								
	Fifth	5	25											0 3	0 4	1 6	2 7	3 8	5 11	7 13	11 17								
	Sixth	5	30											0 3	1 5	1 6	3 7	4 9	7 12	10 15	14 20								
	Seventh	5	35											1 3	1 5	2 7	4 8	6 10	9 14	13 17	18 22								
J	First	8	8	↓	↓	↓	↓	↓	↓	*	↑	↓	# 2	# 2	# 3	# 3	# 4	# 4	0 5	0 6	↑	↑	↑	↑	↑	↑	↑	↑	↑
	Second	8	16										# 2	# 3	# 3	0 4	0 5	1 6	1 7	3 9									
	Third	8	24										0 2	0 3	0 4	0 5	1 6	2 8	3 9	6 12									
	Fourth	8	32										0 3	0 4	0 5	1 6	2 7	3 10	5 12	8 15									
	Fifth	8	40										0 3	0 4	1 6	2 7	3 8	5 11	7 13	11 17									
	Sixth	8	48										0 3	1 5	1 6	3 7	4 9	7 12	10 15	14 20									
	Seventh	8	56										1 3	1 5	2 7	4 8	6 10	9 14	13 17	18 22									
K	First	13	13	↓	↓	↓	↓	↓	*	↑	↓	# 2	# 2	# 3	# 3	# 4	# 4	0 5	0 6	↑	↑	↑	↑	↑	↑	↑	↑	↑	↑
	Second	13	26									# 2	# 3	# 3	0 4	0 5	1 6	1 7	3 9										
	Third	13	39									0 2	0 3	0 4	0 5	1 6	2 8	3 9	6 12										
	Fourth	13	52									0 3	0 4	0 5	1 6	2 7	3 10	5 12	8 15										
	Fifth	13	65									0 3	0 4	1 6	2 7	3 8	5 11	7 13	11 17										
	Sixth	13	78									0 3	1 5	1 6	3 7	4 9	7 12	10 15	14 20										
	Seventh	13	91									1 3	1 5	2 7	4 8	6 10	9 14	13 17	18 22										

↓ = Use first sampling plan below arrow (refer to continuation of table on following page, when necessary). If sample size equals or exceeds lot or batch size, do 100 percent inspection.

↑ = Use first sampling plan above arrow.

* = Use corresponding single sampling plan (or alternatively, use multiple sampling plan below, where available).

++ = Use corresponding double sampling plan (or alternatively, use multiple sampling plan below, where available).

Ac = Acceptance number.

Re = Rejection number.

\# = Acceptance not permitted at this sample size.

† = If, after the final sample, the acceptance number has been exceeded, but the rejection number has not been reached, accept the lot but reinstate normal inspection (see 10.1.4).

(See 9.4 and 9.5)

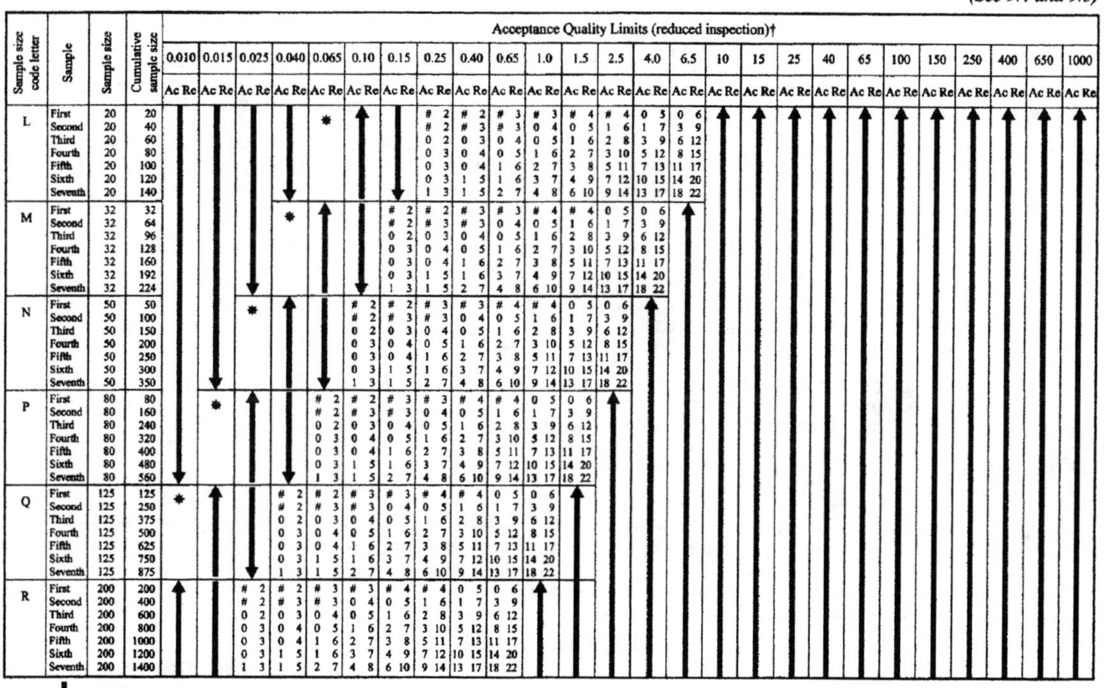

Sample size code letter	Sample	Sample size	Cumulative sample size	Acceptance Quality Limits (reduced inspection)†																									
				0.010	0.015	0.025	0.040	0.065	0.10	0.15	0.25	0.40	0.65	1.0	1.5	2.5	4.0	6.5	10	15	25	40	65	100	150	250	400	650	1000
				Ac Re	Ac Re	Ac Re	Ac Re	Ac Re	Ac Re	Ac Re	Ac Re	Ac Re	Ac Re	Ac Re	Ac Re	Ac Re	Ac Re	Ac Re	Ac Re	Ac Re	Ac Re	Ac Re	Ac Re	Ac Re	Ac Re	Ac Re	Ac Re	Ac Re	Ac Re
L	First	20	20				↓	*	↑	↓	# 2	# 2	# 3	# 3	# 4	# 4	0 5	0 6	↑	↑	↑	↑	↑	↑	↑	↑	↑	↑	↑
	Second	20	40								# 2	# 3	# 3	0 4	0 5	1 6	1 7	3 9											
	Third	20	60								0 2	0 3	0 4	0 5	1 6	2 8	3 9	6 12											
	Fourth	20	80								0 3	0 4	0 5	1 6	2 7	3 10	5 12	8 15											
	Fifth	20	100								0 3	0 4	1 6	2 7	3 8	5 11	7 13	11 17											
	Sixth	20	120								0 3	1 5	1 6	3 7	4 9	7 12	10 15	14 20											
	Seventh	20	140								1 3	1 5	2 7	4 8	6 10	9 14	13 17	18 22											
M	First	32	32			↓	*	↑	↓	# 2	# 2	# 3	# 3	# 4	# 4	0 5	0 6	↑											
	Second	32	64							# 2	# 3	# 3	0 4	0 5	1 6	1 7	3 9												
	Third	32	96							0 2	0 3	0 4	0 5	1 6	2 8	3 9	6 12												
	Fourth	32	128							0 3	0 4	0 5	1 6	2 7	3 10	5 12	8 15												
	Fifth	32	160							0 3	0 4	1 6	2 7	3 8	5 11	7 13	11 17												
	Sixth	32	192							0 3	1 5	1 6	3 7	4 9	7 12	10 15	14 20												
	Seventh	32	224							1 3	1 5	2 7	4 8	6 10	9 14	13 17	18 22												
N	First	50	50		↓	*	↑	↓	# 2	# 2	# 3	# 3	# 4	# 4	0 5	0 6	↑												
	Second	50	100						# 2	# 3	# 3	0 4	0 5	1 6	1 7	3 9													
	Third	50	150						0 2	0 3	0 4	0 5	1 6	2 8	3 9	6 12													
	Fourth	50	200						0 3	0 4	0 5	1 6	2 7	3 10	5 12	8 15													
	Fifth	50	250						0 3	0 4	1 6	2 7	3 8	5 11	7 13	11 17													
	Sixth	50	300						0 3	1 5	1 6	3 7	4 9	7 12	10 15	14 20													
	Seventh	50	350						1 3	1 5	2 7	4 8	6 10	9 14	13 17	18 22													
P	First	80	80	↓	*	↑	↓	# 2	# 2	# 3	# 3	# 4	# 4	0 5	0 6	↑													
	Second	80	160					# 2	# 3	# 3	0 4	0 5	1 6	1 7	3 9														
	Third	80	240					0 2	0 3	0 4	0 5	1 6	2 8	3 9	6 12														
	Fourth	80	320					0 3	0 4	0 5	1 6	2 7	3 10	5 12	8 15														
	Fifth	80	400					0 3	0 4	1 6	2 7	3 8	5 11	7 13	11 17														
	Sixth	80	480					0 3	1 5	1 6	3 7	4 9	7 12	10 15	14 20														
	Seventh	80	560					1 3	1 5	2 7	4 8	6 10	9 14	13 17	18 22														
Q	First	125	125	*	↑	↓	# 2	# 2	# 3	# 3	# 4	# 4	0 5	0 6	↑														
	Second	125	250				# 2	# 3	# 3	0 4	0 5	1 6	1 7	3 9															
	Third	125	375				0 2	0 3	0 4	0 5	1 6	2 8	3 9	6 12															
	Fourth	125	500				0 3	0 4	0 5	1 6	2 7	3 10	5 12	8 15															
	Fifth	125	625				0 3	0 4	1 6	2 7	3 8	5 11	7 13	11 17															
	Sixth	125	750				0 3	1 5	1 6	3 7	4 9	7 12	10 15	14 20															
	Seventh	125	875				1 3	1 5	2 7	4 8	6 10	9 14	13 17	18 22															
R	First	200	200	↑		# 2	# 2	# 3	# 3	# 4	# 4	0 5	0 6	↑															
	Second	200	400			# 2	# 3	# 3	0 4	0 5	1 6	1 7	3 9																
	Third	200	600			0 2	0 3	0 4	0 5	1 6	2 8	3 9	6 12																
	Fourth	200	800			0 3	0 4	0 5	1 6	2 7	3 10	5 12	8 15																
	Fifth	200	1000			0 3	0 4	1 6	2 7	3 8	5 11	7 13	11 17																
	Sixth	200	1200			0 3	1 5	1 6	3 7	4 9	7 12	10 15	14 20																
	Seventh	200	1400			1 3	1 5	2 7	4 8	6 10	9 14	13 17	18 22																

↓ = Use first sampling plan below arrow. If sample size equals or exceeds lot or batch size, do 100 percent inspection.

↑ = Use first sampling plan above arrow.

Ac = Acceptance number.

Re = Rejection number.

\# = Acceptance not permitted at this sample size.

† = If, after the final sample, the acceptance number has been exceeded, but the rejection number has not been reached, accept the lot, but reinstate normal inspection (see 10.1.4).

Table V(A) Factors for Determining Approximate Values for Average Outgoing Quality Limits for Normal Inspection (Single Sampling)

AOQL NORMAL PLANS

(see 11.4)

Code letter	Sample size	Acceptance quality limits																									
		0.010	0.015	0.025	0.040	0.065	0.10	0.15	0.25	0.40	0.65	1.0	1.5	2.5	4.0	6.5	10	15	25	40	65	100	150	250	400	650	1000
A	2															18			42	69	97	160	220	330	470	730	1100
B	3														12			28	46	63	110	150	220	310	490	720	1100
C	5													7.4			17	27	39	63	90	130	190	290	430	660	
D	8												4.6			11	17	24	40	56	82	120	180	270	410		
E	13											2.8			6.5	11	15	24	34	50	72	110	170	250			
F	20										1.8			4.2	6.9	9.7	16	22	33	47	73						
G	32									1.2			2.6	4.3	6.1	9.9	14	21	29	46							
H	50								0.74			1.7	2.7	3.9	6.3	9.0	13	19	29								
J	80							0.46			1.1	1.7	2.4	4.0	5.6	8.2	12	18									
K	125						0.29			0.67	1.1	1.6	2.5	3.6	5.2	7.5	12										
L	200					0.18			0.42	0.69	0.97	1.6	2.2	3.3	4.7	7.3											
M	315				0.12			0.27	0.44	0.62	1.00	1.4	2.1	3.0	4.7												
N	500			0.074			0.17	0.27	0.39	0.63	0.90	1.3	1.9	2.9													
P	800		0.046			0.11	0.17	0.24	0.40	0.56	0.82	1.2	1.8														
Q	1250	0.029			0.067	0.11	0.16	0.25	0.36	0.52	0.75	1.2															
R	2000			0.042	0.069	0.097	0.16	0.22	0.33	0.47	0.73																

Note: For a more accurate AOQL, the above values must be multiplied by $\left(1 - \frac{\text{sample size}}{\text{lot of batch size}}\right)$ (see 11.4)

Table V(B) Factors for Determining Approximate Values for Average Outgoing Quality Limits for Tightened Inspection (Single Sampling)

AOQL TIGHTENED PLANS

Code letter	Sample size	Acceptance quality limits																									
		0.010	0.015	0.025	0.040	0.065	0.10	0.15	0.25	0.40	0.65	1.0	1.5	2.5	4.0	6.5	10	15	25	40	65	100	150	250	400	650	1000
A	2																			42	69	97	160	260	400	620	970
B	3															12			28	46	65	110	170	270	410	650	1100
C	5														7.4			17	27	39	63	100	160	250	390	610	
D	8													4.6			11	17	24	40	64	99	160	240	380		
E	13												2.8			6.5	11	15	24	40	61	95	150	240			
F	20											1.8			4.2	6.9	9.7	16	26	40	62						
G	32										1.2			2.6	4.3	6.1	9.9	16	25	39							
H	50									0.74			1.7	2.7	3.9	6.3	10	16	25								
J	80								0.46			1.1	1.7	2.4	4.0	6.4	9.9	16									
K	125							0.29			0.67	1.1	1.6	2.5	4.1	6.4	9.9										
L	200						0.18			0.42	0.69	0.97	1.6	2.6	4.0	6.2											
M	315					0.12			0.27	0.44	0.62	1.0	1.6	2.5	3.9												
N	500				0.074			0.17	0.27	0.39	0.63	1.0	1.6	2.5													
P	800			0.046			0.11	0.17	0.24	0.40	0.64	0.99	1.6														
Q	1250		0.029			0.067	0.11	0.16	0.25	0.41	0.64	0.99															
R	2000	0.018			0.042	0.069	0.097	0.16	0.26	0.40	0.62																
S	3150			0.027																							

Note: For a more accurate AOQL, the above values must be multiplied by $\left(1 - \frac{\text{sample size}}{\text{lot of batch size}}\right)$ (see 11.4)

Table VI(A) Limiting Quality (in Percent Nonconforming) for Which $P_a = 10\%$ (for Normal Inspection, single Sampling)

LQ(Nonconforming Units) 10% PLANS

Code letter	Sample size	Acceptance quality limits															
		0.010	0.015	0.025	0.040	0.065	0.10	0.15	0.25	0.40	0.65	1.0	1.5	2.5	4.0	6.5	10
A	2															68	
B	3														54		
C	5													37			58
D	8												25			41	54
E	13											16			27	36	44
F	20										11			18	25	30	42
G	32									6.9			12	16	20	27	34
H	50								4.5			7.6	10	13	18	22	29
J	80							2.8			4.8	6.5	8.2	11	14	19	24
K	125						1.8			3.1	4.3	5.4	7.4	9.4	12	16	23
L	200					1.2			2.0	2.7	3.3	4.6	5.9	7.7	10	14	
M	315				0.73			1.2	1.7	2.1	2.9	3.7	4.9	6.4	9.0		
N	500			0.46			0.78	1.1	1.3	1.9	2.4	3.1	4.0	5.6			
P	800		0.29			0.49	0.67	0.84	1.2	1.5	1.9	2.5	3.5				
Q	1250	0.18			0.31	0.43	0.53	0.74	0.94	1.2	1.6	2.3					
R	2000			0.20	0.27	0.33	0.46	0.59	0.77	1.0	1.4						

Table VI(B) Limiting Quality (in Nonconformities per Hundred Units) for Which $P_a = 10\%$ (for Normal Inspection, Single Sampling)

LQ (Nonconformities) 10% PLANS

Code letter	Sample size	Acceptance quality limits																									
		0.010	0.015	0.025	0.040	0.065	0.10	0.15	0.25	0.40	0.65	1.0	1.5	2.5	4.0	6.5	10	15	25	40	65	100	150	250	400	650	1000
A	2															120			200	270	330	460	590	770	1000	1400	1900
B	3														77			130	180	220	310	390	510	670	940	1300	1800
C	5													46			78	110	130	190	240	310	400	560	770	1100	
D	8												29			49	67	84	120	150	190	250	350	480	670		
E	13											18			30	41	51	71	91	120	160	220	300	410			
F	20										12			20	27	33	46	59	77	100	140						
G	32									7.2			12	17	21	29	37	48	63	88							
H	50								4.6			7.8	11	13	19	24	31	40	56								
J	80							2.9			4.9	6.7	8.4	12	15	19	25	35									
K	125						1.8			3.1	4.3	5.4	7.4	9.4	12	16	23										
L	200					1.2			2.0	2.7	3.3	4.6	5.9	7.7	10	14											
M	315				0.73			1.2	1.7	2.1	2.9	3.7	4.9	6.4	9.0												
N	500			0.46			0.78	1.1	1.3	1.9	2.4	3.1	4.0	5.6													
P	800		0.29			0.49	0.67	0.84	1.2	1.5	1.9	2.5	3.5														
Q	1250	0.18			0.31	0.43	0.53	0.74	0.94	1.2	1.6	2.3															
R	2000			0.20	0.27	0.33	0.46	0.59	0.77	1.0	1.4																

Table VIII Limit Numbers for Reduced Inspection

LIMIT NUMBERS

(see (4.7.3))

Number of sample units from last 10 lots or batches	Acceptance quality limits																									
	0.010	0.015	0.025	0.040	0.065	0.10	0.15	0.25	0.40	0.65	1.0	1.5	2.5	4.0	6.5	10	15	25	40	65	100	150	250	400	650	1000
20–29	*	*	*	*	*	*	*	*	*	*	*	*	*	*	*	0	0	2	4	8	14	22	40	68	115	181
30–49	*	*	*	*	*	*	*	*	*	*	*	*	*	*	0	0	1	3	7	13	22	36	63	105	178	277
50–79	*	*	*	*	*	*	*	*	*	*	*	*	*	0	0	2	3	7	14	25	40	63	110	181	301	
80–129	*	*	*	*	*	*	*	*	*	*	*	*	0	0	2	4	7	14	24	42	68	105	181	297		
130–199	*	*	*	*	*	*	*	*	*	*	*	0	0	2	4	7	13	25	42	72	115	177	301	490		
200–319	*	*	*	*	*	*	*	*	*	*	0	0	2	4	8	14	22	40	68	115	181	277	471			
320–499	*	*	*	*	*	*	*	*	*	0	0	1	4	8	14	24	39	68	113	189						
500–799	*	*	*	*	*	*	*	*	0	0	2	3	7	14	25	40	63	110	181							
800–1249	*	*	*	*	*	*	*	0	0	2	4	7	14	24	42	68	105	181								
1250–1999	*	*	*	*	*	*	0	0	2	4	7	13	24	49	69	110	169									
2000–3149	*	*	*	*	*	0	0	2	4	8	14	22	40	68	115	181										
3150–4999	*	*	*	*	0	0	1	4	8	14	24	38	67	111	186											
5000–7999	*	*	*	0	0	2	3	7	14	25	40	63	110	181												
8000–12499	*	*	0	0	2	4	7	14	24	42	68	105	181													
12 500–19999	*	0	0	2	4	7	13	24	40	69	110	169														
20 000–31499	0	0	2	4	8	14	22	40	68	115	181															
31 500 & Over	0	1	4	8	14	24	38	67	111	186																

* = Denotes that the number of sample units from the last 10 lots or batches is not sufficient for reduced inspection for this AQL. In this instance, more than 10 lots or batches may be used for the calculation, provided that the lots or batches used are the most recent ones in sequence, that they have all been on normal inspection, and that none has been rejected while on original inspection.

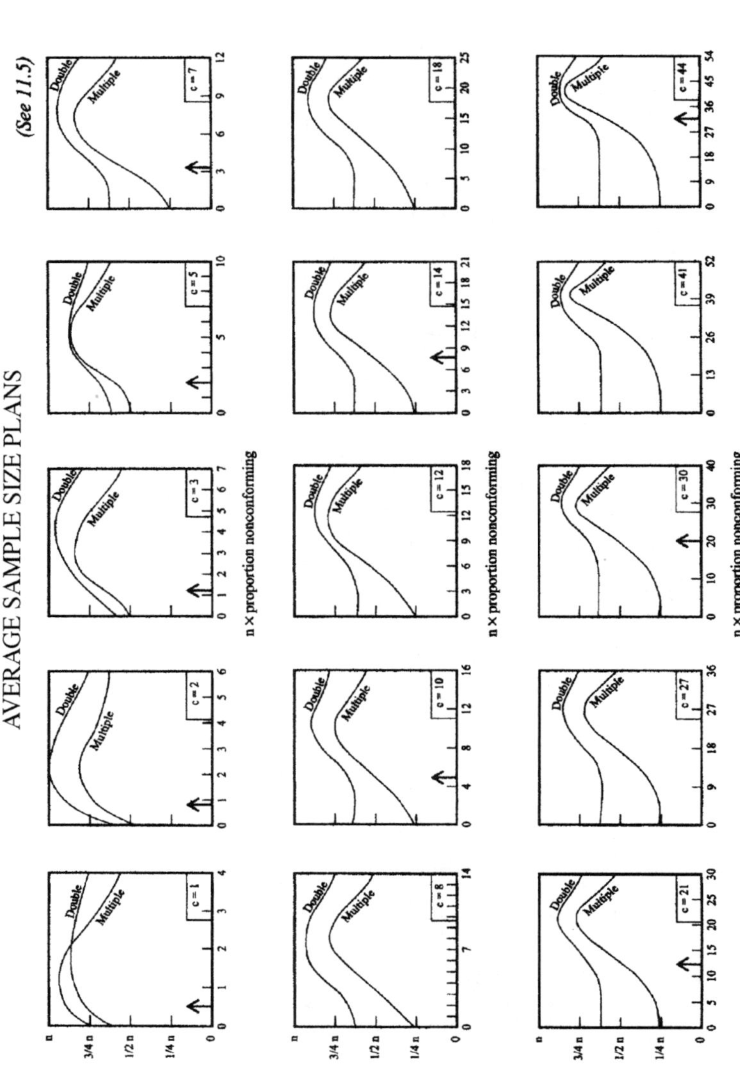

n = Equivalent single sample size

c = Single sample acceptance number

↑ = Reference point, shows performance at AQL for normal inspection

Figure IX Average sample size curves for double and multiple sampling plans (normal and tightened inspection).

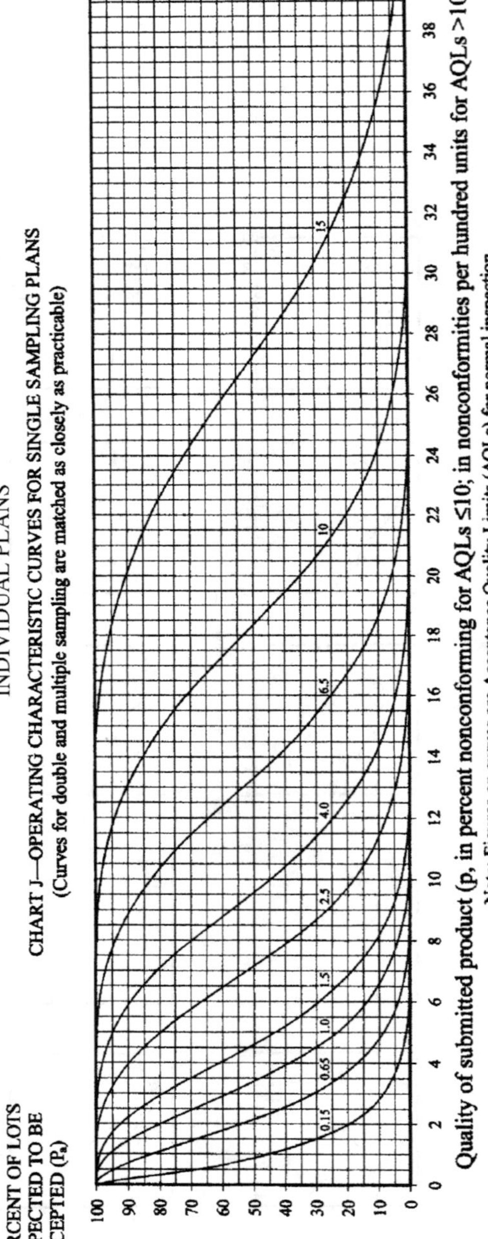
INDIVIDUAL PLANS
CHART J—OPERATING CHARACTERISTIC CURVES FOR SINGLE SAMPLING PLANS
(Curves for double and multiple sampling are matched as closely as practicable)
PERCENT OF LOTS EXPECTED TO BE ACCEPTED (P_a)
100 90 80 70 60 50 40 30 20 10 0
0.15 0.65 1.0 1.5 2.5 4.0 6.5 10 15
0 2 4 6 8 10 12 14 16 18 20 22 24 26 28 30 32 34 36 38 40
Quality of submitted product (p, in percent nonconforming for AQLs ≤10; in nonconformities per hundred units for AQLs >10)
Note: Figures on curves are Acceptance Quality Limits (AQLs) for normal inspection.

Figure X-J-a Tables for sample size code letter: J

Table X-J-b Tabulated Values for Operating Characteristic Curves for Single Sampling Plans

P_a	Acceptance Quality Limits (normal inspection)																					
	0.15	0.65	1.0	1.5	2.5	4.0	X	6.5	X	10	0.15	0.65	1.0	1.5	2.5	4.0	X	6.5	X	10	X	15
	p (in percent nonconforming)										p (in nonconformities per hundred units)											
99.0	0.0126	0.187	0.550	1.04	2.28	3.73	4.51	6.17	7.93	9.76	0.0126	0.186	0.545	1.03	2.23	3.63	4.38	5.96	7.62	9.35	12.9	15.7
95.0	0.0641	0.446	1.03	1.73	3.32	5.07	6.00	7.91	9.89	11.9	0.064	0.444	1.02	1.71	3.27	4.98	5.87	7.71	9.61	11.6	15.6	18.6
90.0	0.132	0.667	1.39	2.20	3.99	5.91	6.90	8.95	11.0	13.2	0.132	0.665	1.38	2.18	3.94	5.82	6.79	8.78	10.8	12.9	17.1	20.3
75.0	0.359	1.201	2.16	3.18	5.30	7.50	8.61	10.9	13.2	15.5	0.360	1.20	2.16	3.17	5.27	7.45	8.55	10.8	13.0	15.3	19.9	23.4
50.0	0.863	2.09	3.33	4.57	7.06	9.55	10.8	13.3	15.8	18.3	0.866	2.10	3.34	4.59	7.09	9.59	10.8	13.3	15.8	18.3	23.3	27.1
25.0	1.72	3.33	4.84	6.30	9.14	11.9	13.3	16.0	18.6	21.3	1.73	3.37	4.90	6.39	9.28	12.1	13.5	16.3	19.0	21.7	27.2	31.2
10.0	2.84	4.78	6.52	8.16	11.3	14.3	15.7	18.6	21.4	24.2	2.88	4.86	6.65	8.35	11.6	14.7	16.2	19.3	22.2	25.2	30.9	35.2
5.0	3.68	5.79	7.66	9.41	12.7	15.8	17.3	20.3	23.2	26.0	3.74	5.93	7.87	9.69	13.1	16.4	18.0	21.2	24.3	27.4	33.4	37.8
1.0	5.59	8.01	10.1	12.0	15.6	18.9	20.5	23.6	26.6	29.5	5.76	8.30	10.5	12.6	16.4	20.0	21.8	25.2	28.5	31.8	38.2	42.9
	0.25	1.0	1.5	2.5	4.0	X	6.5	X	10	X	0.25	1.0	1.5	2.5	4.0	X	6.5	X	10	X	15	X
	Acceptance Quality Limits (tightened inspection)																					

Note: Binomial distribution used for percent nonconforming computations; Poisson for nonconformities per hundred units.

Table X-J-2 Sampling Plans for Sample Size Code Letter: J

Type of sampling plan	Cumulative sample size	Acceptance quality limits (normal inspection)																	Cumulative sample size
		Less than 0.15	0.15	0.25	X	0.40	0.65	1.0	1.5	2.5	4.0	X	6.5	X	10	X	15	Higher than 15	
		Ac Re	Ac Re	Ac Re	Ac Re	Ac Re	Ac Re	Ac Re	Ac Re	Ac Re	Ac Re	Ac Re	Ac Re	Ac Re	Ac Re	Ac Re	Ac Re	Ac Re	
Single	80	∇	0 1	Use Code Letter H	Use Code Letter L	Use Code Letter K	1 2	2 3	3 4	5 6	7 8	8 9	10 11	12 13	14 15	18 19	21 22	Δ	80
Double	50	∇	*				0 2	0 3	1 4	2 5	3 7	3 7	5 9	6 10	7 11	9 14	11 16	Δ	50
	100						1 2	3 4	4 5	6 7	8 9	11 12	12 13	15 16	18 19	23 24	26 27		100
Multiple	20	∇	*				# 2	# 2	# 3	# 4	0 4	0 4	0 5	0 6	1 7	1 8	2 9	Δ	20
	40						# 2	0 3	0 3	1 5	1 6	2 7	3 8	3 9	4 10	6 12	7 14		40
	60						0 2	0 3	1 4	2 6	3 8	4 9	6 10	7 12	8 13	11 17	13 19		60
	80						0 3	1 4	2 5	3 7	5 10	6 11	8 13	10 15	12 17	16 22	19 25		80
	100						1 3	2 4	3 6	5 8	7 11	9 12	11 15	14 17	17 20	22 25	25 29		100
	120						1 3	3 5	4 6	7 9	10 12	12 14	14 17	18 20	21 23	27 29	31 33		120
	140						2 3	4 5	6 7	9 10	13 14	14 15	18 19	21 22	25 26	32 33	37 38		140
		Less than 0.25	0.25	X	0.40	0.65	1.0	1.5	2.5	4.0	X	6.5	X	10	X	15	X	Higher than 15	
		Acceptance quality limits (tightened inspection)																	

Δ = Use next preceding sample size code letter for which acceptance and rejection numbers are available.

∇ = Use next subsequent sample size code letter for which acceptance and rejection numbers are available.

Ac = Acceptance number.

Re = Rejection number.

* = Use single sampling plan above (or alternatively use code letter M).

= Acceptance not permitted at this sample size.

NOTE: "End of the Partial reproduction of ANSI/ ASQ Z1.4-2003."

When there is reason for avoiding more than a limiting percentage of nonconforming units (or nonconformities) in a lot or batch, Tables VI and VII may be useful for fixing minimum sample sizes to be associated with the AQL and inspection level specified for the inspection of a series of lots or batches. For example, if an LQ of 5% is desired for individual lots with an associated P_a of 10% or less, then if an AQL of 1.5% is designated for inspection of a series of lots or batches. Table VI indicates that the minimum sample size must be that given by code letter M.

Where there is interest in a limiting *process level*. Tables XII and XIII, which give LQ values and ANSI Z1.4 scheme performance may be used in a similar way to fix minimum sample sizes.

In the case of an isolated lot, it is preferable for the customer to adapt a sampling plan with a small consumer's risk. The ideal method of calculating the sample size and risk is by use of the hypergeometric probability function. ASQC Q3-1988 contains sampling plans that have been calculated on this basis and therefore provide a more accurate set of tables for these situations.

Let us explain this approach further. Suppose that we decide that we want an AOQL of 1.0% as protection. For any given lot size, there is a large number of single and double sampling plans, each one of which has an AOQL of 1.0% approximately. Which of these should we take? We first must decide whether to use single or double sampling. Let us say it is to be double sampling. Now then if the average fraction defective has been running at p-bar = 0.005 = 0.5%, we must sort through all double sampling plans with AOQLs of 1% to find which one has the minimum ATI when the incoming quality level is p-bar = 0.005 and for a given lot size N. Using this plan on product at p-bar, we will (1) have AOQL protection at 1.0% and (2) obtain this protection through minimum ATI as long as the incoming quality is at p-bar. But if p-bar changes, then probably some other plan will provide the specified AOQL protection and minimum ATI. A change in lot size may also require a different plan to minimize the ATI while supplying the required protection.

This aim is useful in many cases. But note that it is quite distinct from the ABC Standard.

10.2 THE DODGE–ROMIG SAMPLING TABLES

The earliest program of sampling plans was developed in the Bell System in the 1920s and was published by Dodge and Romig. A revised edition was published in 1959 (Dodge and Romig, 1959). These plans have by no means outlived their usefulness.

10.2.1. Aim

The general aim of the Dodge–Romig Tables is, first, to provide the consumer with desired protection, either a specified AOQL or a specified LTPD, lot tolerance percent defective, with 10% risk, that is, a specified p_o. Then secondly, the specified consumer protection is provided on a minimum overall inspection cost, that is, on a minimum average total inspection (ATI). This minimum ATI is attainable as a function of the incoming fraction nonconforming, p.

10.2.2. Description

There are four different sets of tables, being the combinations of single (S) vs. double (D) and lot tolerance percent defective (L) vs. average outgoing quality limit (A). Thus, the four sets are SL, DL, SA, and DA. The available LTPDs are (in percents) 0.5, 1.0, 2.0, 3.0, 4.0, 5.0, 7.0 and 10.0. The AOQLs available are (in percents) 0.1, 0.25, 0.50, 0.75, 1.0, 2.0, 2.5, 3.0, 4.0, 5.0, 7.0, 10.0. Therefore a DA-2 table means double sampling plans with an AOQL = 2.0%. In each such table, there are rows of classes of lot size up to 100,000, some 20 classes. Then there are six columns of classes of p-bar in each table.

To enter the Dodge–Romig tables to find a plan, one uses the following:

1. Decide on single or double sampling.
2. Decide on LTPD or AOQL type of protection.
3. Specify the amount of the type of protection chosen in 2.
4. The information in 1, 2, 3 locates the table to use.
5. In this table, locate the row class containing the given lot size.
6. Find the column class of p-bars which contains our observed p-bar. If p-bar is unknown, use the largest class of p-bars.
7. From 5 and 6, we find a cell in the table which lists for single sampling: the sample size, n, and acceptance number called c. Or for double sampling, samples n_1 and n_2 (which differ) and the acceptance number c_1 for the first sample and acceptance number c_2 for the total nonconforming in the two samples.
8. Also in each cell is given the amount of the opposite kind of consumer protection. Thus in the AOQL tables the cells list the LTPD, called p_t, for the plan.
9. The plans are listed for inspection for nonconforming units (defectives). If, however, inspection is for nonconformities (defects), use average nonconformities per 100 units, (instead of p-bar) in percent. Then we count nonconformities in samples, rather than nonconforming units.

10.3 OTHER SAMPLING INSPECTION PLANS

There were a number of good sampling plans as forerunners of the ABC Standard. Many of their best features were carried into the ABC Standard (see Dodge, 1969, 1970). We mention again the collection of sampling plans indexed on p_{95} and p_{10}, whose usefulness lies in isolated lots or infrequently submitted lots (see Ref. 4 in Chapter 9).

Another plan which has its place in some applications was published by H. C. Hamaker and associates in the Phillips Review (Einhoven, Netherlands) in 1949 and 1950

Table 10.1 Combinations of *f* and *i* for Given AOQL Protection for Continuous Sampling Plan CSP-1. Values of i to Qualify are in Body of Table

Sample 1 in	f	AOQL (%) 0.05	0.10	0.20	0.30	0.50	1.00	2.00	3.00	5.00
5	0.200	1430	720	360	240	142	72	35	24	14
8	0.125	1950	960	480	320	193	96	48	32	19
10	0.100	2210	1100	550	370	220	110	55	31	21
15	0.0667	2700	1320	660	440	270	132	67	44	26
20	0.050	3080	1500	750	510	310	150	75	50	29

(three papers). The plans were indexed on the "point of indifference", namely p_{50} and the slope of the OC curve at that point. Steep slopes of course call for sharp discrimination and larger sample sizes. The general objective is to equalize producer and consumer risks.

Still another type of sampling inspection by attributes is pure "sequential sampling", wherein, after each piece is inspected for nonconformities, we can (1) accept the lot or process, (2) reject it, or (3) ask for another piece to inspect. No limit is set on the sample size which varies. The objective is to achieve a given OC curve for protection, on a minimum ASN.

During the 1970s, many companies began to require the use of $c = 0$ plans taken for the most part from the ABC Standard then in use (MIL STD 105D or ANSI/ASQ Z1.4). While this set of plans gave management a secure feeling of quality (no nonconforming units in the sample were condoned) they tend to often reject lots that are of quite good quality. For example, for a lot of 1500 units we would use a code letter of J in Table 10.1 of the ANSI/ASQ Z1.4. This gives us a sample size of 80 and an AQL of 0.15%. This AQL is the one specified for $c = 0$.

10.4 CONTINUOUS SAMPLING PLANS

A rather different type of sampling plan from those we have been studying will now be discussed. In many processes, units of product come along slowly enough so that each unit may be inspected for nonconforming units. That is, it is feasible to inspect 100%, and it is being done in this manner to control quality. In such a situation, it may be feasible to sample inspect at least some of the time in order to save inspection time. The continuous sampling plan which is now to be discussed does this, with adequate safeguards, in order to deliver an AOQL type of protection.

Let us first give the conditions for which the plan was designed:

1. There are no natural lots submitted for inspection and a decision. Instead, there is a continuous flow of product such as on a conveyor line.
2. There are discrete units of product, each of which is either good or nonconforming, i.e., has none or has few nonconformities.
3. Production is not so rapid as to make 100% inspection infeasible.
4. The quality of production has an acceptably low fraction nonconforming for most of the time.

Continuous sampling plans for a particular nonconformity or for a class of nonconformities call for alternating periods of 100% sorting and of sampling inspection. The relative amount of time spent on each depends upon the quality level of the nonconformities or nonconforming units in question. If quality is excellent, most of the time is spent on sampling, or if relatively poor most of time is at 100% sorting.

10.4.1. The Original CSP-1 Plan

The first continuous sampling plan was developed by Dodge (1943, 1947). The procedure is as follows:

1. At the beginning, inspect for the nonconformities in question, 100% of the units consecutively, as produced, and continue such inspection until i units in succession are found to be free from the nonconformities.

2. When i consecutive good units are found, discontinue 100% inspection, and inspect only a fraction f of the units, selecting the sample units one at a time from the flow of units, in such a way as to insure an unbiased sample.
3. Whenever under sampling a defective is found, revert immediately to 100% inspection of succeeding units, continuing until again i consecutive units are found free from nonconformities, as in step 1, when sampling inspection is resumed.
4. Correct or replace with good units all nonconforming units to found.

It is thus to be seen that there are two constants governing the plan, namely i and f. The larger the i is the more difficult it is to qualify for sampling and thus the greater the protection supplied against relatively poor quality. Similarly, the larger f is the higher the proportion of units sampled, and thus the more difficult it is to stay on sampling, for any given incoming p_o. How do we find i and f? First, we must specify our desired protection. This is in terms of the AOQL, i.e., the worst long-run average quality outgoing, no matter what quality level p_o come in. Therefore, we first decide on a desired AOQL amount of protection. We might then specify f and find I or specify I and find f, from the charts in Dodge (1943, 1947). The former is the usual approach. Accordingly, we present in Table 10.1 some combinations. If others are desired, see the original sources or a reproduction in Ref. 2 of Chapter 9.

From the table, we see that in order to obtain AOQL consumer protection of 1.00%, we will use $I = 110$ if we specify an f of 0.100, or one in each 10 units. So we will need 110 in succession all good in order to qualify for sampling. Once on sampling, we will choose one at random out of each '10. This is best done by use of a table of random numbers, such as our Table D, or an abstract of this table. The main thing to avoid is choosing every 10th unit, because of the possibility of a cycle in the occurrence of nonconformities. Or if we choose a larger sampling proportion $f = 0.200$, or one in each five, then it only takes 12 good units in succession to qualify. It is easily seen that f and i are inversely related for any given AOQL.

10.4.2. Variations from CSP-1

If quality is more or less intermediate, then there may be considerable alternation between 100% and sampling inspection periods. This can be a troublesome factor in adjustment of personnel loads. One way to help is by using inspection personnel for the sampling periods and production personnel for the 100% inspection. One advantage of this is to encourage production to improve its quality performance.

Another way to avoid excessive changing from sorting to sampling and vice versa is to use the CSP-2 plan (Dodge and Torrey, 1951). In this plan, we again have the two constants, I and f, for a given AOQL and the initial qualification for sampling is the same, i.e., I consecutive good units. But when sampling and a nonconforming unit is found, we do not immediately return to 100% inspection of units. Instead, we are alerted and watch the sampled units. If among the next i sampled units a second nonconforming unit is found, we immediately revert to 100% inspection. But if no further nonconforming unit is found among the next i sampled units, we are requalified fully, so to speak. And a single defective does not force us off from sampling, but we would as before watch the next i sampled units for a second defective which would terminate the sampling period.

Now it is obvious that for a given AOQL and f, CSP-2 is more lenient for a fixed qualification number i, because we are less likely to be thrown off of sampling. Therefore, for given AOQL and f, we must use a larger i under CSP-2 than when under CSP-1. In fact, i for CSP-2 is roughly a third more than for CSP-1. This rule can be used on entries of Table 10.16 if you wish to use CSP-2.

A variation of CSP-2 is called CSP-3. It is basically like CSP-2, but when the first nonconforming unit is encountered, we look at all of the next four units, instead of sampling. If these are all good then we continue on sampling, counting these four among the next I, all of which must be good in order to be fully requalified. Thus CSP-3 is essentially a CSP-2 plan; but has added protection against a short run of spotty quality, causing repeated nonconforming units.

10.5 CHAIN SAMPLING PLAN, ChSP-1

Whenever a quality characteristic involves destructive or costly tests so that only a small number of tests, n, can be justified per lot, an acceptance number of Ac $= 0$ is naturally used. But for all sample sizes, n, when Ac $= 0$, the shape of the OC curve is like those of Figure 9.8, i.e., with the concave side everywhere up. This makes it relatively difficult to pass even quite good material. Pa falls rapidly from 1.00 as p_o increases. Moreover, with an n of 5 or 10, the OC curve is rather undiscriminating. For these reasons, the chain sampling plan ChSP-1 was developed to improve the shape of the OC curve, making it more discriminating with an elongated S shape more like that in Figure 9.10 (Dodge, 1955). Further conditions for using ChSP-1, as indicated in the reference, are (1) the product to be inspected consists of a series of lots produced by an essentially continuing process, which under normal conditions can be expected to be of the same quality p_o, and (2) the product comes from a source in which the consumer has confidence.

The procedure is as follows:

1. Take a random sample of n units or specimens and inspect or test each against the requirement specified.
2. Use the acceptance number Ac $= 0$ for nonconforming units; except use Ac $= 1$ if no nonconforming units were found in the immediately preceding i samples of n (i is a constant number specified in the plan and may be 1, 2, 3, . . .).

Thus we specify n and i, and routinely use Ac $= 0$, except that when one nonconforming unit shows up in the sample of n, we must check back over the preceding i lots for the presence of

Table 10.2 Points p_o on the OC Curve for Which Pa = 0.95 and 0.10 for Ordinary Ac = 0 Plans and for ChSP Plans with Same Sample Size n

				Proportion of lots expected to be accepted	
	n	Ac	i	0.95	0.10
Ordinary	4	0	–	0.013	0.44
ChSP-1	4		4	0.031	0.44
	4		2	0.042	0.44
Ordinary	5	0	–	0.010	0.37
ChSP-1	5		4	0.024	0.37
	5		2	0.030	0.37
Ordinary	6	0	–	0.008	0.32
ChSP-1	6		4	0.020	0.32
	6		2	0.027	0.32
Ordinary	10	0	–	0.005	0.21
ChSP-1	10		4	0.012	0.21
	10		2	0.017	0.21

any nonconforming units. If there was one, we reject the present lot (but do not reject any of the previous lots, which are probably out of our hands anyway). There could not be two, for then we would have had an earlier rejection. Thus whenever a defective is found, the next i lots must show perfect samples ($d = 0$) for all of them to be accepted. But if they are, then we can once more tolerate a single nonconforming unit, but would then be alerted and could not have a second nonconforming unit in the next I samples without a rejection. Of course, 2 = 2 in any n gives immediate rejection.

The OC curve has a slightly different meaning than those we have been presenting in this and the preceding chapters. Now the vertical scale is the percentage of lots expected to be accepted for a given process p_o. Table 10.2 gives p_o values for which Pa is 0.95 and 0.10 for several ordinary Ac = 0 plans and for ChSP-1 plans with the same n. Note how in each case p_{95} increases as the number I of previous lots decreases. Thus these ChSP-1 plans give a better chance of acceptance at relatively very good quality levels, while retaining the same consumer protection, p_{10}. This is from the shape of the OC curves being an elongated S rather than being everywhere concave upward.

10.6 SKIP-LOT SAMPLING PLAN, SkSP-1

The skip-lot sampling plan developed by Dodge (1955) is a plan for omitting the testing or inspection of a certain proportion of a series of lots, while still providing specified protection. It is basically a continuous sampling plan like CSP-1, but is applied to lots instead of units or pieces. The lots in the stream are each tested or inspected. This may be on the basis of some measurable characteristic such as chemical content or it may be by a sample inspection decision to accept or reject or even 100% inspection. After a sufficiently long series of consecutive lots, i, have been found to meet requirements, the inspection or testing can go to the performance of this check on only a proportion, f, of the lots. But as soon as a nonconforming lot is found among this proportion, f, of lots, we go back to inspecting or testing every lot, and must requalify. Thus, this SkSP-1 is in reality a CSP-1 plan applied to lots.

Now we are concerned with the proportion of nonconforming lots being put out by the production process. The plan seeks to provide AOQL protection of the following type: For each incoming fraction of nonconforming lots, the plan will provide a lower fraction nonconforming outgoing, that is, an AOQ fraction of nonconforming lots. Considering all the AOQs, there will be a maximum, the AOQL. The plan is indexed on this AOQL (for lots nonconforming).

There are two procedures: A_1, where each nonconforming lot is to be either corrected or replaced by a conforming one, and A_2, where each nonconforming lot is to be rejected and is not to be replaced by a conforming lot. The required number

i to qualify differs by one in the two cases. The following plans are given in Dodge (1959):

	AOQL (%)	f	i Procedure A1	i Procedure A2
Standard plan	2	1/2	14	15
Other plans	3	1/2	9	10
	5	1/2	5	6
	5	1/3	9	10
	5	1/4	12	13

The skip-lot plan seems a reasonable approach to inspection of lots for noncritical nonconformances. But if the nonconformance in question is really critical, it would seem best to at least have a substantial sample checked from every lot. In fact commonly lots are at least 100% inspected for critical nonconformances.

10.7. SUMMARY

We have been discussing a variety of sampling plans in this chapter, each of which is ready to be used, without making any calculations. The objective of the ABC Standard (ANSI/ASQ Z1.4) is to obtain a quality level outgoing which is at least as good as the specified AQL and to accomplish this while using sampling decisions on lots. The other plans discussed were designed to provide some sort of consumer protection while saving on inspection costs. The Dodge–Romig tables enable the consumer to obtain either a specified AOQL or LTPD, and further, to achieve this on a minimum ATI for any given p-bar incoming quality.

The chain sampling, ChSP-1, plan operates on lots being given expensive or destructive tests, so that small sample sizes are imperative. But use of an $\mathrm{Ac} = 0$ gives an undesirable concave-up shape to the OC curve. By using information on previous lots, the shape of the OC curve can be improved, while retaining the small sample size. On the other hand, when the objective is to control the percentage of unsatisfactory lots to a previously set AOQL, under appropriate conditions we can use the skip-lot plan, SkSP-1, to provide this protection while requiring only a proportion f of lots to be tested or inspected.

If concern is with the fraction nonconforming or nonconforming of units of product in continuous production, we can use the continuous plan, CSP-1. This plan provides for maintenance of AOQL protection units. It saves on inspection by providing for sampling after i consecutive units are all good. One hundred percent inspection is reinstated whenever a nonconforming unit is found in the sampled pieces. Plans for CSP-2 and CSP-3 can be used to decrease the frequency of alternation of sampling and sorting 100%.

10.8 PROBLEMS

10.1. Find the following plans from the ANSI/ASQ Z1.4:

	Lot size	Inspection level	S, D, or M	N, T, or R	Nonconforming or nonconformities	AQL (%)
(a)	500	II	Single	Normal	Nonconforming	0.40
(b)	500	II	Single	Tightened	Nonconforming	0.40
(c)	500	II	Single	Reduced	Nonconforming	0.40
(d)	500	I	Single	Normal	Nonconforming	0.40
(e)	2000	II	Double	Normal	Nonconforming	0.65
(f)	2000	II	Double	Normal	Nonconforming	0.065
(g)	2000	II	Double	Normal	Nonconformities	15
(h)	5000	III	Double	Tightened	Nonconforming	0.40
(i)	5000	III	Multiple	Tightened	Nonconforming	0.40
(j)	5000	II	Multiple	Normal	Nonconforming	0.15
(k)	1000	II	Double	Reduced	Nonconforming	2.5
(l)	1000	II	Single	Reduced	Nonconforming	2.5

10.2. Find single sampling plans in the ABC Standard for lot size $N = 500$, inspection level II, with AQL $= 2.5$ (nonconforming) for (a) normal, (b) tightened, and (c) reduced.

Draw the three OC curves on the same axes, labeling the scales. Comment on the curves.

10.3. A Dodge–Romig sampling plan for an AOQL of 1% and lots of 3000 calls for $n_1 = 80$, $n_2 = 170$, $Ac_1 = 0$, $Re_1 = 5$, $Ac_2 = 4$, $Re_2 = 5$. Find Pa when $p_o = 0.04$. Is this compatible with the listed $p_{10} = 0.036$?

10.4. A Dodge–Romig single sampling plan for lots of $N = 3000$ and p-bar $= 0.21$–0.40% is $n = 80$, Ac = 1. It is for an AOQL of 1.0%. Draw the OC and AOQ curves and check this AOQL.

10.5. Find the CSP-1 continuous sampling plan for an AOQL of 1.0% with $f = 0.125$. Explain the meaning of i and f. What is the corresponding CSP-2 plan? Again explain i and f.

10.6. In what sense is the OC curve for the ChSP-1 chain sampling plan $n = 10$, Ac = 0, $i = 2$ more favorable than that for the ordinary sampling plan $n = 10$, Ac = 0?

10.7. Explain in simple terms the operation of a SkSP-l skip sampling plan having AOQL = 2% and $f = 1/2$. What advantages and disadvantages do it have?

REFERENCES

ANSI/ASQ Z1.4–2003, American Society for Quality, Milwaukee, WI.

HF Dodge. *Notes on the evolution* of acceptance sampling plans, Parts I to IV. *J. Quality Techn.* 1: 77–88, 155–162, 225–232, 2: 1–8, 19, 1969, 1970.

HF Dodge, HG Romig. Sampling Inspection Tables. New York: Wiley, 1959.

HF Dodge. A sampling inspection plan for continuous production. Ann. Math. Statist. 14:264–279, 1943.

HF Dodge. Sampling plans for continuous production. Indust. Quality Control 4(3):5–9, 1947.

HF Dodge, MN Torrey. Additional continuous sampling inspection plans. Indust. Quality Control 1(5):1–12, 1951.

Military Standard MIL- STD-1235(ORD). *Single and Multi-level Continuous Sampling Procedures and Tables for Inspection by Attributes.* Department of the Army, 1962. (No longer in print).

HF Dodge. Chain sampling inspection plan. Indust. Quality Control 11(4):10–13, 1955.

HF Dodge. Skip-lot sampling plan. Indust. Quality Control 11(5): 3–5, 1955.

11

Sampling by Variables

In the preceding two chapters, the decision for acceptance or rejection of a lot or process was based upon attributes, that is, counts of the number of nonconforming units or of nonconformities on the units in samples. Such an approach is always possible. For, if the quality characteristic is a measurement, such as a dimension, a weight of contents or a tensile strength, this can always be converted to an attribute by comparison to a maximum or a minimum limit, or to two limits. Thus, the measurement either lies inside limits or it does not. On the other hand, many attribute characteristics are not capable of being converted to a measurement.

But when the quality characteristic is a measurement, there is the possibility of basing the decision on the measurements on units within the sample. This has the possibility of making sound decisions on smaller samples, because the measurements provide more information than just whether the measurement lies within or outside of limits. For example, it will tell how "nonconforming" the unit was. But there are three "prices" to pay for the smaller sample size for a comparable power of discrimination: (1) It usually takes more time to make and record measurements than to merely find out whether the unit is in outside of limits, (2) knowledge of the shape or type of measurement distribution is essential and (3) decisions by measurements are somewhat more complicated and commonly require at least some simple calculations. The second of these is unfortunately often overlooked and does require a substantial amount of previous data.

But properly supported, decisions based upon samples of measurements can be very effective. Basically, the decision is whether the lot or process seems to have a satisfactory distribution.

11.1. KNOWLEDGE OF DISTRIBUTION TYPE

As we have just been pointing out, if we are to take full advantage of our observed sample measurements, we need to know what shape or type the distribution takes. This is quite distinct from the parameters: the population average μ and standard deviation σ.

If we are making a decision on a process, we may well know from past experience that the distribution is approximately normal. This is often the case, especially where conditions have been controlled, and where the process has not been reset while the product in question was produced. In fact, the sampling decision may be whether the process level is in need of resetting or adjusting.

On the other hand, when we are examining a lot a vendor has sent us, we must be more careful, and a normal distribution is not so readily assumed. For one thing, the producer may have run the process at several distinct levels, Thus, if we were to measure up the entire lot and tabulate a frequency table, such as Table 2.2, we might find several humps of high frequencies, or possibly a lopsided unsymmetrical distribution. Such is often the case when conditions have not been well controlled. Another type of abnormality is that in which the producer has been having trouble meeting our specification limits and thus has had the output sorted, more or less perfectly, to specification limits. Such a situation can cause a frequency distribution that stops abruptly at the ends, with one or both tails chopped off by inspection, human or mechanical. A third cause of nonnormality of distribution is where there is a physical limitation in one direction. Out-of-round, out-of-square, dimensions cannot be below zero nor can eccentricity (distance between center lines). These give rise to distributions tailing out on the high side. Similar tendencies to unsymmetrical distributions may occur in percent impurity, content weights in containers, strengths, and weights of castings or gloss readings. For such measurements, we cannot soundly use methods which are designed under the assumption that the distribution is normal. Unless otherwise stated, the methods to be discussed all assume a normal distribution of measurements. Such an assumption is likely to be safe if the process is running under homogeneous or controlled conditions. But it is desirable to tabulate at least 100 measurements, or better 200, to check the shape.

11.2. GENERAL AIM: TO JUDGE WHETHER DISTRIBUTION IS SATISFACTORY

The general aim in acceptance sampling by variables is to use measurements on a sample of units from a lot or process in order to judge whether the population distribution seems to be satisfactory. By a satisfactory distribution, we usually mean that the percentage of units having measurements outside the specification limit or limits is acceptably small, perhaps 1%, for example. The two cases give rise to "one-tail" and "two-tail" tests, i.e., where concern is with only one tail of the distribution or with both tails. In the former case are strengths subject to a minimum, or blowing times of fuses under marginal overload, subject to a maximum limit. In the latter are most dimensions, that is, those which have two specification limits ($\pm$ limits). Also, for example, hardness limits may be in both directions.

If in addition to knowing that the distribution is approximately normal, we also know the population or process standard deviation from past data, we are indeed in a strong position. And we can set up simple acceptance criteria for the process or lot, based on the mean, x-bar, only.

11.3. DECISIONS ON LOT MEAN, KNOWN σ, NORMAL DISTRIBUTION

It may seem strange that we might know the lot or process standard deviation, σ, and yet fail to know the mean, μ. But this can well happen. The process level is subject to many factors in practice which may not affect the variability. Or, to put it in control chart terms, the R chart may well show good control and have a quite constant R-bar, while the x-bar chart shows lack of control. In such a case, we may estimate σ by R/d_2 This is σ for individual x's, σ_x. It tells about how far away from μ we must expect an individual x to lie. And in fact an x may rarely be as much as $3\sigma_x$ away from μ.

11.3.1. Review of Distributions of x and of x-bar

Let us suppose that we have a normal distribution of individual x's with average μ and standard deviation σ_x. Then we recall from Chapter 1 that the distribution of sample averages will also be normal, and the average of x-bar's is the same as that for the x's, namely

$$\mu_{x\text{-bar}} = E(x\text{-bar}) = \mu \tag{11.1}$$

and the standard deviation of the x-bar's is

$$\sigma_{x\text{-bar}} = \sigma_x/\sqrt{n} \tag{11.2}$$

that is, a typical x-bar may be expected to lie closer to μ than we expect an x, by a factor of $\sqrt{n}$.

To explain further, let us say that for a dimension we have, $\mu = 0.1720$ in. and $\sigma = 0.0002$ in. Then in Figure 11.1 we show the distribution of x. Now for sample means from samples of $n = 4$ dimensions, we have by (11.2)

$$\sigma_{x\text{-bar}} = 0.0002 \text{ in.}/\sqrt{4} = 0.0001 \text{ in.}$$

while the average for the x-bar's is the same as for the x's, namely 0.1720 in. Thus in Figure 11.1b, the x-bar's are more

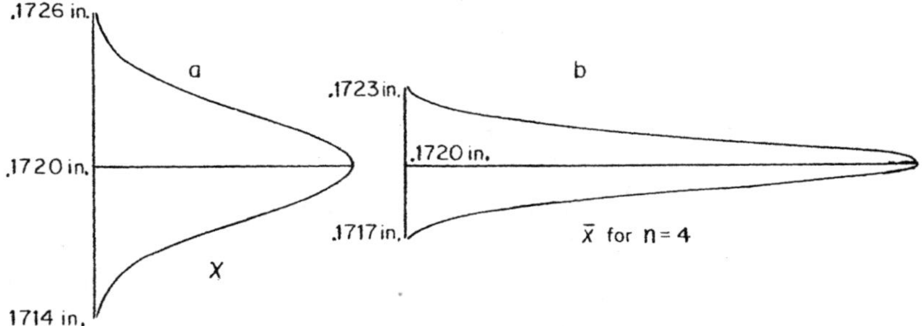

Figure 11.1 Comparison of distribution of x and of $\bar{x}$ for $n = 4$ when $\mu = 0.1720$ in. and $\sigma_x = 0.0002$ in. Normal distribution of x is assumed.

closely clustered around μ than are the x's. If n were 1arger, $\sigma_{x\text{-bar}}$ would be even smaller. (The relative frequencies for x-bar's are twice as great as those for x's so that the total areas enclosed by the curves are the same.)

We shall make constant use of these facts in the following examples of sampling plans. Be careful to make sure whether you are concerned with the distribution of individual x's or with that for averages, x-bar's.

11.3.2. A Specific Example

Suppose that we are to sample test a lot of parts for tensile strength. This being an expensive and destructive test, we are interested in a small sample size. Past results indicate the strengths to be normally distributed, with a standard deviation running at about 3000 psi. The specified minimum tensile strength is 70,000 psi. The main question with each lot is whether the lot mean is safely above 70,000 psi. Accordingly, we plan to make tensile tests on a sample of n parts and to make our decision from the sample mean strength x.

Now what do we mean by "safely above 70,000 psi?" A reasonable possibility is to take as an acceptable lot mean strength the value $\mu = \text{lower specification} + 3\sigma_x = 70{,}000 + 3(3000) = 79{,}000$. If $\mu = 79{,}000$, what percentage of parts will have strengths below 70,000? To answer this we make use of (7.12);

$$Z = \frac{x - \mu}{\sigma_x} = \frac{70{,}000 - 79{,}000}{3000} = -3.00$$

Then we use Table A to find the probability of a strength below 70,000. It is $0.0013 = 0.13\%$ or about one out of 770 parts. This would seem to be small enough to regard $1 = 79{,}000$ psi as clearly safe, depending, of course, upon the criticality of the part.

Next suppose we set $\mu =$ lower specification $+ 2\sigma_x, = 70{,}000 + 2(3000) = 76{,}000$ as an undesirable lot mean strength. What percentage would be below 70,000 if $\mu = 76{,}000$ psi? Again use (7.12):

$$Z = \frac{x - \mu}{\sigma_x} = \frac{70{,}000 - 76{,}000}{3000} = -2.00$$

which by Table A gives $0.0228 = 2.28\%$ or one in 44, which might well be undesirable.

Thus, we wish our sampling plan to distinguish between lots having $\mu_1 =$ 76,000 psi and lots having $\mu_2 = 79{,}000$ psi. As we have already mentioned, we will use x-bar to decide about the lot. Now if x-bar were to be above 79,000, we might well feel safe and accept the lot, whereas if x-bar is below 76,000, we would undoubtedly wish to reject the lot. But if x-bar lies between 76,000 and 79,000, what would we wish to do? Somewhere between them there must be a cut-off point, say K, so that

x-bar $> K$ Accept lot

x-bar $< K$ Reject lot

We need to find a reasonable K value, and not only that, but also the sample size n.

In order to find K and n for the test, we must make our decision making more precise by deciding upon risks as follows: let the small Greek alpha, α, stand for the probability of erroneous rejection if μ is actually at μ_2 the acceptable level. Likewise let the small Greek beta, β, stand for the probability of erroneous acceptance if μ is actually at μ_1, the rejectable level. The α and β are called "risks of wrong decisions" and can be set at will.

Summarizing:

$$\alpha = P \text{ (rejection, given } \mu \text{ is acceptable at } \mu_2)$$

$$\beta = P \text{ (acceptance, given } \mu \text{ is rejectable at } \mu_1)$$

Usually, risks are set as desired and then the sample size is determined. If n seems too large to be economic, then one or both risks must be increased and/or the difference between μ_1 and μ_2 increased.

Let us take risks of 0.10, that is, for example, if $\mu = 79{,}000$ psi there will be 1 chance in 10 of erroneous rejection, and a 90% chance of acceptance. We may now proceed to find n and K. To do this we note first that n must be large enough to make the x-bar distributions sufficiently narrow, as shown in Figure 11.2. (The respective distributions of the x's are far wider, of course.)

In order to find n and K, consider first the distribution of the x-bar's, centered at $\mu = \mu_2 = 79{,}000$ psi. We shall need to find the standard normal curve value z for which the

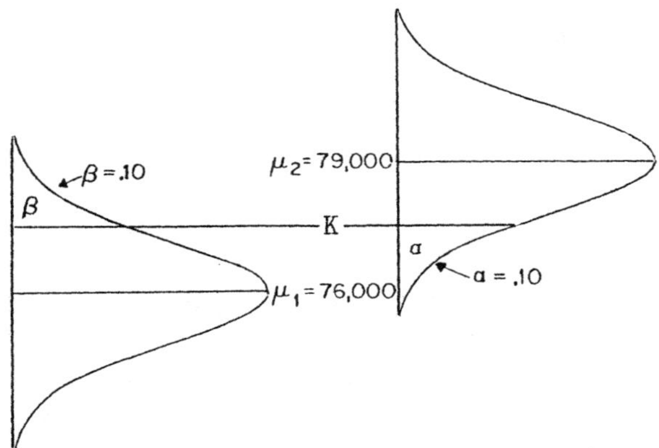

Figure 11.2 Two distributions of averages $\bar{x}$, one acceptable at $\mu_2 = 79{,}000$ psi, the other rejectable at $\mu_1 = 76{,}000$ psi. Also shown are the respective risks, α of rejection at μ_2, β of acceptance at μ_1. Each set to be 0.10.

lower tail is 0.10. For this we look in the first part of Table A (where the z's are negative) for 0.10. It is not there exactly, but we do find

$$z = -1.280 \qquad p = 0.1003$$
$$z = \qquad\qquad p = 0.1000$$
$$z = -1.290 \qquad p = 0.0985$$

Interpolating for the required z, we must go 3/18 of the way from -1.280 toward -1.290 or -1.280 $-(3/18)(0.010) =$ -1.282. We are now in a position to use (7.12) for the x-bar distribution:

$$Z = \frac{x-\mu}{\sigma_x} \text{ becomes } Z = \frac{x-\mu}{\sigma_{x\text{-bar}}} \text{ or } -1.282 = \frac{K - 79{,}000}{3000/\sqrt{n}}$$

Thus, we have arrived at an equation in two unknowns n and K. We need another, which comes from the distribution at μ_1.

This time we have an upper tail with a positive z value having 0.10 probability above it or 0.90 below. We could look for 0.9000 in Table A and interpolate to find $z = +1.282$. Or we could use the symmetry of the normal curve and merely change the sign of our other z. Then

$$+1.282 = \frac{K - 76{,}000}{3000/\sqrt{n}} \text{ or } +1.282 = (K - 76{,}000)\frac{\sqrt{n}}{3000}$$
$$-1.282 = \frac{K - 79{,}000}{3000/\sqrt{n}} \text{ or } -1.282 = (K - 79{,}000)\frac{\sqrt{n}}{3000}$$

Subtracting yields $2.564 = \sqrt{n}$, and squaring both sides gives $n = 6.57$

But since we can only make six or seven tests we choose $= 7$.

Now to find K, we could substitute into either equation. But since we rounded n up to the whole number 7, the other equation (not substituted into) will not be exactly satisfied. Or since the two risks were taken equal, we can merely take K half-way between μ_1 and μ_2, that is:

$$K = 77{,}500$$

Our sampling plan is then to take $n = 7$ tensile tests and find x-bar:

$$x\text{-bar} \geq 77{,}000 \qquad \text{Accept lot}$$
$$x\text{-bar} \leq 77{,}000 \qquad \text{Reject lot}$$

The actual risks will be just a bit less than 0.10. For we have

$$z = \frac{77{,}500 - 79{,}000}{3000/\sqrt{7}} = -1.32$$

giving $\alpha = 0.0934 = \beta$.

Now if seven tests are too expensive, then we may lower μ_1 below 76,000 or raise μ_2 above 79,000 or increase one or both of the risks, α and β. That is, we must sacrifice some of the sharpness of discrimination in order to make n smaller. Or if we can afford a larger n we can make a sharper discrimination, subject to smaller risks.

We may sketch the OC curve P_a vs. μ by drawing a smooth curve through the points (0.09, 76,000), (0.50, 77,500), (0.91, 79,000) such that P_a approaches zero below 76,000, and one above 79,000.

11.3.3. One-Specification Limit Test

Now let us generalize the specific example just discussed. For this, we will use the notation

$$z_p = \text{standard normal curve } z \text{ with probability } p \text{ above it} \tag{11.3}$$

Thus, $z_{0.10}$ has an upper tail of 0.10, and is +1.282, $-z_{0.10}$ by symmetry has a lower tail of 0.10, and is −1.282, z_α has an upper tail of 0.10, and so on.

Case 1. Acceptable mean $\mu >$ rejectable mean μ_1 risk of rejection if $\mu = \mu_2$ is α, risk of acceptance if $\mu = \mu_1$ is β. The equal ions then are

$$-z_\alpha = \frac{K - \mu_2}{\sigma/\sqrt{n}}, \qquad z_\beta = \frac{K - \mu_1}{\sigma/\sqrt{n}} \tag{11.4}$$

in which α, β, μ_1, μ_2 and σ known, and n and K are to be found. z_α and z_β of course are found in Table A. For n we have then

$$n = \frac{(z_\alpha + z_\beta)^2 \sigma^2}{(\mu_2 - \mu_1)^2} \tag{11.5}$$

which we round up to the first whole number. Then K may be found from either equation of (11.4).

The OC curve goes through the three points (β, μ_1), (0.5, K) and ($1-\alpha$, μ_2).

Case 2. Rejectable mean $\mu_2 >$ acceptable mean μ_1, risk of rejection if $\mu = \mu_1$ is α, risk of acceptance if $\mu = \mu_2$ is β. The equations then are:

$$z_\alpha = \frac{K - \mu_1}{\sigma/\sqrt{n}}, \qquad -z_\beta = \frac{K - \mu_2}{\sigma/\sqrt{n}} \tag{11.6}$$

The solution for n is still (11.5), and then we may substitute into either of (11.6) to find K.

An OC curve may be sketched through the points ($1-\alpha$, μ_1), (0.5, K), (β, μ_2).

11.3.4. A Two-Way Example

When there are two specification limits for a measurement, we have a problem of protecting against the lot or process mean μ being too high or too low. For example, a lot of parts is to be tested for hardness on the Rockwell C scale, to specifications of 60 and 70. Past experience has shown that lot distributions are approximately normal with $\sigma = 1.5$ points. The main thing to test for is the lot mean. We pick as the acceptable mean $\mu_o = 65$. For the two rejectable lot means, we may well come in from the specification limits $2\sigma = 3$, giving $\mu = 60 + 3 = 63$ and $\mu = 70 - 3 = 67$. If $\mu = 67$, what proportion of parts will be outside specifications? None will be below 60 which is 4.67σ below 67, but some will be above 70.

Use

$$z = \frac{70 - 67}{1.5} = +2$$

The probability below $z = +2$ is 0.9112, leaving 0.0228 above. Likewise if $\mu = 63$, there will be 0.0228 below 60, that is, about 2% too soft.

Next we need to set risks of wrong decisions. As usual, we let α be the risk of rejecting an acceptable lot, namely one with $\mu = \mu_o = 65$. Suppose we set this at $\alpha = 0.01$. Next, β is the risk of accepting a rejectable lot, i.e., one with $\mu = 63$ or 67, say, μ_1 and μ_2. The value for β will depend upon how critical the hardness is for the part in question. Suppose we set $\beta = 0.10$.

The form of the test is to set two limits around μ_o, between which the average sample hardness, x-bar, is expected to lie, say, $\mu_o \pm k$. If x-bar lies within we accept the lot, if outside we reject the lot. Our job is to find n and k, subject to the risks α and β, and for the critical means μ_1, μ_o, and μ_2.

See Figure 11.3 which pictures the distributions of x-bar at the three critical means, to provide the risks $7 = 0.01$ and $\beta = 0.10$. Notice that $\alpha = 0.01$ is split into two parts, 0.005 above $\mu_o + k$ and 0.005 below $\mu_o - k$. On the other hand, if $\mu = \mu_2 = 67$, then all of $\beta = 0.10$ lies below $\mu_o + k$ and similarly if $\mu = \mu_1 = 63$, all of $\beta = 0.10$ will lie above $\mu_o - k$. Now let us set up appropriate equations to find n and k. Remembering that we have distributions for x-bar's, we have as the standard deviation, $\sigma_{x\text{-bar}} = \sigma_x/\sqrt{n}$. The two distributions, which are cut by the line $\mu_o + k = 65 + k$, have means 65 and 67 and thus

$$z_{\alpha/2} = \frac{(65 + k) - 65}{1.5/\sqrt{n}} = z_{0.005}$$

$$-z_{\beta} = \frac{(65 + k) - 67}{1.5/\sqrt{n}} = z_{0.10}$$

Using Table A and interpolating we find $z_{0.005} = 2.575$, $z_{0.10} = 1.282$. Therefore, we have

$$\frac{k}{1.5/\sqrt{n}} = 2.575$$

$$\frac{k - 2}{1.5/\sqrt{n}} = -1.282$$

$$\sqrt{n}\,\frac{k}{1.5} = 2.575$$

$$\sqrt{n}\,\frac{k - 2}{1.5} = -1.282$$

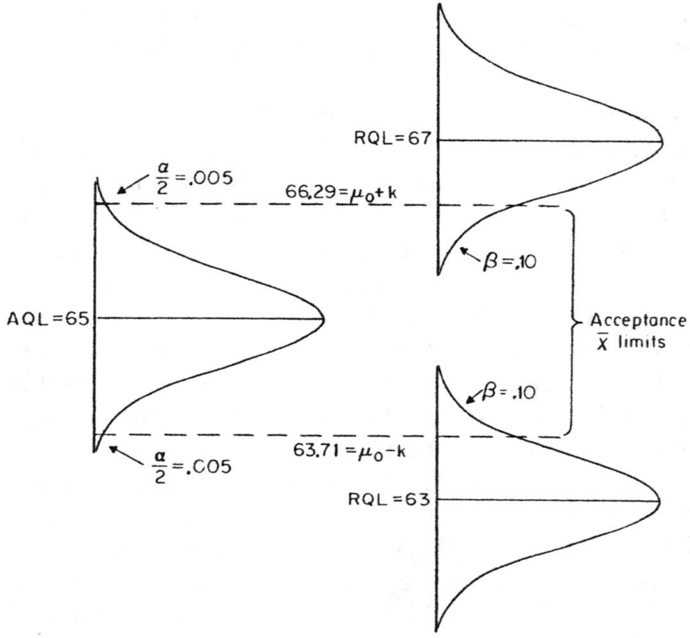

Figure 11.3 Two-way test for the mean $\bar{x}$ for the hardness example. AQL $=\mu_o=65$, RQL $=\mu_o \pm 2=63$, 67. Also shown are the areas for α and β risks and the acceptance limits for $\bar{x}$ at $\mu_o \pm k=65 \pm 1.29$.

By subtracting

$$\sqrt{n}\frac{2}{1.5} = 3.857$$
$$\sqrt{n} = 2.89$$
$$n = 8.37 \text{ or } 9$$

We may now substitute $n=9$ into either equation to find k. (They will give slightly differing k's because of using $n=9$ instead of 8.37.) Suppose we take the first equation

$$\sqrt{9}\frac{k}{1.5} = 2.575 \quad \text{or} \quad k = 1.29$$

There are also two equations available from the cut-off line at $\mu_o - k$, but by the symmetry they prove to be identical to the two we just solved. Thus, our plan is to take $n = 9$ hardness readings and find x-bar from them. Then

x-bar between 65 ± 1.29	Accept lot or process
x-bar outside 65 ± 1.29	Reject lot or process

An OC curve of P_a vs. μ might be sketched.

11.3.5. Two-Specification Limit Test

Knowing σ, we can compare it with the specified tolerance, $U - L$. If $U - L$ is, say, 8σ or more we may best use two one-way tests, one near L and one near U. But if $U - L$ is less than 8σ or so, then we probably will do best with a two-way test such as just given. In the previous example, $U - L = 6.67\sigma$.

The general set-up then is that we have a desired nominal or acceptable mean μ_o, and equally spaced on opposite sides of μ_o and two rejectable means μ_1 and μ_2, that is, $\mu_o - \mu_1 = \mu_2 - \mu_o$. Also we have to set risks: α of rejection if $\mu = \mu_o$ and β of acceptance if $\mu_o = \mu_1$ or μ_2. Then the equations at $\mu_o + k$ are

$$z_{\alpha/2} = \frac{\mu_o + k - \mu_o}{\sigma/\sqrt{n}} \qquad -z_\beta = \frac{\mu_o + k - \mu_2}{\sigma/\sqrt{n}} \tag{11.7}$$

The solution for n is

$$n = (z_{\alpha/2} + z_\beta)^2 \left[\frac{\sigma}{\mu_2 - \mu_o}\right]^2 \tag{11.8}$$

Then rounding (11.8) up to the next whole number, we can substitute n into either of (11.7) to find k. Then for sample means x-bar:

x-bar between $\mu_o \pm k$	Accept lot or process
x-bar outside $\mu_o \pm k$	Reject lot or process

As in the example, we could give two more equations at $\mu_o - k$, but they prove to be identical to (11.7).

An OC curve may be sketched through the five points (β, μ_1), $(0.5, \mu_o - k)$, $(1 - \alpha, \mu_o)$, $0.5, \mu_o + k)$, (β, μ_2).

11.3.6. Controlling the Percentage Nonconforming

We could take the following one-way approach to control the percentage outside of limits, say, at an upper specification limit U: (1) Set μ_1 at such a level that there is a tolerably small proportion, p_1, above U, i.e., call $\mu_1 = U - z_{p1}\sigma$. (2) Set μ_2 at such a higher level that there is a rejectable proportion p_2 above U, i.e., call $\mu_2 = U - z_{p2}\sigma$. (3) Using these $\mu_1$and μ_2 in Section 11.3.3, case 2, we proceed as there described. Substituting the foregoing values of μ_1 and μ_2 into (11.5), we obtain

$$n = \left(\frac{z_\alpha + z_\beta}{z_{p_1} - z_{p_2}} \right)^2 \tag{11.9}$$

Then find the acceptance criterion K from either of (11.6), namely

$$\begin{aligned} z_\alpha &= \frac{K - z_{p_1}}{\sigma/\sqrt{n}}\sigma \quad \text{or} \quad -z_\beta = \frac{K - z_{p_2}\sigma}{\sigma/\sqrt{n}} \\ K &= U - z_{p_1}\sigma + \frac{z_\alpha \sigma}{\sqrt{n}} \quad \text{or} \quad K = U - z_{p_2}\sigma + \frac{z_\beta \sigma}{\sqrt{n}} \end{aligned} \tag{11.10}$$

In the case of a lower specification L, we use the following equations:

$$K = L + z_{p_1}\sigma - \frac{z_\alpha \sigma}{\sqrt{n}} \quad \text{or} \quad K = L + z_{p_2}\sigma - \frac{z_\beta \sigma}{\sqrt{n}} \tag{11.10a}$$

11.4. DECISIONS ON LOT BY MEASUREMENTS, σ UNKNOWN, NORMAL DISTRIBUTION

We may make the following one-way test under these conditions. The objective is to make a sound decision on the percentage of parts lying beyond the one specification limit, say, an upper limit U. With σ for the lot unknown, we must take account of the wide variety of combinations of μ and σ which would be satisfactory. Thus, for example, suppose $U = 1000$

and that 1% above 1000 is permissible. We first seek $z_{0.01}$ in Table A, finding

Cumulative probability	z
0.9898	2.32
0.9901	2.33

giving $z_{0.01} = 2.32 + (2/3)0.01 = 2.327$. Then if $\sigma = 1$, μ of $1000 - 2.327(1) = 997.673$ or lower would be satisfactory for then p_o would be 101 or less. But if $\sigma = 10$, then μ must be $1000 - 2.327(10) = 976.73$ or less for p_o to be 0.01 or less. The larger the σ is, the farther below U we must have μ.

In order to use the technique being discussed, we now set up the problem much as in Section 9.8, which was using a single sampling test by attributes, that is, n and Ac. Now we will be working with the same sort of requirements, but making the test using x-bar and s from a sample of measurements. (The assumption of normality is vital here.) Specifically, we have

$$\text{Acceptable quality level} = \text{AQL} = p_1$$
$$\text{Rejectable quality leve1} = \text{RQL} = p_2$$
$$P(\text{reject if } p_1) = \alpha$$
$$P(\text{accept if } p_2) = \beta$$

We thus have two critical fractions defective for parts beyond U and corresponding risks of wrong decisions, α and β, all four of which may be set at will.

The test takes the form:

1. Take n observations and find x-bar and s.
2. Then $x\text{-bar} + ks \leq U$ Accept
   $x\text{-bar} + ks > U$ Reject

The multiple k of the sample standard deviation s, provides a "buffer" so that x-bar does not come too close to U. How close is safe depends upon both k and s. Our task, therefore, is to find n and k for the desired control of fraction nonconforming, p_o. Specifically, we have the following formulas from Statistical Research Group, Columbia

University (1947):

$$k = \frac{z_\alpha z_{p_2} + z_\beta z_{p_1}}{z_\alpha + z_\beta} \tag{11.11}$$

$$n = \frac{k^2 + 2}{2}\left[\frac{z_\alpha + z_\beta}{z_{p_1} - z_{p_2}}\right]^2 \tag{11.12}$$

In general, using (11.12) does not give a whole number for n, so we round upward to the next nearest integer for the sample size.

If our concern is with a lower specification limit L, we still use (11.11) and (11.12) then the decisions

$$\begin{aligned} &x\text{-bar} - ks \geq L \quad \text{Accept} \\ &x\text{-bar} - ks < L \quad \text{Reject} \end{aligned}$$

Let us take as an example

$$\begin{aligned} &\text{AQL} = p_1 = 0.01 \quad \alpha = 0.05 \\ &\text{RQL} = p_2 = 0.05 \quad \beta = 0.10 \end{aligned}$$

From Table A, we have $z_{0.01} = 2.327$, $z_{0.05} = 1.645$, $z_{0.10} = 1.282$, and then (11.11) gives

$$k = \frac{1.645(1.645) + 1/282(2.327)}{1.645 + 1.282} = 1.944$$

$$n = \frac{1.994^2 + 2}{2}\left[\frac{1.645 + 1.282}{2.327 - 1.645}\right]^2 = 53.2 \text{ or } 54$$

So we take 54 measurements and find z-bar and s. Then

$$\begin{aligned} &x\text{-bar} + 1.944s \leq U \quad \text{Accept} \\ &x\text{-bar} + 1944 > U \quad \text{Reject} \end{aligned}$$

We might compare this with the pure attribute sample size, using Section 9.8: $p_2/p_1 = 5 \ = R_o$, then Ac $= 3$, $np_2 = 6.68$ and $n = 134$. Or if we knew σ and use it, Section 11.3.6 gives by (11.9) $n = 18.4$ or 19. Note the large gain achieved by knowing σ. (Also there is no need to calculate s in order to make the test!)

11.5. SINGLE-SAMPLE TEST ON VARIABILITY

This test is a convenient one for deciding whether a lot or process standard deviation, σ, is acceptable. It is easily applied. It can be used for making a decision on the variability of a lot. An important application is that of testing whether a production process is capable of meeting some tolerance. Say the tolerance $T = U - L$ is given. Then if σ for the process is about one-eighth of T, we would say that the process is fully capable of meeting T, even with set-up error and some drift of the mean. But if σ is as much as one-fifth or even one-sixth of T, the process will have to be closely set up and the average tightly controlled to meet specifications. So we would have to distinguish between $\sigma_1 = T/8$ ($C_p = 1.33$) and $\sigma_2 = T/5$ ($C_p = 0.83$).

Another useful application is for an acceptance test on a gauge or measurement technique. We might be able to set an acceptable standard deviation of measurement error σ_1 and a rejectable standard deviation of measurement error σ_2. The objective is to accept most of the time $(1 - \alpha)$ if $\sigma_e = \sigma_2$ and to reject most of the time $(1 - \beta)$ if $\sigma_e = \sigma_2$. This is where σ_e is the true standard deviation for repeated measurements or analyses on the same or on homogeneous material. Then we have the following four test quantities to be set for the test:

$$\text{AQL} = \sigma_1 \qquad \alpha = P\,(\text{reject if } \sigma = \sigma_1)$$
$$\text{RQL} = \sigma_2 \qquad \beta = P\,(\text{accept if } \sigma = \sigma_2)$$

Table F can be easily used if we agree to use equal risks $\alpha = \beta$, and to let them take one of the four values 0.10, 0.05, 0.02, or 0.01.

There are two steps to take once the four test quantities are set. First, we find the quotient σ_2/σ_1 which describes the discrimination ratio. Then we choose whichever of columns (2)–(5) corresponds to our chosen $\alpha = \beta$. We now seek in that column the desired σ_2/σ_1, taking the nearest entry less than or equal to our σ_2/σ_1. Then for this entry, we find in column

(1) the sample size, n. The second step is to find in this row and the appropriate column (6)–(9) the multiplier for σ_1^2 to obtain K. Then the plan is to find s^2 for n observations and

$$s^2 \leq K \quad \text{Accept}$$

$$s^2 > K \quad \text{Reject}$$

Risk, α, is preserved but β may be slightly decreased by this test.

As an example, suppose a lathe is to be used under certain conditions to meet a tolerance of 0.0004 in. We would like to approve the process as fully capable if $\sigma = \sigma_1 = 0.00004$ in., would like to reject if $\sigma = \sigma_2 = 0.00007$ in. Suppose further that we set $\alpha = \beta = 0.10$. Then in column (2) we look for $\sigma_2/\sigma_1 = 1.75$. The sample size is $n = 13$. To find the test criterion, K, look in column (6) opposite $n = 13$, finding 1.55

$$K = 1.55(0.00004\,\text{in.})^2 = 0.248(0.0001\,\text{in.})^2$$

Now run off 13 parts under homogeneous conditions, measure them and calculate s^2. If $s^2 \leq K$ approve the lathe for the job, but if $s^2 > K$, regard the lathe as not capable, under present conditions.

A final word might be said about the OC curve. We have, of course, $P_a = 1 - \alpha$ when $\sigma = \sigma_1$ and $P_a = \beta$ at $\sigma = \sigma_2$. Also as σ decreases below σ_1, P_a approaches one, and as σ increases above σ_2, P_a approaches zero. Also as σ goes from σ_1 to σ_2 P_a smoothly decreases from $1 - \alpha$ down to β. These facts are enough for a sketch of the OC curve. Further information on the OC curve and the derivation of the method are available in Burr (1976).

11.6. DESCRIPTION OF ANSI/ASQ Z1.9

The ANSI/ASQ Z1.9 (1980) is an outgrowth of research beginning in 1939 (Ramer, 1939) and continuing through World War II when the Military Standard 105 was developed. As its title implies, this standard was for controlling the percentage beyond one or two specification limits, through the use of measurement statistics x, s, and R. Since concern is with the tail or tails of the distribution of the individual x's, the assumption of normality is quite crucial, and is probably not sufficiently emphasized in the Standard, ANSI/ASQ Z1.9 and other references. This assumption may also be given insufficient attention in practice.

Now let us proceed with a description of the Z1.9 plans.

1. All are single sampling plans, making decisions on the basis of n measurements, rather than upon counts of the number outside of specifications. All assume that the distribution of individual measurements, x, in the lot is normal.
2. Nevertheless, the plans are concerned with the percentage beyond a single specification limit, L or U, or outside of two limits, L and U. These are one-way and two-way plans, respectively.
3. Plans are available for two general cases: σ known—Section D in the standard using σ for variability; σ unknown—Section B using sample s or Section C using sample range.
4. Plans are to control the lot fraction nonconforming. They include normal, tightened, and reduced plans.
5. There are two forms, 1 and 2. (This is unfortunate because it makes the standard more bulky and complicated than necessary.) Decisions on a lot would be the same under either form. Since Form 2 is used whenever a two-way decision (both L and U given) is made, we recommend using Form 2 throughout, i.e., for both one- and two-way decisions.
6. OC curves for P_a vs. lot fraction nonconforming are provided for normal and tightened inspection plans. They are in the Standard's Table A.3, a page for each sample size code letter, listed for normal inspection; but also correct for tightened inspection by finding what normal inspection plan corresponds to the desired tightened plan.
7. To start to find a plan, specify an acceptable quality level, AQL, in percent nonconforming. Table A.1 converts this to one of the index AQL's, e.g., .65%.

Then for inspection levels I–V (smaller to larger sample sizes, n) and lot size, N, Table A.2 gives a sample size code letter, B, C, ... ,Q. Use level IV if none is specified. Thus, we have a code letter and an AQL in percent.

8. Now choose Section B, C, or D according to knowledge of σ and preference for R or s, as in our step 3.
9. Form 2—Using sample standard deviation, s: Section B in the Standard.

   a. One-way, e.g., an upper specification, U.

      i. For the AQL and code letter from 7, find in Table B.2 the sample size, n, (first column) and the maximum allowed estimated percent nonconforming, M, in the column for our AQL. (This M will be larger than the AQL, so that lots with $p_o = \text{AQL}$ will have a high P_a.) For a random sample of n measurements, x, find x-bar and s.
      ii. Form the "quality index"

      $$Q_U = (U - x\text{-bar})/s \qquad (11.13)$$

      (This is much like (7.12) $z = (x - \mu)/\sigma$ for obtaining the percentage beyond some limit, x, but instead of μ and σ, it uses x-bar and s.) From Q_U and n, enter Table B.5 to find the estimated percentage above U, say P_U.
      iii. Decision

      $$\begin{array}{ll} P_U \le M & \text{Accept lot} \\ P_U > M & \text{Reject lot} \end{array} \qquad (11.14)$$

      iv. If protection is relative to a lower specification L, use step (i) as it is, but in step (ii), use

      $$Q_L = (x\text{-bar} - L)/s \qquad (11.15)$$

      Then use Table B.5 to find p_L and make

the decision by

$$
\begin{aligned}
&P_L \leq M \quad \text{Accept lot} \\
&P_L > M \quad \text{Reject lot}
\end{aligned}
\tag{11.16}
$$

v. Tightened plans are found in Table B.3 by entering from the AQL's at the bottom, rather than those at the top as in normal plans. Reduced plans are found in Table B.4.

b. Two-way plans, L and U both given

   i. Same as step a.i.
   ii. Form Q_L by (11.15) and Q_U by (11.13) and use Table B.5 to find estimated percentages p_L and p_U.
   iii. Decision

$$
\begin{aligned}
&p_L + p_U \leq M \qquad \text{Accept lot} \\
&p_L + p_U > M \qquad \text{Reject lot}
\end{aligned}
\tag{11.17}
$$

   iv. Same as step a.v.

c. Switching rules are based in considerable part upon the estimated fraction defective for a series of lots, that is, upon $P\text{-bar}_U + P\text{-bar}_L$, usually for 10 lots. The rules are similar to those in ANSI/ASQ Z1/4.

10. Form 2—Using sample ranges R: Section C in the standard.

   a. If $n = 3$, 4, 5, or 7, use R. Or if $n = 10, 15, \ldots$, break up the sample into sub-samples of five each (first five measurements made, second five, and so on) finding range for each and use R.
   b. Proceed as in 9a and 9b, except using R or R-bar and Tables C.e–C.5 and

$$
\begin{aligned}
Q_U &= c(U - x\text{-bar})/R\text{-bar} \\
Q_L &= c(x\text{-bar} - L)/R\text{-bar}
\end{aligned}
\tag{11.18}
$$

   Which provide estimated percentages p_U and p_L from Table C.5. Tables C.3 and C.4 give the c

values for each n; they are like d_2 values for control charts.

11. Form 2—Using known population σ: Section D in the standard.

    a. Procedure very similar to step 9, but use Tables D.3–D.5, the first two providing n, M and υ-quantity. Then use

    $$Q_U = \upsilon(U - x\text{-bar})/R\text{-bar}$$
    $$Q_L = \upsilon(x\text{-bar} - L)/R\text{-bar} \qquad (11.19)$$

    And find estimates p_U and p_L from Table D.5

12. If there is real doubt as to the normality of the distribution of the x's, then one possibility is to accept only by Z1.9. But if the variables approach using Z1.9 would call for rejection, then do not reject yet, but continue onward, counting the number of pieces outside of specifications, until the sample size n for an appropriate attribute plan is completed. Such an attribute plan might be from the Z1.4 Standard. This approach is called "variables-attributes sampling". In particular, it protects a producer who may have a process running outside of L to U, but who has sorted his product carefully to these limits.
13. If the shape of the distribution of x's is not normal, but is known and is unsymmetrical, then one might use Zimmer and Burr (1963). If the longer tail in a skewed curve is toward U, one can use a smaller n than for the normal, but if the short tail is toward U a larger n is needed.

11.7. CHECKING A PROCESS SETTING

Our two suggested plans for checking the level μ of a process follow plans given in Burr (1976). The typical way to keep a production process "on the beam" is to use a measurement control chart. However, it may occur that from past experience it is found that the process maintains reasonably good control and that σ_x is known. It may then be desirable to merely check the process level from time to time, and in particular to check the initial setting of the process. There are two cases to consider. There is first the case of two specification limits L and U for which the tolerance $T = U - L$ is only six or seven σ_x's i.e., $C_p = 1.0$–1.2. In this case, then, we must maintain μ quite close to the nominal (middle of the specification range $(U - L)/2$) in order to avoid pieces out of specification. Secondly there are cases where the tolerance, T, is eight or more times σ_x, i.e., $C_p \geq 1.3$. We may then use a one-way check at each specification. Or, of course, there may be just one limit, a minimum, L, or a maximum, U, but not both.

11.7.1. Two-way Check of Level

We set the safe or desired level at the nominal $(L + U)/2 = \mu_o$, and then two unsafe or undesirable levels equally spaced on opposite sides of μ_o as follows:

Safe process level, μ_o:

$$P(\text{approval if } \mu = \mu_o) = 0.942 \cong 0.95$$

Unsafe process levels $\mu_o \pm 1\sigma_x$:

$$P(\text{reject if } \mu = \mu_o \pm \sigma_x) = 0.897 \cong 0.90$$

Plan to give such control of risks of wrong decisions is:

take $n = 10$ measurements and find x-bar

Then x-bar between $\mu_o - 0.6\sigma_x$ and $\mu_o + 0.6\sigma_x$, approve setting

x-bar outside, reject and reset the process

11.7.2 One-way Check of Level

For a check as to whether μ is too close to an upper specification, U, we regard $\mu = U - 3\sigma_x$ as a safe level, for then (from Table A) there will only be 0.0013 of the pieces above U. Likewise, we regard $\mu = U - 2\sigma_x$ as an unsafe level for then (by Table A) there will be 0.0228 of the pieces above U. A practical one-way check with approximate 0.10 risks is the following.

Take $n = 7$ measurements and find x-bar. Then

x-bar $\leq U - 2.5\sigma_x$ Approve setting
x-bar $> U - 2.5\sigma_x$ Reject setting

$$P(\text{accept if } \mu = U - 3\,\sigma_x) = 0.907$$
$$P(\text{accept if } \mu = U - 2.5\,\sigma_x) = 0.500$$
$$P(\text{reject if } \mu = U - 2\,\sigma_x) = 0.907$$

If μ is to be checked near the lower specification, L, use

Take $n = 7$ measurements and find x-bar. Then

x-bar $\geq U + 2.5\sigma_x$ Approve setting
x-bar $< U + 2.5\sigma_x$ Reject setting

The risks on this plan are symmetrical to those on the check at U.

11.8. SUMMARY

The sampling plans in this chapter have been using measurements to determine whether the lot or process distribution is acceptable. The basic assumption has been that the distribution was normal. Thus, we were specifically concerned, in the decision, with μ and/or σ. In some cases, we assumed that σ was known from previous records, whereas in other cases we assumed that σ was unknown. Tests can be made directly on μ and/or σ as in Section 11.3 and 11.5. Or the test can ascertain whether the combination of μ and σ is such as to give an acceptably small percentage of pieces outside of specifications, as in Sections 11.3.6 and 11.4.

It is recommended that control chart records be maintained in a series of lots, even though the decision on each individual lot is made using the present chapter. This will determine the control or lack of it, and in particular may permit accurate estimation of σ, so that known σ plans may be used instead of unknown σ plans, thus saving on the sample size and calculations.

ANSI/ASQ Z1.9 is an effective system of integrated plans, providing normal, tightened, and reduced inspection for protection against one or two specification limits. It is for control of the fraction or percent nonconforming. The plans assume normality, but one hedge, if this is not a safe assumption, is to use variables-attributes, permitting quick acceptance via measurements, but only rejecting via attributes.

Simple checks on process average level were also given.

11.9. PROBLEMS

11.1. The time of blow for fuses under a marginal circuit condition is subject to a maximum specification limit of 150 sec $= U$. Experience has shown that $\sigma_x = 25$ sec. Regarding $\mu_1 = 100$ sec as the AQL and $\mu_2 = 125$ sec as the RQL, and setting respective risks of $\alpha = 0.10$ and $\beta = 0.10$, find an appropriate sampling plan. Sketch the OC curve for P_a vs. μ, labeling axes.

11.2. The "Scott value" for material for a battery is subject to a maximum specification of 26 and σ_x is known from experience to be 0.56. Assume normality. Set up a sampling plan so that if $\mu = \mu_1 = 24.5$, the probability of rejection is 0.05, whereas if $\mu = \mu_1 = 25.0$, the probability of acceptance is 0.05. Sketch the OC curve P_a vs. μ, labeling the axes.

11.3. For the weight of contents of a package of a food product, the minimum specification for weights is 500 g. From past experience σ_x, is known to be 0.6 g. Taking $\mu = \mu_1 = 501$ g as the RQL, and $\mu = \mu_2 = 502$ g as the AQL, and using respective risks α and β both at 0.05, find a sampling plan. Assume normal distribution of the x's. Sketch the OC curve, P_a vs. μ, labeling axes.

11.4. For the length of a spring under a compressive force of 50 kg, there is set minimum limit of 10.0 cm. Assume that $\sigma_x = 0.2$ cm, and that the distribution of the x's is normal.

Set $\mu = \mu_1 = 10.4$ cm as the RQL and $\mu = \mu_2 = 10.6$ cm as the AQL, and take risks of $\beta = \alpha = 0.05$. Find a sampling plan for the test and sketch the OC curve P_a vs. μ, labeling axes.

11.5. For doubled thickness of rubber gaskets for metal tops for food jars, σ_x was known to be 0.002 in., and the distribution of x's to be normal. Consider $\mu = 0.106$ in. = AQL and $\mu = 0.103$ in. or 0.109 in. as RQL's. Set $\alpha = b = 0.01$. (Actually, if μ is at either RQL, the rubber is used in another application.) Set up an appropriate sampling plan. Sketch the OC curve, P_a vs. μ labeling axes.

11.6. A small part has diameter specifications of 0.1100 and 0.1102 in., each lot being produced under a single setup. Experience has shown a normal distribution of diameters with σ_x, $= 0.00003$ in. Taking $\mu = \mu_o = 0.1101$ in. = AQL and $\mu = \mu_o \pm \mu_o = 0.00003$ in. as the RQL's and setting α and β risks at 0.05, find a sampling plan. Assume normality. Sketch the OC curve, P_a vs. μ, labeling axes.

11.7. Devise a test of variability for lots having $\sigma_1 = 1.715$, $\sigma_2 = 3.43$, $\alpha = \beta = 0.10$, assuming normality. Sketch the OC curve P_a vs. σ, labeling axes. Test once each on distributions A and C of Table 7.3, by experimentation.

11.8. Muzzle velocity of target ammunition is to be tested for variability. Take $\sigma_1 = 5$ ft/sec, and use risks $\alpha = \beta = 0.05$. Set up a sampling plan to make the test. Sketch the OC curve, P_a vs. σ, labeling axes.

11.9. A new gauge is being considered. If the standard deviation, σ_1, on homogeneous material is 0.00002 in., the gauge should be approved, whereas if at $\sigma_2 = 0.00003$ in., it should be rejected. Use risks of $\alpha = \beta = 0.10$ and find a sampling plan.

11.10. An analytical technique is supposed to have a standard deviation error of measurement (on homogeneous material) of 2 ppm. Set $\sigma_1 = 1.5$ ppm and $\sigma_2 = 2.5$ ppm and both risks at 0.05. Determine an appropriate test for the proposed technique.

11.11. For fuses as in Problem 11.1, considering σ_x as unknown, devise a sampling plan for measurements so that if $p_o = p_1 = 0.005$, $P_a = 0.95$ while if $p_o = p_2 = 0.02$, $P_a = 0.05$.

11.12. For the contents weights in Problem 11.3, considering σ_x as unknown, devise a sampling plan for measurements so that if $p_o = p_1 = 0.005$, $P_a = 0.95$ while if $p_o = p_2 = 0.05$, $P_a = 0.05$.

11.13. Verify the risks given in the one-way check of a process setting in Section 11.7.2.

11.14. Verify the risks given in the two-way check of a process setting in Section 11.7.1.

Assuming that ANSI/ASQ Z1.9 is available, find a sampling plan for the following requirements, assuming normal distribution of the x's.

11.15. Lot size, $N = 1000$, inspection level IV, one specification, U, σ unknown, normal inspection, using sample s, AQL = 0.65%.

11.16. Lot size, $N = 3000$, inspection level IV, two specifications, σ unknown, normal inspection using sample R's, AQL = 0.40%.

REFERENCES

ANSI/ASQ Z1.9-1980. *Sampling Procedures and Tables for Inspection by Variables for Percent Nonconforming*. ANSI & ASQ.

IW Burr. Statistical Quality Control Methods. New York: Dekker, 1976.

HG Romig. Allowable Average in Sampling Inspection. PhD thesis, Columbia University, New York 1939.

Statistical Research Group, Columbia University. Techniques of Statistical Analysis. New York: McGraw-Hill, 1947.

WJ Zimmer, W Burr. Variable sampling plans based on non-normal populations. Indust. Quality Control. 20 (1), 18–26, 1963.

12

Tolerances for Mating Parts and Assemblies

Nearly all individual piece parts are manufactured to be assembled with other piece parts built to match them. For example, a shaft is made for an assembly within a bearing. It must be possible for the shaft to go into the bearing. That is, assuming perfectly round pieces, x_1, the inside diameter of the bearing, must be larger than x_2, the outside diameter of the shaft in order to permit assembly. On the other hand, the clearance in diameters, $x_1 - x_2$, should not be too large or the fit will be too loose. Therefore, we have a problem as to what tolerances to set for the bearing and the shaft to permit assembly in a very high percentage of cases, but to avoid fits which are too loose. The objective is to accomplish this with random assembly which means picking at random the bearing and shaft to be assembled. One other way sometimes used is to measure all bearing and all shafts and selectively assemble the large diameter bearings and shafts together and the small diameter shafts and bearing together. This has disadvantages: (1) it is expensive and time consuming, and (2) when we need to replace a bearing we have trouble.

In general terms, then, our problem is that given the distributions of component dimensions or characteristics, what is the distribution of the assembly characteristic? And how may we put statistical laws to work in helping us?

12.1. AN EXAMPLE OF BEARING AND SHAFT

Let us suppose that we have a process that will produce shafts about 2 cm in diameter with a standard deviation, $\sigma_x = 0.0003$ cm and another process for bearings with $\sigma_y = 0.0004$ cm. Moreover, these are achievable without sorting and with good control of averages. Then we might well set 3σ specification limits for *shafts* of $2.0009 \pm 0.0009 = 2.0000, 2.0018$ cm. Next, so as not to overlap causing interference, we might set limits for the *bearings* of 2.0020, $2.0044 = 2.0032 \pm 0.0012$ cm. Thus our two nominal diameters are 2.0032 and 2.0009 cm.

Since these σ's are achieved without sorting and the averages are in control, we will also assume the distributions to be normal. Let us set

y = inside diameter of a bearing,
x = outside diameter of a shaft, and
$w = y - x$ = diametral clearance of a random pair.

Then we have

$\sigma_y = 0.0004$ cm = standard deviation of bearing diameters, and
$\sigma_x = 0.0003$ cm = standard deviation of shaft diameters.

and we also set the processes to run at

$\mu_y = 2.0032$ cm = mean diameter of bearings, and
$\mu_x = 2.0009$ cm = mean diameter of shafts.

Now then we can be assured that the clearance, w, will virtually always lie between the following two extremes:

$$\min_w = \min_y - \max_x = 2.0020 - 2.0018 = 0.0002\,\text{cm}$$

$$\max_w = \max_y - \min_x = 2.0044 - 2.0000 = 0.0044\,\text{cm}$$

This is a common way of thinking among engineers and has been for a very long time. There is nothing really wrong about it, but there is a way to do better by using the laws of statistics and taking advantage of compensating errors. For one thing, what is the chance of a clearance of only 0.0002 cm? This could occur with a 1-in-1000 minimum bearing and 1-in-1000 maximum shaft (for 3σ extremes). This gives a 1-in-1,000,000 chance for such a small clearance. (This is a rough analogy of what is operating, not an accurate approach.)

Now given:

$$W = y - x \tag{12.1}$$

and μ_y, μ_x, σ_y, σ_x and that x and y are independent (as in random assembly), it can then be proved that

$$\mu_w = \mu_y - \mu_x \tag{12.2}$$

$$\sigma_w = \sqrt{\sigma_y^2 + \sigma_x^2} \tag{12.3}$$

Take a good look at (12.3). σ_w is not the sum $\sigma_y + \sigma_x$ as was assumed implicitly in the foregoing way of setting tolerances for y and x. σ_w, is, in fact, much less than $\sigma_y + \sigma_x$ since the square of $\sigma_y + \sigma_x$, i.e., $(\sigma_y + \sigma_x)^2 = \sigma_y^2 + 2\sigma_x\sigma_y + \sigma_x^2$, not $\sigma_y^2 + \sigma_x^2$. Or to be numerical, σ_w is not $\sigma_y + \sigma_x$ which is $0.0004 + 0.0003 = 0.0007$ cm. Instead

$$\sigma_w = \sqrt{0.0004^2 + 0.0003^2} = 0.0005\,\text{cm}$$

Therefore, using (12.2) and (12.3), we have for diametral clearance w in (12.1) the following for our two processes:

$$\mu_w = \mu_y - \mu_x = 2.0032 - 2.0009 = 0.0023\,\text{cm}$$
$$\sigma_w = \sqrt{\sigma_y^2 + \sigma_x^2} = \sqrt{0.0004^2 + 0.0003^2} = 0.0005\,\text{cm}$$

Therefore, with the production processes set as assumed, we are actually meeting limits for w in a $\pm 3\sigma$ sense of

$$\mu_w \pm 3\sigma_w = 0.0023 \pm 3(0.0005) = 0.0023 \pm 0.0015$$
$$= 0.0008, 0.0038$$

These limits are considerably closer than 0.0002, 0.0044 cm as found from the purely additive basis on σ's, or tolerances.

Just how rare would a diametral clearance w of 0.0002 cm or less be. Using our μ_w, σ_w in (7.12)

$$z = \frac{0.0002 - \mu_w}{\sigma_w} = \frac{0.0002 - 0.0023}{0.0005} = -4.2$$

Using a larger table than our Table A, we find a probability for z of -4.2 or less to be 0.000013.

Therefore if we wish to have an actual minimum diametral clearance, w, of 0.0002 cm, while retaining x limits at 2.0009 ± 0.0009 cm, we can make use of $\sigma_w = 0.0005$ cm as follows:

$$\mu_w = 0.0002 + 3(0.0005) = 0.0017$$

But $\mu_w = \mu_y - \mu_x$, and so substituting what we know, $0.0017 = \mu_y - 2.0009$ giving $\mu_y = 2.0026$. Therefore, we can set limits for the bearings at

$$\mu_y \pm 3\sigma_y = 2.0026 \pm 3(0.0004) = 2.0026 \pm 0.0012$$
$$= 2.0014, 2.0038$$

This will give 3σ limits for clearance of $0.0017 \pm 0.0015 = 0.0002$, 0.0032, thus avoiding the looser fits up to 0.0044 cm, originally in mind.

See Figure 12.1, which pictures the two approaches: traditional and statistical. The distribution Ic shown for clearances is only approximately correct for the two uniform or rectangular distributions for shaft and bearing, which fill the tolerance interval. If the processes give $\pm 3\sigma$ normal curve in meeting of tolerances as shown, then the clearances behave as pictured in IIc and IIIc. Distribution IIc is safely away from zero. But we can use overlapping tolerances for shaft and bearing as shown, provided we meet them with $\pm 3\sigma$ normal curve distributions, and will then have IIIc, which is still safely above zero, but has a more favorable maximum clearance of 0.0032 cm.

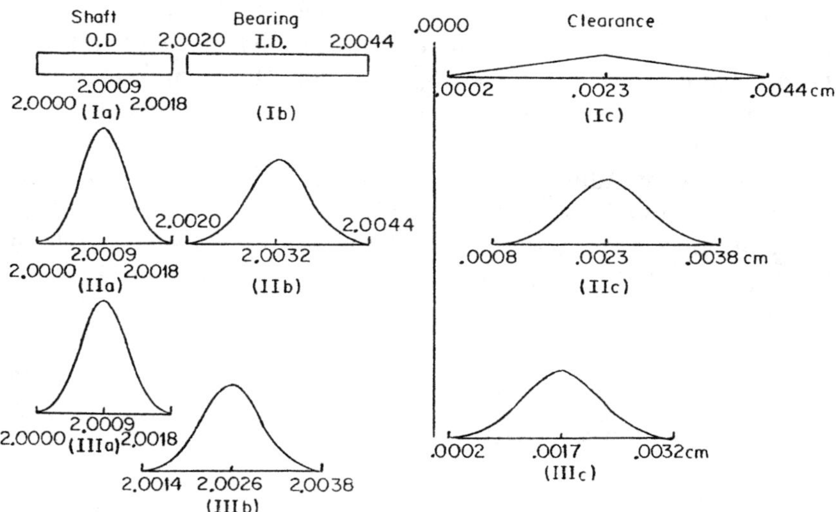

Figure 12.1 Three tolerance-meeting sets of distributions for a shaft and bearing. Section I is for the traditional purely additive tolerances. The two component distributions Ia and Ib are uniform between the limits and give approximately the clearance distribution shown. II and III are for $\pm 3\sigma$ normal-curve meeting of tolerances. IIa and IIb give IIc for clearances. The overlapping tolerance distributions of III still give IIIc for clearances, with extremely small chance for a bearing smaller than a shaft in random assembly.

It is worth mentioning with regard to this example that the assumption we used of independence of bearing and shaft may not fully apply if there is drift due to tool wear. Thus shaft outside diameters may tend to gradually increase, and bearing inside diameters may tend to gradually decrease. And so, initially the clearance may be relatively large and later on relatively small. We have assumed that such tool-wear drifting has been held under control by resetting as needed.

12.2. AN EXAMPLE OF AN ADDITIVE COMBINATION

Assemblies where component characteristics are added are very common in industry, perhaps more so than subtractive combinations such as discussed in Sec. 12.1 on clearances. Examples are assemblies on a shaft, resistances in a series circuit, thickness of two pieces bolted together, thickness after two coatings, and contents or impurity in a mixture of two or more liquids.

Let us consider a hypothetical example on resistances in a circuit. Suppose we connect three resistors in a series. Then the component resistances may be called w, x, and y, and the resistances of the wire and connections, z. Then for the total resistance, we have

$$R = w + x + y + z \tag{12.4}$$

Let us illustrate with

$$\mu_w = 800, \quad \mu_x = 300, \quad \mu_y = 195, \quad \mu_z = 4$$
$$\sigma_w = 6, \quad \sigma_x = 3, \quad \sigma_y = 3, \quad \sigma_z = 0.5 \text{ ohms}$$

We have for (12.4)

$$\mu_r = \mu_w + \mu_x + \mu_y + \mu_z \tag{12.5}$$

which in the example gives

$$\mu_r = 1299 \text{ ohms}$$

Next, if these resistances are independent, which seems to be quite a safe assumption here, then we also have

$$\sigma_r = \sqrt{\sigma_w^2 + \sigma_x^2 + \sigma_y^2 + \sigma_z^2} \tag{12.6}$$

This gives here

$$\sigma_r = \sqrt{6^2 + 3^2 + 3^2 + 0.5^2} = 7.37$$

That is, just a little more than σ_w alone. Therefore, as long as the averages are maintained at the levels shown, we can expect that nearly all circuit resistances will lie between

$$\mu_r \pm 3\sigma_r = 1299 \pm 3(7.37) = 1299 \pm 22 = 1277, 1321$$

Now let us suppose that instead of working with the normal distributions with averages and standard deviations, which we considered known, we only used specification limits at $\mu \pm 3\sigma$, that is

$$800 \pm 3(6) = 782, 818$$

$$300 \pm 3(3) = 291, 309$$

$$195 \pm 3(3) = 186, 204$$

$$4 \pm 3(0.5) = 2.5, 5.5$$

Many would now add the four low specification limits and the four high ones. This gives 1261.5, 1336.5. Now are these reasonable limits? Of course, if the four component resistances all lie inside their respective limits, then the total resistance for the circuit will indeed lie within 1261.5–1336.5. But even when we have uniform distributions of resistances for the components, between limits as in Figure 12.1 (Ia) and (Ib), the distribution of circuit resistances will not be uniform between 1261.5 and 1336.5. Instead, there will be considerable tapering out to these limits, making extreme values rarer. But if these limits are met with normal curve distributions, then as we have seen, practically all total resistances will lie between 1277 and 1321. The width of this band is 44 as against 75 for the purely additive approach. Thus if we can be assured of a normal curve ($\pm 3\sigma$) meeting of specifications, then we could set specification limits of, say, 1275 and 1325 rather than 1261 and 1337.

It may also be stated that even when the distributions of component characteristics are not normal, the distribution of the combined or assembly characteristic tends toward a normal distribution with μ and σ like (12.5) and (12.6). (This fact is from the Central Limit Theorem.)

Thus we have another example in which the tolerance limits for the characteristic of the assembly are narrower than might have been expected from the limits for the components as a result of compensating errors or variations.

12.3. GENERAL FORMULAS

Let us set down the general formulas for the kinds of relationships, we have been studying. Let us define

$$y = x_1 \pm x_2 \pm \cdots \pm x_k \tag{12.7}$$

which we call an additive–subtractive relation. The signs could be positive or negative throughout or in any combination. There are k characteristics of component parts affecting the assembly characteristic, y. Then we have

$$\mu_y = \mu_1 \pm \mu_2 \pm \cdots \pm \mu_k \tag{12.8}$$

For the next formula, we need to assume that the various component characteristics, x_1, $x_2, \ldots, x_k$ act independently, i.e., that whatever value one x takes has no influence upon or relation to the other x's in the assembly or combination. If the production processes are in control or if assembly is done randomly, this is likely to be a safe assumption. Then

$$\sigma_y = \sqrt{\sigma_1^2 + \sigma_2^2 + \cdots + \sigma_k^2} \tag{12.9}$$

Note that although there are plus and minus signs in (12.7) and (12.8), all signs under the radical in (12.9) are plus. Also we mention that σ_y is much less than the simple sum $\sigma_1 + \sigma_2 + \cdots + \sigma_k$ (as assumed in purely additive tolerances).

Finally, we may say that if the component characteristics, $x_1, x_2, \ldots, x_k$ have normal distributions, then y will also. Moreover, even if the x's are not normally distributed, y still tends to be, especially as k increases. (This could be made more exact by quoting the Central Limit Theorem in one of its several forms.)

The author consulted on one job where 10 different dimensions had a formula for clearance like (12.7). A wider tolerance was being requested on one dimension, so as to eliminate one refined threading operation. Process capabilities were well known. Analysis showed that, given the wider tolerance, there would still be only one chance in 10,000 of inability to assemble the parts. The saving on this one order was $300,000.

In another case, analysis showed that there were 38 dimensions on the shaft of an auto generator which could affect end-play via a relation like (12.7). Thus there was at least some possibility, through using statistical tolerancing, of having limits $\sqrt{38} = 6.2$ times as wide as in additive tolerancing, provided averages are well controlled. Many tolerances were accordingly widened.

12.4. SETTING REALISTIC TOLERANCES

Basic to the setting of realistic tolerances is knowledge of the capabilities of the processes which will produce the components. To find the process capability, we need to have achieved reasonable process control, for otherwise we simply do not know what the process can do. Then we can use (6.10) or (6.25), that is, R-bar$/d_2$ or s-bar$/c_4$ to estimate

σ for each component. Now such a σ is for a short time interval. It is reasonable to add a bit to this short-term σ, in order to estimate the long- term σ, which will contain some setup error and some drifting of the average. We might use

$$\sigma_{(\text{long term})} = (4/3)\sigma_{(\text{short term})} \tag{12.10}$$

It is the $\sigma_{(\text{long term})}$'s which we should substitute into (12.9) to find σ_y. Now setting μ_y at the desired nominal for the assembly, we can set the component nominals μ_i's in (12.8) to yield the desired σ_y. Then we use the $\sigma_{(\text{long term})}$'s of the components to substitute into (12.9) to find σ_y. Then the processes will nearly always (99.7% or so) meet specification limits on the assembly of

$$\mu_y \pm 3\sigma_y \quad \text{(limits for assembly)} \tag{12.11}$$

If these limits are satisfactory, then we have realistic specification limits and need only exercise the controls over components that we have been using. But if the limits in (12.11) are not narrow enough, we will need to make at least one of the components more narrow. The place to start looking is the component with the largest $\sigma_{(\text{long term})}$. There are two ways to proceed: (1) Exercise tight control over the process average and thus bring down $\sigma_{(\text{long term})}$ to close to $\sigma_{(\text{short term})}$ and, therefore, cut (12.9) narrowing (12.11). (2) To arbitrarily set specification limits on a component narrower than $\mu \pm 3\sigma$ $_{(\text{short term})}$ which is to be achieved through 100% sorting and possibly rework. Which component or components to restrict thus is an engineering and economic decision. But since such sorting does not leave a normal distribution, we can conservatively use $\sigma = (U - L)/4$, instead of the normal curve where $\sigma = (U - L)/6$ for $\sigma_{(\text{short term})}$. In this latter case, we will have the expensive inspection job to do with possible reworking as well.

Of course, there is also a possibility: (3) To decide to widen the specification limits to be at those in (12.11) which the processes can economically supply to the assembly. In many cases where specifications are involved discussion with the customer is required.

These general considerations can also be used in telling whether two particular lots of product will satisfactorily assemble. Material review boards can use this technique for analysis and decision.

12.5 RELATIONS OTHER THAN ADDITIVE-SUBTRACTIVE

The majority of functional relationships between assembly and components would seem to take the form of (12.7). But there are a great many other functions that might possibly occur, such as

$$y = x_1 \cdot x_2 \quad \text{or} \quad y = x_1/x_2.$$

Such can be handled. See Burr (1976). μ_y and σ_y are found by algebra and calculus.

12.6. SUMMARY

This chapter has provided an introduction to the way in which dimensions and other component characteristics combine into the characteristic of an assembly. Specifically discussed was the case where the relation is additive–subtractive (12.7) and in which there is independence or random assembly. Then the tolerances which we can set for the components are usually considerably greater than we can with purely additive tolerancing, sometimes called “worst case” tolerancing. But to reap the benefits, we must exercise reasonable control over the various process averages. The distribution of the assembly characteristic tends to be normal. An approach to practical tolerancing was given in Sec. 12.4 where several alternatives were suggested. These can often eliminate much inspection, rework, and scrapping, if implemented. The methods can also be used for decisions on lots of mating parts for objective decisions by material review boards.

If the function for the assembly characteristic is not additive–subtractive, there are other methods which can be used. See Burr (1976) .

12.7. PROBLEMS

12.1. To illustrate the approach to normality as the number of components increases, perform this experiment on dice totals. For a single die, the faces may show 1, 2, 3, 4, 5, 6 with equal probability, that is, 1/6 each. This is a flat-topped uniform distribution. Now throw three dice and count the total of the faces showing. The total can be anywhere from 3 to 18 inclusive, but the extremes are much less likely than, say, 10 or 11. Throw the three dice 108 times, tabulating directly into a frequency table. Compare with the theoretical frequency distribution given in the answers.

12.2. Consider shafts with specification limits 2.0002 and 2.0019 in. The matching bearings also carry specification limits of 2.0009 and 2.0033 in. Note the sizable overlapping of the two sets of limits. Suppose that these limits are met in a $\pm 3\sigma$ sense with normal distributions. Make a guess as to the proportion of random choices of bearing and shaft, which will have negative clearance, that is, will not go together. To illustrate, throw three dice for the number of 0.0001 in. above 2.0000 in. for a shaft, for example, 12 total is 2.0012 in. Likewise, throw six dice (or three, twice) for the inside diameter of a bearing. The difference then is the diametral clearance. You can tabulate the differences, for example, $20 - 12 = 8$. Obtain 50 clearances by dice throws. For three-dice totals, $\mu = 10.5$, $\sigma = 2.96$, and for six-dice totals $\mu = 21$, $\sigma = 4.18$. Use (12.7)–(12.9) to find μ and σ for the clearance, then use (2.11) and Table A to approximate the percentage of clearances below zero. Is this less than you expected?

12.3. A pin shows good control on OD = outside diameter with y-double bar $= 0.21440$, R-bar $= 0.00032$ in. for samples of $n = 5$. A mating collar also shows good control on ID, with x-double bar $= 0.21503$, R-bar $= 0.00040$ in. and $n = 5$. Estimate the process standard deviations for each. Let the diametral clearance be $w = x - y$, and estimate μ_w and σ_w. Assuming normal distributions, what percentage of pairs of pin and collar will have clearance less than 0.0001 in.?

12.4. In manufacture of lamps, the clearance $w = x - y$ is of interest, where $x =$ distance from rim of base to the glass, and $y =$ distance from rim of base to top of inner shell. For x's, large samples gave estimated $\mu = 1.0499$, $\sigma = 0.0428$ in., whereas for y's $\mu = 0.9079$, $\sigma = 0.0120$ in. Good normality was present. Describe the distribution of $w = x - y$. What percentage of w's are negative in random assembly?

12.5. A brass washer and a mica washer are to be assembled one on top of the other. For the former $\mu = 0.1155$, $\sigma = 0.00045$ in., whereas for the latter $\mu = 0.0832$, $\sigma = 0.00180$ in. Find μ and σ for the combined thickness. What assumptions were needed? Assuming normality, set $\pm 3u$ limits for the combined thickness.

12.6. A general concept in interpreting measurements of material is the following: We let $w =$ the observed measurement, $x =$ the "true" measurement, and $y =$ the error of measurement (which may be + or −). Then $w = x + y$. Since x and y are independent we may use (12.9) with $k = 2$. Also if the measurement is "unbiased" then $\mu_y = 0$. Suppose $\mu_x = 115.0$, $\sigma_x = 2.0$, $\sigma_y = 0.5$ mm. Assuming that x and y are normally distributed, describe the distribution of the observed measurement w. How influential was σ_y? (Take note that although $x = w - y$, we could not use $\sigma_x = [\sigma_w{}^2 + \sigma_y{}^2]^{1/2}$ because w and y are not independent. In fact, this equation would give more variation in true dimension than in the measured dimension!)

12.7. At one time 200 ohm resistors, made to = 0.75%, that is, to 200 ± 1.5 ohms were quite costly, due in part to the necessity of sorting out those outside the limits. But production was able to manufacture 100 ohm resistors to ±1% without sorting and with good normality of distribution. Take $\mu = 100$, $\sigma = 0.33$ ohm. Now assembling at random two 100 ohm resistors in series gives what distribution for total resistance? Will such a distribution meet limits of 200 ± 1.5 ohms? Take note that we should check on the independence of resistances in 100 ohm resistors, because wire diameter might cause nonindependence. (A company made quite large savings using this approach.)

12.8. Five pieces are assembled on a shaft, two of the first kind with $\mu = 2.0140$, $\sigma = 0.0014$, two of a second kind with $\mu = 3.2061$, $\sigma = 0.0012$, and one of a third kind with

$\mu = 3.8402$, $\sigma = 0.0022$ (all in inches). For total length w, use $w = x_1 + x_2 + \cdots + x_5$, since random assembly is to be used. What are μ and σ for length w, and what $\pm 3\sigma$ limits can it meet? Are these limits narrower than we would find using the five maximum limits $2.0140 + 3(0.0014)$, etc., and the five minimum limits?

12.9. In packaged weight control, it is not very convenient to obtain directly the weight of the net contents. Instead, we may proceed indirectly by weighing empty containers x, covers y, and total filled packages z. Letting the net weight of contents be w, we thus have $w = z - x - y$. Now we cannot find σ_w by use of (12.9), because z, x and y are not independent. Instead, we can let $z = x + y + w$, in which x, y, and w are likely to be quite independent. Then we use (12.9) for the relation between σ^2's. Knowing three of these, we can find σ_w^2 and σ_w. However, we can use (12.8) on either form of (12.7) to find μ_w. Now, given the actual observed data z: $\mu = 189.5$, $\sigma = 3.5$; for y, $\mu = 3.1$, $\sigma = 1.1$ (in pounds) find μ and σ for w. Assuming normality estimate the percentage of contents weights below the minimum specification of 165 lb.

12.10. The following is an example of aid to a materials review board. In a plant manufacturing soap and cosmetics, a question arose as to whether to use a lot of caps for cologne bottles. Accordingly tests were made on the distribution of cap strength (the torque which would break the cap) and the distribution of the torque which the capping machine would apply to a cap. They found in inch-pounds, respectively,

$$x\text{-bar} = 11.1, \quad s_x = 2.80, \quad y\text{-bar} = 8.0, \quad s_y = 1.51$$

Assuming independence (very safe here) and normality, estimate the average and standard deviation of the margin of safety, $w = x - y$. Estimate the percentage of caps that can be expected to break by a negative w value.

REFERENCE

IW Burr. Statistical Quality Control Methods. New York: Dekker, 1976.

13

Studying Relationships Between Variables by Linear Correlation and Regression

We now take up a useful method of analyzing the relationship between two variables. We make simultaneous observations on the two variables x and y, say, x_1 and y_1, then later on another pair x_2 and y_2, and so on to x_n, and y_n. Each such pair is at the same time or place, or on the same material. Then we seek to study the relationship between the two variables. For example, can we estimate or predict y from x? Are they closely related, loosely related, or unrelated? One very simple way to gain some insight into the relation is to make a "scatter diagram" that is, to plot each pair of x and y on a graph. Thus with a horizontal x-axis and a vertical y-axis, we first plot y_1 against x_1 then y_2 against x_2, and so on. Then a relationship may or may not emerge. Sometimes such a scatter diagram is all that we need in a study.

13.1. TWO GENERAL PROBLEMS

In the first problem, we are especially interested in the estimation or prediction of y from x. For example, we may be using a standard analytical technique and wish to see whether we can accurately estimate its result from that of an alternative less expensive or less time-consuming analysis. Or, we may compare two gauges, or wish to see whether two physical properties such as hardness and tensile strength are closely enough related to predict the latter from the former. Such problems frequently arise in industry.

Another general problem is to study a collection of "input" variables to see which is most closely related to an "output" or quality variable. That variable or those variables most closely related are the ones to work on in trying to improve the process and obtain better quality. The degree or strength of relationships is thus the key to the study.

These two problems respectively emphasize "regression" and "correlation", although there is not really any hard and fast distinction between these two concepts.

In this book, we shall only consider quite basic techniques. For those who wish to go more deeply into the subject, Burr (1974) and Draper and smith (1966) are recommended.

13.2. FIRST EXAMPLE—ESTIMATION

In the manufacture of piston rings, C-shaped castings are precisely ground and machined to specifications. A processing of the completed rings called "ferroxing" was being used. It used high-temperature steam to form a blue oxide coating to protect the rings against rust. The measurement in question was the force necessary to close the ring to the specified gap and is called "tension". is tension must not be too great, or it will cause excessive wear, nor too little for then the piston may leak pressure and the cylinder lose power. During ferroxing, strains are relaxed and the tension is increased. Three different forms of suspension during the ferroxing were being studied. The data we shall use are from a T-bar suspension with the gaps up. We give here 20 rings out of the 300 used in the study. The tension of each ring was measured before and after ferroxing. The correlating feature is that x_1 was the tension before for the first ring and y_1 the tension after for the same ring, and so on. Then the question is as to how we may predict y from x, and how accurate is the prediction, i.e., to how much error is it subject?

We shall first show a method of calculation in Table 13.1, which may seem cumbersome, but does help the reader to see what is happening. Then by coding and other methods analogous to (2.6) and (2.5) the calculations can be simplified. Now although practically every plant with an electronic computer will have a program available for linear regression, the reader

Table 13.1 Pounds "Tension" of 20 Piston Rings, Before and After "Ferroxing", Respectively, x_i and y_i for the ith Ring

x	y	$x - x$-bar	$y - y$-bar	$(x - x\text{-bar})^2$	$(y - y\text{-bar})^2$	$(x - x\text{-bar})(y - y\text{-bar})$
5.2	5.7	+0.185	+0.135	0.034225	0.018225	0.024915
5.8	6.4	+0.785	+0.835	0.616225	0.697225	0.655475
4.8	5.4	−0.215	−0.165	0.046225	0.027225	0.035475
6.4	7.0	+1.385	+1.435	1.918225	2.059225	1.987415
5.2	5.8	+0.185	+0.235	0.034225	0.055225	0.043415
4.6	5.1	−0.415	−0.465	0.172225	0.216225	0.192975
5.7	6.3	+0.685	+0.735	0.469225	0.540225	0.503475
5.9	6.6	+0.885	+1.035	0.783225	1.071225	0.915975
4.6	5.2	−0.415	−0.365	0.172225	0.133225	0.151475
4.9	5.3	−0.115	−0.265	0.013225	0.070225	0.030475
4.2	4.9	−0.815	−0.665	0.664225	0.442225	0.541975
4.2	4.8	−0.815	−0.765	0.664225	0.585225	0.623475
4.3	4.8	−0.715	−0.765	0.511225	0.585225	0.546975
5.5	6.0	+0.485	+0.435	0.235225	0.189225	0.210975
4.8	5.3	−0.215	−0.265	0.046225	0.070225	0.056975
5.6	6.2	+0.585	+0.635	0.342225	0.403225	0.371475
4.7	5.1	−0.315	−0.465	0.099225	0.216225	0.146475
4.7	5.1	−0.315	−0.465	0.099225	0.216225	0.146475
4.7	5.2	−0.315	−0.365	0.099225	0.133225	0.114975
4.5	5.1	−0.515	−0.465	0.265225	0.216225	0.239475
100.3	111.3	0.000	0.000	7.285500	7.945500	7.540500

is strongly advised to carry through some problems by the methods to be shown. In the first, two columns of Table 13.1 are listed the before and after ferroxing tension (in pounds), respectively, for x and y.

Figure 13.1 shows the y's plotted against the respective x's. The relationship seems to be quite close, i.e., given a before ferroxing x, we can predict y with considerable confidence. For example, if $x = 5.4$, we would estimate by eye that y would be close to 5.9. Moreover, it looks as though a straight line might be made to fit the points quite well, i.e., the relationship is "linear".

Definition 13.1 (Linear relation). If a plot of y vs. x seems to follow a straight line, even if quite loosely, then we call the relation linear.

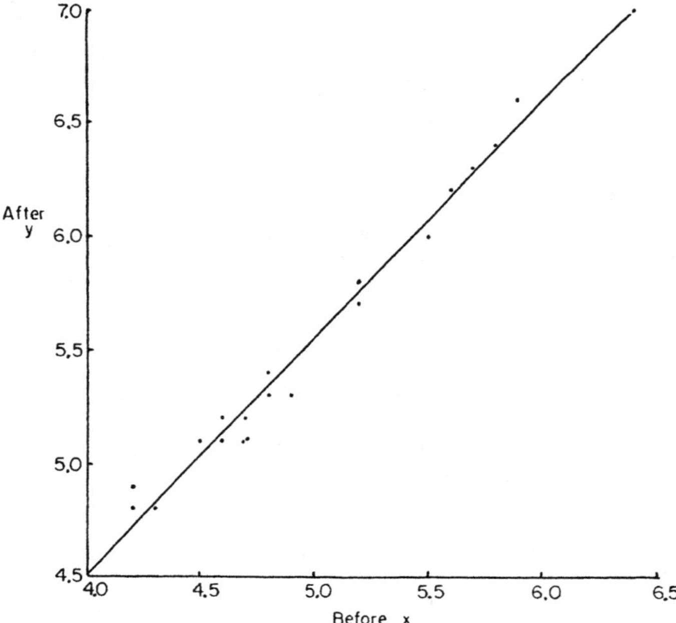

Figure 13.1 Observations of pounds tension for 20 piston rings of x before vs. y after ferroxing. The best-fitting line, y on x, is also shown.

Definition 13.2 (Nonlinear relation). If a plot of y vs. x does not seem to follow a straight line but instead some curve, then we call the relation nonlinear.

Since Figure 13.1 shows such a linear trend, our object is to fit a straight line to the data given in Table 13.1. In fact, we wish to find the equation of the best fitting straight line. For estimating y from x, we choose to use that line for which the sum of the squares of the distances from the line to the points, measured vertically, is as small as it can be for the given data. This is called the "least squares fitted line". It can be proved that the following method gives this line.

The reader may well recall that the general equation of a straight line takes the form

$$y = b_o + b_1 x \tag{13.1}$$

where b_o is the "intercept", i.e., the value of y when x is 0, and where b_1 is the "slope", i.e., the amount of change in y when x increases by one unit. Then our problem becomes that of finding what values to use for b_o and b_1 so as to fit the given data as well as possible (in the least-squares sense)

$$x - \text{bar} = 100.3/20 = 5.015\text{lb}, \quad y - \text{bar} = 111.3/20 = 5.565\text{lb}$$

Next we find the respective deviations ($x - x$-bar) and ($y - y$-bar) in the next two columns. The totals of these deviations are 0 (except possibly for round-off errors). This provides a good check on our deviations. Next we find the squares of the deviations or at least their sums $\sum(x - x\text{-bar})^2$, $\sum(y - y\text{-bar})^2$. Such sums of squares are readily accumulated on many desk calculators, without writing each individual squared deviation. Finally, we obtain in the last column the products of the x deviation ($x - x$-bar) by the y deviation ($y - y$-bar). For example, $(+0.185)\ (+0.135) = 0.024975$. Signs must be carefully watched, but in our problem both deviations carried the same sign in all pairs, so that the products were all positive.

Now for our "best" slope, we have

$$b_1 = \frac{\sum (x - \bar{x})(y - \bar{y})}{\sum (x - \bar{x})^2} \tag{13.2}$$

For our example, we, therefore, have

$$b_1 = \frac{7.540500}{7.285500} = 1.0350$$

Next we need the intercept, b_o, which is to be found by

$$b_o = \bar{y} - b_1 x \quad \text{(intercept of "best" line)} \tag{13.3}$$

This gives here

$$b_o = 5.565 - 1.0350\ (5.015) = 0.374$$

so that our estimation equation of y from x is

$$\hat{y} = 1.035 + 0.374$$

where y-hat, i.e., $\hat{y}$, designates the estimated y.

The estimating equation y-hat $= b_o + b_1 x$ is always satisfied by the pair (x-bar, y-bar), so that the best fitting straight line of y on x passes through the point of averages (x-bar, y-bar) and, of course, has slope b_1. Here this point is (5.015, 5.565). In order to draw the best-fitting line, it is desirable to find two more points on the line and plot all three. We find (4.0, 4.514) and (6.4, 6.998). These three lie on a line and supply a check in drawing the line.

Next we would like to know something about how well we are able to estimate y from x. One measure is the "coefficient of linear correlation", called r. It may be defined by

$$r = \frac{\sum (x - \bar{x})(y - \bar{y})}{\sqrt{\sum (x - \bar{x})^2 \sum (y - \bar{y})^2}} \quad \text{(linear correlation coefficient)} \tag{13.4}$$

This coefficient measures the relative closeness of fit of the points around the line. As may be seen by comparing (13.2) and (13.4), r and the slope b_1 have the same sign because the numerators are identical and both denominators are always positive. The larger in size that r is, the relatively better the straight-line fits. Perhaps the best way to use r in this connection is through its square. Thus

$$\begin{aligned} r^2 &= \text{coefficient of determination} \\ &= \text{proportion of } \sum(y - y\text{-bar})^2 \\ &\quad \text{which is explained by the line} \end{aligned}$$

The total variation of the y's around their average is $\sum(y - y\text{-bar})^2$, and r tells us what part of this can be related to x via the best-fitting straight line.

Now let us use these formulas in our example. We have

$$r = \frac{7.5405}{\sqrt{7.2855(7.9455)}} 0.99108$$

which indicates a strong linear relationship (as we see in Figure 13.1). Also

$$r^2 = 0.99108^2 = 0.98224$$

Therefore, over 98% of the total variation in y is linearly related to x via the best-fitting line, for the before and after tensions. This indicates excellent predictability.

Let us consider briefly another way to describe the predictability. Since by (13.5) r^2 is the proportion of $\sum(y - y\text{-bar})^2$ that is explained by x via the straight line, then $1 - r^2$ is the unexplained proportion or

$$(1 - r^2) \sum(y - y\text{-bar})^2 = \sum[y(\text{observed}) - y(\text{estimated})]^2$$

that is the sum of the squares of the distances from the line to the observed points, measured vertically. This sum is best divided by $n - 2$, after which the square root is taken. We call this result the "standard error of the estimate". It gives us an only-to-be expected distance of the y's from the line, within the data at hand. Calling this $s_{y \cdot x}$, we thus have

$$s_{y \cdot x} = \sqrt{\frac{(1 - r^2)\sum(y - \bar{y})}{n - 2}} \quad \text{(standard error of the estimate)} \tag{13.6}$$

In our example, this becomes

$$s_{y\cdot x} = \sqrt{\frac{(1 - 0.99108^2)\sum(7.9455)}{20 - 2}} = 0.0885$$

This means that within our 20 points, when we estimate y from x using y-hat$=0.374 + 1.035x$, we can expect the observed y to be off from the estimated y by 0.0885, on the average. However, when we go outside our original 20 data points, i.e., extrapolate, we can expect to miss hitting y by a slightly larger amount especially toward the extreme x's as a result of errors in our obtained b_o and b_1. We can make a comparison of $s_{y\cdot x}$ with s_y as follows:

If we do not know x, our best guess at a random y value is y, whereas if we know x and have a linear relation, we use the regression equation for an estimate of y. The average error in the latter estimate is $s_{y\cdot x} = 0.0885$, whereas for the former estimate, the average error is

$$s_y = \sqrt{\frac{\sum(y - \bar{y})^2}{n - 1}} = \sqrt{\frac{7.9455}{19}} = 0.647$$

Therefore, knowledge of x cuts the estimating error down from 0.647 to 0.0885.

This $s_{y\cdot x}$ thus gives us a useful picture of the accuracy with which we may estimate y from x using the best-fitting line. We may sometimes be off by $2s_{y\cdot x}$ but very seldom by $3s_{y\cdot x}$.

One further use of the best-fitting line is in predicting a lot mean y-bar, i.e., the average tension after ferroxing from a lot with the mean, x-bar, i.e., the average tension before ferroxing. This we do by (using the carat symbol for the estimate)

$$\hat{y} = 0.374 + 1.035\bar{x}$$

Hence we can tell what average tension to expect after ferroxing if we estimate the lot mean tension by x-bar before ferroxing.

This story on piston rings is a good example of a case in which reliable estimations of y from x may be made because the correlation coefficient, *r*, was high. The results of the full study compared three forms of ring suspension during ferroxing and proved useful in choosing one method and in making predictions on lots and controlling the production process to meet specifications.

13.3. SECOND EXAMPLE—CORRELATION

Our second example illustrates the second general type of application mentioned in Sec. 13.1. Correlation can be used to sift through the various "input" variables to find which are most closely related to an important "output" variable. In this case, the output variable was tons rejected for bad surface on an open-earth heat of steel. Altogether in the original study, there were about 20 or more input variables and 130 heats of steel. (Careful records are made in steel plants on each heat of steel.) Now in such process industries as steel making, plastics, foundries, chemical and food production, we must not expect large correlations for at least two reasons: (1) there are so many factors involved, which may influence the quality produced, and (2) our predecessors have probably already discovered the strong relations of really influential input variables. (Another possible trouble is inaccurate measurement of the output variable itself, i.e., our measurements are not highly reproducible.)

Therefore, our problem becomes that of trying to decide which of many rather loose or poor relationships are the best or the least poor within the current practice and specifications. Which input variables might it be desirable to control more closely than we have in the past? In this way, correlation coefficients may be used as a criterion of choice.

In this approach, however, the reader is urged to draw scatter diagrams of many, if not all, pairs of points, because some relationships may be nonlinear, which may tend to make the correlation coefficient, *r*, rather meaningless. Sometimes *r* may be very small even though there is a fairly distinct curvilinear relationship. In many statistical software packages, there is a "casement plot" which will provide the reader with a plot of all combinations of the variables (two at a time).

Now see Table 13.2 in which we give a small portion of the original data, namely the 20 heats of steel and only five of the input or independent variables. (These variables are not "independent" in any mathematical or statistical sense, but they are subject to control.) We have plotted five scatter diagrams in Figure 13.2 for the five variables vs. the quality variable—tons rejected. None of the graphs show much strength of relation. In particular, the heat with 67 tons rejected seems to be quite a disturbing factor to the relations. It may well have been largely caused by some other variable or condition than those given here.

Table 13.2 Data on Five Input and One Ouput Variables for 20 Heats of Steel, a Portion of the Original Data Comprising 130 Heats and About 20 Input Variables

Heat number	0.01% Ladle manganese	0.01% Ladle carbon	Number of bags of coke	Pouring temperature (°F)	Average sealing time (sec)	Tons rejected for bad surface
1	47	28	12	2870	31	38
2	45	25	2	2860	26	0
3	81	20	0	2855	115	2
4	51	21	4	2840	67	12
5	87	19	4	2855	14	67
6	60	22	4	2845	75	12
7	51	22	8	2850	120	55
8	49	27	8	2878	46	33
9	51	25	10	2845	70	26
10	84	16	2	2875	45	16
11	88	21	4	2875	31	12
12	44	17	5	2870	25	11
13	47	28	14	2870	27	8
14	50	25	0	2840	120	3
15	48	22	8	2860	127	6
16	53	18	2	2880	23	3
17	46	20	1	2840	44	1
18	43	19	1	2890	98	7
19	46	26	6	2860	44	0
20	48	14	5	2860	40	0

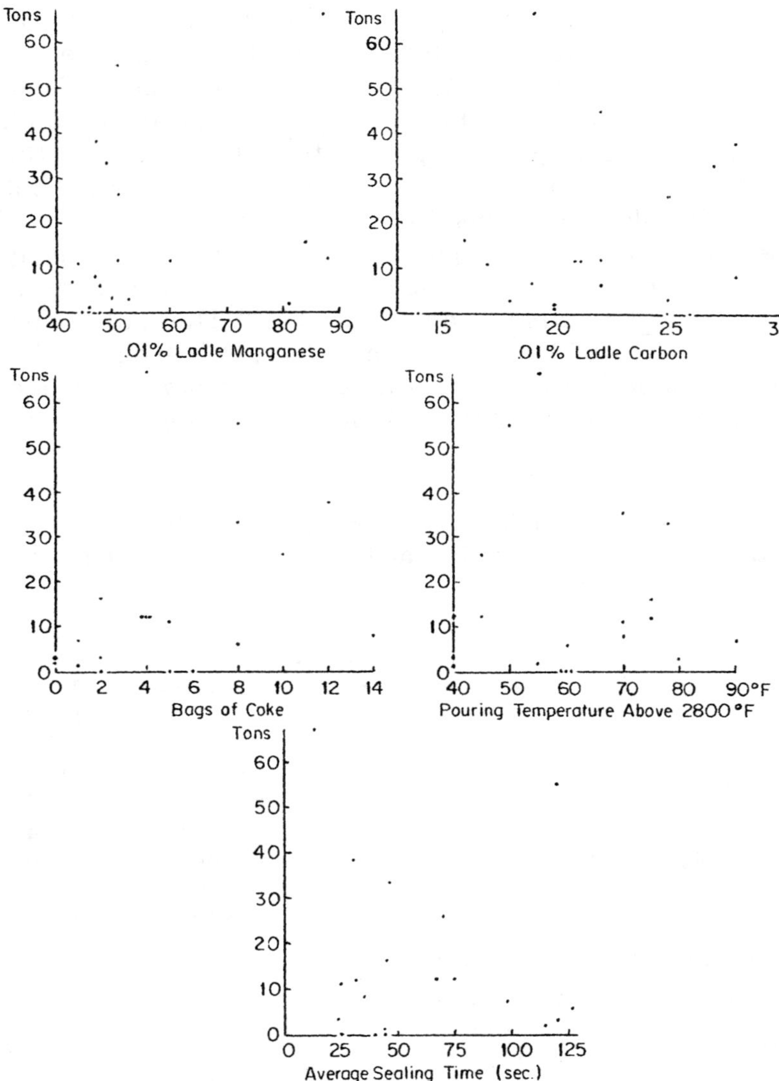

Figure 13.2 Tons of steel rejected for bad surface in heat vs. five input variables. Data from Table 13.2, being a small portion from the original study.

Let us again illustrate the calculations for r by use of deviations of the variables (see Table 13.3). There we see the same five column headings as in Table 13.1. But in Table 13.1, every product ($x - x$-bar) by ($y - y$-bar) was positive and, moreover, the relatively large positive deviations in the respective columns fell together as did the negatives. That is, no large deviation was "wasted" on a relatively small deviation of the same or opposite sign. This gave a relatively large positive sum for ($x - x$-bar) ($y - y$-bar)'s of about the same size as $\sum(x - x\text{-bar})^2$ and $\sum(y - y\text{-bar})^2$. Thus (13.4) gave a large positive r of 0.99. But in Table 13.3, we see quite a different situation in the building up of $\sum(x - x\text{-bar})$ ($y - y$-bar) with the products both plus and minus occurring. Note how

Table 13.3 Calculation Table for Correlation Between y = Tons Rejected for Bad Surface in Heat and x = 0.01% Ladle Manganese. Use Made of Deviations of Variables

x	y	$x - \bar{x}$	$y - \bar{y}$	$(x - \bar{x})^2$	$(y - \bar{y})^2$	$(x - \bar{x})(y - \bar{y})$
47	38	−9	+22.4	81	501.76	−201.6
45	0	−11	−15.6	121	243.36	+171.6
81	2	+25	−13.6	625	184.96	−340.0
51	12	−5	−3.6	25	12.96	+18.0
87	67	+31	+51.4	961	2641.96	+1593.4
60	12	+4	−3.6	16	12.96	−14.4
51	55	−5	+39.4	25	1552.36	−197.0
49	33	−7	+17.4	49	302.76	−121.8
51	26	−5	+10.4	25	108.16	−52.0
84	16	+28	+0.4	784	0.16	+11.2
88	12	+32	−3.6	1024	12.96	−115.2
44	11	−12	−4.6	144	21.16	+55.2
47	8	−9	−7.6	81	57.76	+68.4
50	3	−6	−12.6	36	158.76	+75.6
48	6	−8	−9.6	64	92.16	+76.8
53	3	−3	−12.6	9	158.76	+37.8
46	1	−10	−14.6	100	213.16	+146.0
43	7	−13	−8.6	169	73.96	+111.8
46	0	−10	−15.6	100	243.36	+156.0
48	0	−8	−15.6	64	243.36	+124.8
1119	312	−1	0	4503	6836.80	+1604.6

the algebraic sum of the column is much less than that for either sum of squares of the deviations. Now we use (13.4) finding

$$r = \frac{+1604.6}{\sqrt{4503(6836.80)}} = \frac{1604.6}{5548.5} = 0.28919$$

Thus we have a rather low positive correlation. Only $0.28919^2 = 0.08363$ of the variation, $\sum(y - y\text{-bar})^2$, among these 20 points is linearly related to percent of ladle manganese. And yet within the 130 heats in the original study, percent of manganese was one of the variables showing at least some relation to tons rejected for bad surface condition. Prediction of y from this x is, of course, relatively poor. But at least

$$b_1 = \frac{\sum(x - \bar{x})(y - \bar{y})}{\sum(x - \bar{x})^2} = \frac{+1604.6}{4503} = +0.356$$

was positive, indicating that higher rejection tended to go with higher manganese.

In the whole study, by means of scatter diagrams and correlation coefficients, four input variables were selected: percent manganese, percent sulfur, bags of coke, and sealing time (time delay before capping the ingot mold to kill the action). The actions taken were as follows:

1. higher rejection tended to go with higher sulfur (b_1 positive); so further attempt was made to hold down the percent of sulfur;
2. rejection vs. sealing time for 130 heats exhibited a curved relationship with low sealing times giving fair results while the best results were at 2 min or more sealing time; therefore sealing times were set at about 2 min; and
3. the percent manganese and bags of coke seemed somewhat related. A cross-tabulation was made for 356 heats. The average percent rejection was found for all combinations of the two variables in question. Classes of percent manganese were 0.30–0.39, 0.40–

0.49, . . . ,0.80 and over, and for bags of coke: 0, 1 and 2, 3 and 4, 5 to 7, 8–10 and 11 and over. If manganese was 0.30–0.39, bags of coke were no problem but as percent manganese increased, the number of bags of coke should be held down to three or four at most if at all possible.

These practice changes when incorporated led to very considerable improvement in surface quality.

13.4. SIMPLIFYING THE CALCULATIONS

In looking over Table 13.1, the reader may well be feeling unhappy and saying to him or herself "This is not for me!" So you should be well motivated by now for some simplifications. Of course, one approach is to use a computer with the appropriate software having correlation/regression analysis capability. This is readily obtainable in many statistical packages and some of the spread sheet packages. One can, however, go overboard in using such programs, accepting the results at full face value without making the scatter diagrams. The latter may reveal a nonlinear trend which could render the linear regression calculations inefficient or even downright misleading. Or there may be one or more really off-trend or isolated points which could well "muddy" the whole picture and (the mathematically inclined computer software being ignorant of this circumstance). So be sure to make scatter diagrams for variables whose relations are being studied.

Another point is that if the correlation is high, it is possible for the calculational methods to be discussed next, to lead to very "heavy cancelation", and thereby to create problems. For this reason, some programmers prefer to use formulas like (13.2) and (13.4) instead of the formulas now discussed.

As we have seen in Tables 13.1 and 13.3, one first finds x-bar and y-bar, then subracts them from each of the x's and y's to find the deviations, $x - x$-bar, $y - y$-bar. Then their sums of squares and the algebraic sum of the cross-products must be found. Instead of all this work, we can work directly with our x and y values to accumulate five sums, $\sum x$, $\sum y$, $\sum x^2$, $\sum y^2$ and $\sum xy$. For example, in Table 13.1, the last sum would be $\sum xy = 5.2(5.7) + 5.8(6.4) + \cdots + 4.5(5.1)$.

Now even with motor-driven desk calculators, it is often possible to accumulate all five totals in just one time through the data. (This is done by making use of $(x+y)^2 = x^2 + 2xy + y^2$, keeping x and y well separated so that the accumulating sums $\sum x$, $2\sum xy$, and $\sum y^2$ do not run together.)

Then we form the following:

$$\text{n}\sum x^2 - \left(\sum x\right)^2, \quad \text{n}\sum y - \left(\sum y\right)^2,$$
$$\text{n}\sum xy - \left(\sum x\right)\left(\sum y\right)$$

and use these results to find r, b_1, $s_{y\cdot x}$ and also b_o

$$r = \frac{n\sum xy - (\sum x)(\sum y)}{\sqrt{\left[n\sum x^2 - \sum(x)^2\right]\left[n\sum y^2 - \sum(y)^2\right]}} \tag{13.7}$$

$$b_1 = \frac{n\sum xy - \sum(x)(\sum y)}{n\sum x^2 - \sum(x)^2} \tag{13.8}$$

$$s_{y\cdot x} = \sqrt{\frac{(1-r^2)\left[n\sum y^2 - (\sum y)^2\right]}{n(n-2)}} \tag{13.9}$$

But note the possibility of very heavy cancelation.

For example, for pouring temperatures in Table 13.2, we have small relative variation. We find $\sum x = 57{,}218$ and $\sum y = 163{,}699{,}134$. But if instead we use $v = x - 2800$, we find $\sum v = 1218$ and $\sum v^2 = 78{,}334$. These give, respectively,

$$\begin{array}{rr} N\sum x^2 = 3,273,982,680 & n\sum v^2 = 1,566,680 \\ \underline{(\sum x)^2 = 3,273,899,524} & \underline{(\sum v)^2 = 1,483,524} \\ 83,156 & 83,156 \end{array}$$

Note the very heavy cancelation for the x's showing why full precision must be carried.

For the first example of Sec. 13.2, we have the following results:

$$\sum x = 100.3, \ \sum y = 111.3, \ \sum x^2 = 510.29, \ \sum y^2 = 627.33$$

$$\sum xy = 565.71, \ n\sum x^2 - \left(\sum x\right)^2 = 145.71$$

$$n\sum y^2 = 158.91, \ n\sum xy - \left(\sum x\right)\left(\sum y\right) = 150.81$$

$$r = \frac{150.81}{\sqrt{145.71(158.91)}} = 0.99108$$

$$b_1 = \frac{150.81}{145.71} = 10350$$

$$s_{y.x} = \sqrt{\frac{(1-0.99108^2)(158.91)}{20(18)}} = .0885$$

$$b_o = \bar{y} - b_1\bar{x} = 5.565 - 1.0350(5.105) = 0.374$$

All of these results agree with those found before. Note that we save time (and possibly errors) in avoiding all subtractions for $x - x$-bar and $y - y$-bar. Moreover, if n is not as simple as 20, we would have to decide as to how much precision to carry in x-bar and y-bar. Formulas (13.7) through (13.9) require no rounding off until practically the end of the calculation.

Another aid to calculation is to use "coding" as in Sec. 2.6. For example, in Table 13.2, the pouring temperatures were all 2840°F to 2890°F. We could subtract from each temperature the same amount, 2800°F or 2840°F so as to avoid any negative numbers. Then the coded numbers would be much simpler to use. Also we avoid heavy cancelation in $n\sum x^2 - (\sum x)^2$. We can use the following to code and simplify x and y as in (2.6):

$$v = \frac{x - x_o}{d}, \quad w = \frac{y - y_o}{e} \quad (d,\ e \text{ positive numbers}) \tag{13.10}$$

Using such coding leaves r unchanged, that is,

$$r_{vw} = r_{xy} \tag{13.11}$$

but we do need to decode the other quantities. Thus in line with (2.9) and (2.10)

$$\bar{x} = x_o + d\bar{v}, \quad \bar{y} = y_o + e\bar{w}, \quad b_{y.x} = b_{w \cdot v}\left(\frac{e}{d}\right), \quad s_{y \cdot x} = s_{w \cdot v}(e) \tag{13.12}$$

The simplest is when we merely subtract x_o and y_o, then we need not bother with d and e since they would each be one.

Of course, if a properly programmed computer is used, coding is unnecessary.

13.5. INTERPRETATIONS AND PRECAUTIONS

Again, let us emphasize the desirability of making scatter diagrams when studying relationships, preferably for all pairs of points (Casement Plot), but in any case, a substantial portion of them. This will help in visualizing the relation. It may bring to light a nonlinear tendency. It is possible for r to be 0, even though there is a strong nonlinear trend with excellent predictability. Or there may be isolated points way off from the general trend, or a point which is along the trend but very far away from the rest of the points. Such are evidences of nonhomogeneity. The author saw another type of nonhomogeneity in a steel-making case. The r was about $+0.2$ but the scatter diagram revealed two patches of points in each of which the correlation coefficient was about -0.5, i.e., as the value of x increased, y tended to decrease within each patch. It resulted from two different ways of calculating the quality variable.

Another aspect of correlation and regression is the sampling error. Would we obtain the same values for r, b_o, b_1, and $s_{y \cdot x}$ if we were to take a larger sample of points (x, y), or another sample of the same number? Quite obviously not. Formulas are available; see, for example, Burr (1974) and Draper and Smith (1966). Such errors, as is usual in statistics, vary inversely as the square root of the sample size, n, so that to cut an error in half we need to make n four times as large.

We shall include one test for the significance of r. Consider first the correlation between two variables, s and y, which are completely independent. Therefore, the true correlation coefficient, ρ, is 0 for the population of pairs. Nevertheless, if we gather a sample of, say, 10 pairs (x,y), then calculate r, it will not be zero. (The author has only seen one correlation coefficient of zero in a long career.) Suppose that the true correlation coefficient, ρ, is 0, then how far can the observed sample correlation coefficient, r, be off from 0? It depends upon the sample size, n. A good test as to whether ρ could be 0 and still yield an r as large in size as we did observe is the following one:

$$t = \frac{r\sqrt{n-2}}{\sqrt{1-r^2}} \tag{13.13}$$

Now if we have at least 30 pairs ($n = 30$) and if $\rho = /'0$, then only once in about 20 times will t be larger in size than 2.0. Thus if $\rho = 0$ in fact, then the probabi1ity of $|t| > 2.0$ is about 0.05. Therefore, whenever (13.13) does yield a t bigger in size than 2.0, we can quite reliably conclude that ρ is not 0 and that there really is some linear correlation present. (But remember to look at the scatter plot.)

Let us check on the two examples given. In the second $r = 0.28919$, $n = 20$. This is given by (13.13)

$$t = \frac{0.28919\sqrt{20-2}}{\sqrt{1-0.28919^2}} = 1.28$$

Thus such a correlation as was observed could perfectly well come from an uncorrelated population, and we thus have no reliable evidence that ρ is not 0. However, in the actual application, n was 130 and the r coefficient was clearly significant.

For the first example, $r = 0.99108$, $n = 20$. Therefore,

$$t = \frac{0.99108\sqrt{20-2}}{\sqrt{1-0.99108^2}} = 31.6$$

which is far above 2.0. So that even with only 20 pairs (x,y), we have extremely reliable evidence that ρ is not 0, but rather some strong positive value. Of course, with the 300 pairs in the original study, there was a much higher t, and actually r was high at 0.986. Little use cannot be made of nonsignificant correlations, but if significant, an r as low as ± 0.3 may well prove useful. (Please note that if one were to look at the plot of these data giving rise to the correlation coefficient, it would be foolish to try to distinguish which of a myriad of possible y's belong to a given x value.) For example, a friend of the author's performed a correlation study on a number of factors which might be able to predict the success or failure of a student in an engineering university to graduate. The result of his study of 5000 students yielded a correlation coefficient of 0.35 which tested to be significantly different than 0. He got his PhD but when he was asked whether he could predict the success or failure of a single student, he replied that the variability ($s_{y \cdot x}$) was too large to do that!

Next we must discuss cause and effect vs. correlation. A large-sized correlation coefficient does not necessarily mean that either variable "causes" the other to vary as observed. There often is some common cause affecting both variables. For example, if you heat a block of steel, measuring accurately length and width from time to time, a close linear relation will be observed. But neither "causes" the other to vary. Both increase because of the temperature rise.

But if a significant correlation is observed there would seem in general to be some causative factor at work, in practical cases. And we would do well to study carefully those input variables showing the largest r's in size vs. the quality variable or variables. High r's can also be well used for predictive purposes, even if there is no cause-and-effect relation present. Of course, if there is a causative relation, so much the better. In all of this, engineering knowledge and judgment are invaluable.

Following the preceding points, we also mention that two or more input factors may have a joint effect not suspected by studying the separate relation of either to the dependent variable. Such effects can be studied by "multiple regression and correlation", and by "analysis of variance" in designed experiments. In multiple regression, an objective is to develop an estimation equation involving two or more independent variables (see Burr, 1974; Draper and Smith, 1966).

13.6. SOME APPLICATIONS

The bore hardness of a certain cylinder block was subject to specification limits. Normally the hardness on the Rockwell B scale ran from 88 to 96. The problem was that in order to measure the bore hardness the block had to be cut in half, destroying the block. It was conjectured that hardness measured on the top deck (a nondestructive test) might be sufficiently related for prediction. This hardness was measured by Brinell readings. We call the respective hardnesses, x and y, each being actually an average of five readings on a single block. A study was made for 25 blocks yielding $r = 0.79$. This provided an estimation equation which was sufficiently accurate to make bore hardness measurement unnecessary, although an occasional check measurement was still run. Thus, considerable time and money were saved.

In another somewhat similar study in the meat packing industry, a study was made of percent fat content y vs. percent moisture content x of a type of meat. Both x and y are subject to specification limits, but y is a much messier and more time-consuming measurement. In the scatter diagram, the author saw, that r was at least 0.99, a better correlation than the author might have expected between repeat measurement of y alone! Further analysis was made, but the scatter diagram was almost sufficient by itself. This enabled the company to very considerably decrease the making of percent fat content measurements and still to meet specifications on both characteristics.

In an effort to control the thickness of fiberglass after recovery, y (the quality specification for usage), the machine thickness, x, was studied as a predictor. Fifty pairs of measurements were taken, yielding $r = 0.9963$, or 99.26%

(0.9963^2) of the variation in y was explained. The estimation equation was $y = 0.868x - 0.072$ in., cutting $s_y = 0.4338$ down to $s_{y \cdot x} = 0.0377$. The close relation and the estimation equation aided in controlling the machine thickness to give desired results on recovered thickness.

A certain large-running order of steel was experiencing considerable trouble in meeting the customer's Rockwell specifications. The latter measurements were made in the customer's plant. By correlating his results with about 20 input variables per heat in the steel plant, three variables were chosen having the largest (least small) correlations with the Rockwell values. The three were all ladle chemistry: carbon, manganese, and residuals (tin and other trace elements). These were studied and a relation developed involving the three variables. This led to a table for the melt foreman, which gave for various combinations of carbon and manganese the maximum and minimum limits for residuals. Thus the heat would have at least a 90% chance to meet the customer's specification range. Nor further trouble was experienced. If the chemistry of a heat did not have a 90% chance of success with this customer, the steel would be rolled on some other order for another customer.

Failure to meet specifications of $90° \pm 0.333°$ on the angle of bend after forming was blamed on variation in the thickness of stock. A correlation study was made on 128 pieces with stock thicknesses running from 0.076 to 0.081 in. giving $s_y = .1115°$, so $\pm 3s_y = \pm 0.3345°$, which is not excessive relative to $90° \pm 0.333°$. But y-bar was not 90° but instead was 89.798°, that is, about $2s_y$ off from 90°, leading to many angles out of limits. Were the two variables related? For the 128 pieces, $r = -0.7078$, which was highly significant. Thickness for x-bar was 0.07818 in. The estimation equation was $y = 6.47714 - 85.43x$. Substituting the desired y-bar $= 0$ (from 90°), we solve for $x = 6.47774/85.43 = 0.0758$ in. Therefore, if the average thickness of stock is decreased to about 0.0758 in., with the same variation in thickness, then y will meet $90° \pm 0.333°$ very well. Note that this approach assumes that the linearity continues on into lower x's. Linearity was good in the 128 pieces.

In a study of the diameters of coil condensers for automobiles, 179 condensers showed a correlation of $r = +0.6591$ for diameter vs. total foil thickness. Other input factors were total paper thickness and number of turns. They too provided good correlations. This study permitted better process control for meeting diameter specifications.

13.7. PROBLEMS

13.1. The data given below are for thicknesses in 0.00001 in. of nonmagnetic coatings of galvanized zinc on 11 pieces of iron and steel. The destructive (stripping) thickness is y; the nondestructive (magnetic) thickness is x:

x	105	120	85	121	115	127	630	155	250	310	443
y	116	132	104	139	114	129	720	174	312	338	465

We find $n = 11$, $\sum x = 2461$, $\sum y = 2743$, $\sum x^2 = 852{,}419$, $\sum y^2 = 1{,}067{,}143$, $\sum xy = 952{,}517$.

a. Plot a scatter diagram and comment.
b. Find r, b_o, b_1 and $s_{y.x}$, write the estimation equation and draw it on the scatter diagram.
c. Is r significant?
d. Estimate y, if $x = 150$.

13.2. The following data are for pounds tension of piston rings before and after ferroxing as in Table 13.1, but with an inverted V bar suspension with the gaps up, first 20 rings out of 300:

x	4.3	5.4	4.9	5.5	4.8	4.4	4.5	4.8	4.9	4.4
y	4.8	5.9	5.5	6.0	5.4	5.1	5.2	5.4	5.5	4.8
x	4.3	4.6	5.0	4.4	4.8	5.9	5.3	5.7	4.8	4.9
y	4.8	5.3	5.6	6.0	5.3	6.6	6.0	6.3	5.6	5.5

a. Plot a scatter diagram and comment.
b. Find r, b_o, b_1, and $s_{y.x}$, write the estimation equation, and draw the line on the scatter diagram.

c. Is r significant?
d. Estimate y, if $x = 4.8$.

For both Problems 13.3 and 13.4, for the respective variables of Table 13.2

a. Plot a scatter diagram and comment.
b. Find r, b_o, b_1, and $s_{y.x}$ write the estimation equation, and draw the line on the scatter diagram.

13.3. Tons rejected y vs. bags of coke x. Estimate y for $x = 2$.

13.4. Tons rejected y vs. sealing time in seconds x. Estimate y for $x = 25$.

13.5. Thirty-five springs were made at each of the following temperatures: 300, 350, 400, ..., 600, so that $n = 245$. The dependent variable was initial tension of spring in pounds. Calculations gave $r = -0.9346$, $b_o = 7.500$, $b_1 = -0.004398$, $\sum(y - y\text{-bar})^2 = 54.26$. Test r for significance. Find $s_{y \cdot x}$ and estimate y for $x = 300$ and 600.

13.6. In a research study of bituminous road mixes, seven samples were tested for x = percent water absorbed vs. y = percent asphalt absorbed. We find $\sum x = 43.18$, $\sum = 36.23$, $\sum x^2 = 285.7292$, $\sum y^2 = 207.5937$, $\sum xy = 242.6675$.

a. Find r and test it for significance.
b. Find b_o and b_1 and write the estimation equation.
c. Find $s_{y \cdot x}$.
d. Estimate y from $x = 4$ and 9. (The range of x's was from 3.96 to 8.57.)

13.7. For 49 pieces of 1/6-in. poplar (wood) about 110×51 in., a study was made of wet, x, vs. dry, y, lengths. This was to study the shrinkage so as to control y to specification needs. The following results were found: $\sum x = 5523$, $\sum y = 5162$ in, and $n\sum x^2 - (\sum x)^2 = 19{,}600$, $n\sum y - (\sum y)^2 = 23{,}672$, $n\sum xy - (\sum x)(\sum y) = 17{,}584$.

a. Find x, y, s_y, b_o, b_1, r, and $s_{y \cdot x}$.
b. Test r for significance.
c. Write the estimation equation, and estimate y for $x = 109$ and 119.

REFERENCES

IW Burr. Applied Statistical Methods. New York: Academic Press, 1974.

N Draper, H Smith. Applied Regression Analysts. New York: Wiley, 1966.

14

A Few Reliability Concepts

14.1. RELIABILITY IN GENERAL

The subject of reliability of product and its performance have come very much to the forefront in industry since 1970s. Although reliability is an integral part of the total quality program of a company or corporation, the reliability field is so broad that it can provide a field of specialization for the individual and is often serviced by a reliability department or division. Probably the field of reliability began to emerge as a discipline with the advent of the air and space age. But the need for reliability is no less important in the automotive, pharmaceutical, and foods industries.

Reliability is, of course, intimately tied in with the various techniques of statistical quality control, which we have been studying in this book. For, in order to obtain reliability, we need to have a product which is produced by well-controlled processes so that their output will function as intended and specified. Thus process control and acceptance sampling are of great importance in securing reliability.

Since the reader is quite likely to work with specialists in reliability, we shall here include some techniques and concepts which are of importance in reliability and which supplement the preceding parts of this book. There are many books available on various aspects of reliability. An excellent book covering may such aspects is reliability: management, methods, and mathematics (Lloyd and Lipow, 1962). A more recent reference for the person who desires an introduction into these aspects of reliability is cited in O'Conner (1985).

14.2. DEFINITIONS OF RELIABILITY

Nearly everyone has some conception of the meaning of "reliability". Equipment or product can be called "reliable" if it can be counted upon to perform satisfactorily its intended function or functions. Now this involves several ingredients. The product is designed and manufactured to be used over some field of application. For example, a chain saw is for cutting wood, green or dry, hard or soft, but not intended for cutting metal which may occur in the wood. Nevertheless, some safeguards can and are built into a chain saw to protect the user against moderate misuse should it occur. Some designs are used to make product not only mistake proof but "damn-fool proof".

Another facet of reliability is the time factor or repetitive usage. A product must be designed and produced so that it will perform its function for at least a minimum length of life, or a minimum number of cycles, such as startings. These guaranteed lengths of life may be short or long, but are an integral part of the picture.

Another concept is that of the contents or composition which is an area much to the forefront today. Products such as foods or pharmaceuticals must contain the prescribed or guaranteed amounts of the contents desired and must contain none or else not over a permissible amount of undesirable contents. Such requirements are controllable by process and testing controls and are a basic part of the reliability picture. Correct labeling also comes into the picture, very especially in pharmaceuticals.

Then too there is the probability facet. In this imperfect world there is usually no way to guarantee in absolute terms the functioning of product. About all we can do is to make the probability of functioning sufficiently high. This was a part of every space trip. And the unreliability came home to the astronauts on several flights.

We may summarize the foregoing with a commonly given definition of reliability.

Definition 14.1 (Reliability). The reliability of product is the probability of its successful functioning under prescribed conditions of usage and for the prescribed minimum time or number of cycles.

There are also uses of the word "reliability", such as reliability of design, reliability of production, proving of reliability by tests, such as life testing, environmental testing, and receiving testing. And then too there is the sample or 100% testing of inventory or stockpile for reliability or functionability. Probability is, however, part of the picture in all of these.

14.3. TIME TO FIRST FAILURE, THE GEOMETRIC DISTRIBUTION

As we have seen, it is desired that product should be designed and built so that it will successfully function a certain minimum number of times without a miss. As the appropriate model, we here present the "geometric distribution".

Conditions for the geometric distribution are

1. At each stage or trial the product functions as required, or it does not.
2. The probability of its functioning at each trial is constantly q_o of failing to function p_o $(p = 1 - q_o)$.
3. Functioning or nonfunctioning on trials is independent.
4. Interest is on the number of trials till the first failure or nonfunctioning.

The first three of these conditions are also assumed for the binomial distribution, so the trials are similar. But here interest centers on the number of trials until the first failure, so that the variable whose probability we want is the sample size, n, not the number of failures in a fixed, specified sample size of n.

Let us, therefore, seek a formula for $P(n)$, the probability that the first failure occurs on the nth trial or piece tested where n may be 1, 2, 3, ...

$$P(n = 1) = p_o \text{ (geometric distribution)} \tag{14.1}$$

So that the first trial results in failure. For the first failure to occur on the second trial, we must have a success followed by a failure. The respective probabilities are q_o and p_o, and since there is independence among our conditions, we may multiply giving

$$P(n = 2) = q_o p_o \text{ (geometric distribution)} \tag{14.2}$$

Next, for the first failure to occur on the third trial, there must have been two consecutive successes, followed by a failure, with probabilities q_o, q_o, p_o. Multiplying gives

$$P(n = 3) = q_o^2 p_o \text{ (geometric distribution)} \tag{14.3}$$

In general, if the first failure occurs on the nth trial, then there had to be $n - 1$ successes followed by a failure, with probabilities $q_o, \ldots, q_o{}^{n-1}$, p_o, there being $n - 1q_o$ values. Thus

$$P(n) = q_o^{n-1} p_o \text{ (geometric distribution)} \tag{14.4}$$

We note that the largest of these is the very first one, $n = 1$, and that each successive probability is q' times as large as the preceding probability (in line with the so-called "geometric progression" from which the geometric distribution gets its name). Of interest is the average number of trials until the first failure and also the standard deviation of the number of trials. For the average, we would need the sum $1P(1) + 2P(2) + 3P(3) + \cdots + n\mathrm{P}(n) + \cdots$. This proves to be

$$E(n) = 1/p_o \text{ (geometric distribution)} \tag{14.5}$$

while we also have

$$\sigma_n = \frac{\sqrt{q_o}}{p_o} \text{ (geometric distribution)} \tag{14.6}$$

Thus for $p_o = 0.01$, the average number of trials until the first failure is $1/0.01$ equals; 100, which seems very natural. The typical departure from this average of 100 is $\sqrt{0.99}/0.01 = 99.5$ or about as large as the average itself. (This relatively large σ_n in relation to $E(n)$ sometimes suggests use of the "negative binomial" distribution which gives the probability of there being n trials until the cth failure (see Lloyd and Lipow, 1962).

The geometric distribution applies equally to consecutive trials on a single piece or to single tests or inspections once to each of a series of pieces. The distribution may be used to estimate the failure rate and for qualification tests.

14.4. LOWER CONFIDENCE LIMIT ON RELIABILITY

The reliability of a piece or assembly is its probability of successful operation under the prescribed conditions and time interval. Thus reliability is analogous to q_o, where $1 - q_o = p_o$ is the failure probability or failure rate.

Now, in practice, we can never obtain the reliability, q_o exactly. The best we can do is to run a series of trials or experiments, and then use the results to obtain an estimate of q_o. Thus suppose we find two failures in 1000 trials. We could then estimate the reliability q_o to be $998/1000 = 0.998$, or the failure rate p_o to be $2/1000 = 0.002$. But if we were to repeat such a series of experiments, we might well obtain different results and estimates. Thus such a "point estimate" is subject to error.

We can determine the amount of error by setting what are commonly called "confidence limits". In reliability testing, however, our chief concern is that the reliability, q_o, be not less than a certain amount, and thus we wish to set a lower confidence limit on the reliability, from the data at hand. This

Table 14.1 Sample Size n of Tests with No Failures to Provide at Least the Specified Minimum Reliability with Listed Confidence Level

	Confidence level			
Minimum reliability	0.90	0.95	0.98	0.99
0.998	1151	1497	1954	2301
0.995	460	598	781	919
0.99	230	299	390	459
0.95	45	59	77	90
0.90	22	29	38	44

takes the form of finding d failures out of n trials, and then stating with some desired confidence or probability that $q_o \geq q_L$. For example, suppose that we observe three failures in 500 trials. Then we can be 90% confident that the reliability is at least 0.9867. This is because, if q_o were 0.9867, the probability of as many as or more successes than 497 in 500 is only 0.10. But we did observe 497 successes. Therefore, we can be 90% confident that q_o is at least as high as 0.9867. Note that the point estimate of reliability q_o is $497/500 = 0.994$, which is much higher.

In Table 14.1, we give a few combinations of confidence and reliability and the required sample size, n, of pieces all to be free from failures in order to give the specified protection.

Thus, for example, if we wish to be 90% confident that an article is 99% reliable, we can achieve this protection by having 230 trials or tests without a single failure. But if we wish to be 99% confident, we must have 459 tests with no failures.

A much larger table is given in Lloyd and Lipow (1962), with many more confidence levels and minimum reliabilities.

We also give in Table 14.2, a set of minimum reliabilities with several confidence coefficients when zero to three failures have been observed in the given number of tests. For example, we are satisfied with 90% confidence and decide to run $n = 500$ tests. We find two failures. Then we are 90% confident that the reliability is at least 0.989.

Table 14.2 Lower Limits to Reliability with Confidence Coefficients 0.90, 0.975, 0.995 When Zero to Three Failures Have Been Observed in n Tests

		Number of failures in n tests			
Number of tests n	Confidence coefficient	0	1	2	3
100	0.90	0.977	0.962	0.948	0.934
	0.975	0.964	0.946	0.930	0.915
	0.995	0.948	0.928	0.911	0.894
150	0.90	0.985	0.974	0.965	0.956
	0.975	0.976	0.963	0.953	0.943
	0.995	0.965	0.951	0.940	0.929
200	0.90	0.989	0.981	0.974	0.967
	0.975	0.982	0.972	0.964	0.957
	0.995	0.974	0.963	0.954	0.946
300	0.90	0.992	0.987	0.982	0.978
	0.975	0.988	0.982	0.976	0.971
	0.995	0.983	0.976	0.970	0.964
500	0.90	0.995	0.992	0.989	0.987
	0.975	0.993	0.989	0.986	0.983
	0.995	0.989	0.985	0.982	0.978
1000	0.90	0.998	0.996	0.995	0.993
	0.975	0.996	0.994	0.993	0.991
	0.995	0.995	0.992	0.991	0.989

A very much larger table is given in a different form in Mainland et al. (1956).

The practical usages of these two tables are quite different. For entries of Table 14.1, we decide to run n tests or trials and hope for no failures. We decide on one of the confidence levels given and a desired minimum reliability to be demonstrated with this confidence. The table then tells us how many tests to run which must be with zero failures. On the other hand, with Table 14.2, we again decide upon a confidence level given, and now choose one of the available sample sizes. After running n tests and observing failures from zero to three, we enter the table to find what reliability we have demonstrated as a minimum.

If while using the approach to Table 14.1, we should encounter a failure, we do not then decide to run enough more tests to complete a sample size in Table 14.2. This would change the probabilities unpredictably.

14.5. THE EXPONENTIAL DISTRIBUTION FOR LENGTH OF LIFE

The reader is no doubt familiar with the way the length of life of a product varies from a very short life for a few up toward a very long life for others. Electric light bulbs are a good example. Some may fail almost immediately, whereas others seemingly go on operating "forever". It is, therefore, desirable to have theoretical distribution models for length of life tests. It is hardly necessary to point out the great need for adequate length of life for all important components of any assembly or product.

The normal distribution for measurements does not commonly have the desired characteristics for the distribution of length of life of product. Instead, a most commonly used model is the so-called "exponential distribution". It is quite a standard distribution for lengths of life, much like the normal distribution is in other fields of application.

For the exponential distribution, we have

$$f(x) = \frac{1}{\mu} e^{-x/\mu} \tag{14.7}$$

where e is the natural logarithmic base, namely 2.71828 . . . , and μ is the theoretical average length of life.

Figure 14.1 shows an example of the exponential distribution. There we see that the density function $f(x)$ is at its maximum when x is 0, and that $f(x)$ steadily decreases as x increases, tailing far out to the few very long lives. The x scale is shown for $\mu = 500$ hr, the average length of life, whereas the vertical scale is in one-over-hours units, and is such as to make the total area under the curve to be one, so as to interpret areas under the curve as probabilities. It may be shown by calculus that

$$\sigma_x = \mu \text{ (exponential distribution)} \tag{14.8}$$

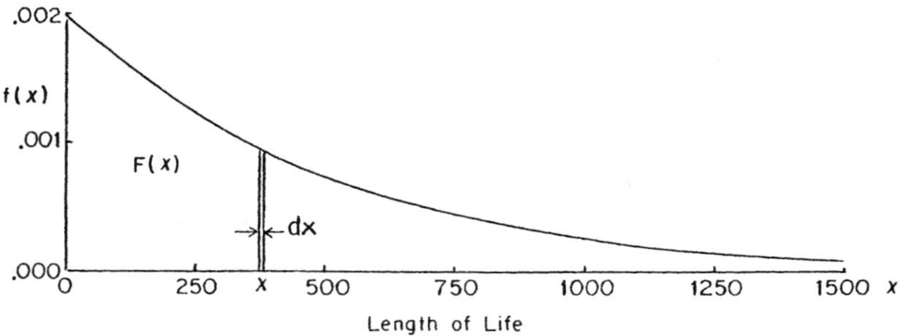

Figure 14.1 An exponential distribution. The scales are drawn for an average length of life of $\mu = 500$ hr. Total area under curve using these scales is one, representing certainly for some length of life to occur.

That is, the standard deviation of the lengths of life is just the same as the mean length of life! Moreover

$$\begin{aligned} F(x) &= 1 - e^{-x/\mu} \\ &= \text{cumulative probability of failure by time } x \end{aligned} \quad (14.9)$$

The value of $F(x)$ tells us what proportion of the pieces put in test or service can be expected to have failed by time x. Note that $F(0) = 0$ and the $F(x)$ increases toward the limit one as x increases without limit.

An important characteristic of distributions of length of life is the "hazard function" $h(x)$. The probability of a unit or piece failing in some short time interval, x to $x + dx$, is the area under the curve (such as in Figure 14.1) between ordinates at these two x values. This may be approximated by $f(x)\,dx$. So if we start with 1,000,000 units, we may expect that approximately 1,000,000 $f(x)\,dx$ will fail between times x and $x + dx$. But how many were there left out of the 1,000,000 at time x? This is the original 1,000,000 minus those already failed, i.e., 1,000,000 $F(x)$, giving the expected number left as $1{,}000{,}000 - 1{,}000{,}000\ F(x) = 1{,}000{,}000[1 - F(x)]$. What then is the failure rate over the interval x to $x + dx$? It is the number failing divided by the number present at time x, i.e.,

$$\frac{1,000,000\, f(x)\,\mathrm{d}x}{1,000,000[1-F(x)]}=\frac{f(x)\,\mathrm{d}x}{1-F(x)}$$

The latter is in general terms for all length-of-life distributions. We define then the "hazard function" in general as

$$h(x)=\frac{f(x)}{1-F(x)} \tag{14.10}$$

For the exponential distribution (14.7) and (14.10), we have

$$\begin{aligned} h(x) &= \frac{\mathrm{e}^{-x/\mu}/\mu}{1-[1-\mathrm{e}^{-x/\mu}]}=\frac{\mathrm{e}^{-x/\mu}/\mu}{\mathrm{e}^{-x/\mu}} \\ &= \frac{1}{\mu}\ (\text{exponential distribution}) \end{aligned} \tag{14.11}$$

Therefore, the hazard function or failure rate function is constant and equal to the reciprocal of the mean length of life. This constant failure rate is a unique property of the exponential distribution. This situation is sometimes called "random failure".

There can well be two additional components to a general hazard function $h(x)$. The first is an even higher initial failure rate than the $1/\mu$ of the exponential distribution. This is sometimes called "infant mortality". Then, after a substantial length of time during which the hazard function remained constant, the hazard function may gradually begin to increase. In some models, $h(x)$ may gradually increase or decrease throughout the range of x's, or $h(x)$ may decrease for early x's and then gradually increase.

The properties of the exponential distribution model make it possible to estimate average length of life and to set tests for determining reliability of life to a required minimum of time.

14.6. RELIABILITY OF COMPLEX EQUIPMENT

We now consider very briefly the difficult problem of securing adequate reliability for a complex assembly. Such an assembly may well consist of several thousand component parts. Of these, some are, of course, more crucial than others. In fact, it may well be possible to divide up the list of parts into: (1) those, the failure of any one of which will probably cause the assembly to fail, and (2) those parts in which failure will very probably not cause the assembly to fail. The former are the "lethal" or vital components which must have very high reliability. Let us see.

Now suppose that there are 500 critical components in the assembly. Let the reliabilities (under specified conditions and length of time) be $R_1, R_2, \ldots, R_{500}$. Now we next make the substantial assumption that failures among these 500 components occur entirely independently. Then the reliability of the system, i.e., the probability that all 500 components function without failure, is the product of the reliabilities of the separate lethal components, that is

$$R(\text{assembly}) = R_1 \cdot R_2 \cdots R_{500} \tag{14.12}$$

Carefully note how very troublesome this product of separate reliabilities is: If just one of these lethal components has a low reliability, then the assembly does also, even if the reliabilities of the other 499 components are virtually one! All must have high reliabilities. (R cannot exceed the lowest R_i.)

As a further analysis of (14.12), let us make the hypothetical assumption that we can make all component reliabilities equally high. How high must they be to provide reliability for the assembly of, say, 0.99? We have from (14.12)

$$R = R_i^{500}$$

or substituting 0.99 for R, we have

$$R = 0.99^{500} \quad \text{or} \quad R_i = 0.99^{1/500}$$

We can use logarithms to solve the last equation for R_i, obtaining $R_i = 0.99998$ (or we could use the binomial expansion).

Now it is possible that many of the component reliabilities are exactly one. But we cannot very well prove this and to establish a single reliability of 0.99998 would require a fantastically large sample size. Thus we have many practical problems.

One approach is to test components under conditions very much more rigorous than the specified conditions for the assembly. If all of a reasonable sample size pass the test of greatly "increased severity", then a very high reliability may be assumed for the specified conditions. Be cautioned, however, that the failure modes under the "increased severity" test are the same as those under normal operating circumstances. The author has had experience, particularly in chemical applications, that different failure modes are observed at substantially increased temperatures.

Another often-used technique for achieving very high reliability is to use "redundancy". Thus suppose we wish to use a relay which will open a circuit under certain conditions with a reliability of $0.995 = q_{o1}$. We may place another design of relay with the same purpose in series with the first one. Let its reliability be $q_{o2} = 0.996$. Now under the critical undesirable conditions if either one or both relays open the circuit their mission is accomplished.

The only way failure of the system of the two can occur is by both failing to open the circuit. If we assume that the failing of the two are independent events, then the probability of failure of the redundant system of the two is by (14.12)

$$p_o = p_{o1}p_{o2} = (1 - q_{o1}(1 - q_{o2}) = 0.005(0.004) = 0.000020$$

so that the reliability of the two in series is 0.999980.

One of the facets of our space program, facilitating very high reliability, has been our great ability at miniaturizing circuits permitting much use of redundancy without undue weight.

The reader may wonder whether there might be as many as 500 lethal components in any system. But consider a large plane. All six descendants of a good friend of the author's were killed in a plane crash which was traced to a single missing cotter pin. Then there was John Glenn, who after his first space flight, was asked for his thoughts; and he is reported to have said that he was calling to mind that each of the components was made by the lowest bidder!

14.7. SUMMARY

Our objective in this chapter has been to give the reader a speaking acquaintance with a few of the techniques with which those in reliability work. In this way, it is hoped that cooperation and communication may be facilitated. Anyone seriously interested in reliability will, of course, wish to study books on the subject, such as Lloyd and Lipow (1962).

But here let us say that one of the absolutely basic building blocks to reliability is control of processes and incoming product. The techniques we have discussed in Chapters 1–13 are thus of great use in obtaining high reliability of the finished product, of whatever type.

14.8. PROBLEMS

For Problems 14.1 and 14.2, find for the respective geometric distributions $P(1)$, $P(2)$, $P(3)$, $P(4)$, μ, and σ:

14.1. $p_o = 0.10$.

14.2. $p_o = 0.05$.

14.3. In an audit of missile systems, we want 90% confidence of a reliability of at least 0.90. What size of random sample of missile systems do we test for no failures to secure the desired confidence?

14.4. We want to prove 0.99 minimum reliability with 98% confidence. What sample size do we require with no failures to establish this degree of protection?

14.5. Suppose that we observe one failure in 200 trials or tests. Find the minimum established limit for reliability with confidence 0.90, 0.975, 0.995, and interpret the results.

14.6. Suppose that we observe one failure in 500 trials or tests. Find the minimum established limit for reliability with confidence 0.90, 0.975, 0.995, and interpret the results. Find also for confidence 0.95.

14.7. If the average length of life for an exponential distribution is 2000 hr, find $h(x)$. What number of the 1200 components remaining on test, not having failed, may be expected to fail in the next 10 hours? (Hint: Let $dx = 10$ hr.)

14.8. Two relays each expected to open under certain dangerous conditions are placed in series in a circuit. Their reliabilities are 0.99 and 0.995 and they act independently. Find the reliability of the system of the two.

14.9. Two relays each expected to close under certain dangerous conditions are placed in parallel in a circuit. Their reliabilities are 0.98 and 0.99. Find the reliability of the system of the two. Assume they act independently.

14.10. An assembly consists of three essential components, A, B and C. The reliabilities for each of these respectively is 0.99, 0.90 and 0.98. Because the reliability of B is so low, the design includes two B components in parallel. Assuming that the components act independently, what is the reliability of the system? (Hint: Draw a diagram of the system before attempting a solution.)

REFERENCES

DK Lloyd, M Lipow. Reliability: Management, Methods and Mathematics. Englewood Cliffs, NJ: Prentice-Hall, 1962.

D Mainland, L Herrera, MI Sutcliffe. Tables for Use with Binomial Samples. Dept. Medical Statistics, New York: New York University, 1956.

PDT O'Conner. Practical Reliability Engineering 2nd ed. New York: John Wiley-Interscience, 1985.

Appendix

Tables of Statistical and Mathematical Functions

Table A Cumulative Probability ≤z, to Four Decimal Places, for Standard Normal Distribution; Units and Tenths for z in Left Column, Hundredths in column Headings

Z	.00	.01	.02	.03	.04	.05	.06	.07	.08	.09
−3.5	.0002	.0002	.0002	.0002	.0002	.0002	.0002	.0002	.0002	.0002
−3.4	.0003	.0003	.0003	.0003	.0003	.0003	.0003	.0003	.0003	.0002
−3.3	.0005	.0005	.0005	.0004	.0004	.0004	.0004	.0004	.0004	.0003
−3.2	.0007	.0007	.0006	.0006	.0006	.0006	.0006	.0005	.0005	.0005
−3.1	.0010	.0009	.0009	.0009	.0008	.0008	.0008	.0008	.0007	.0007
−3.0	.0013	.0013	.0013	.0012	.0012	.0011	.0011	.0011	.0010	.0010
−2.9	.0019	.0018	.0018	.0017	.0016	.0016	.0015	.0015	.0014	.0014
−2.8	.0026	.0025	.0024	.0023	.0023	.0022	.0021	.0021	.0020	.0019
−2.7	.0035	.0034	.0033	.0032	.0031	.0030	.0029	.0028	.0027	.0026
−2.6	.0047	.0045	.0044	.0043	.0041	.0040	.0039	.0038	.0037	.0036
−2.5	.0062	.0060	.0059	.0057	.0055	.0054	.0052	.0051	.0049	.0048
−2.4	.0082	.0080	.0078	.0075	.0073	.0071	.0069	.0068	.0066	.0064
−2.3	.0107	.0104	.0102	.0099	.0096	.0094	.0091	.0089	.0087	.0084
−2.2	.0139	.0136	.0132	.0129	.01258	.0122	.0119	.0116	.0113	.0110
−2.1	.0179	.0174	.0170	.0166	.0162	.0158	.0154	.0150	.0146	.0143
−2.0	.0228	.0222	.0217	.0212	.0207	.0202	.0197	.0192	.0188	.0183
−1.9	.0287	.0281	.0274	.0268	.0262	.0256	.0250	.0244	.0239	.0233
−1.8	.0359	.0351	.0344	.0336	.0329	.0322	.0314	.0307	.0301	.0294
−1.7	.0446	.0436	.0427	.0418	.0409	.0401	.0392	.0384	.0375	.0367
−1.6	.0548	.0537	.0526	.0516	.0505	.0495	.0485	.0475	.0465	.0455
−1.5	.0668	.0655	.0643	.0630	.0618	.0606	.0594	.0582	.0571	.0559
−1.4	.0808	.0793	.0778	.0764	.0749	.0735	.0721	.0708	.0694	.0681
−1.3	.0968	.0951	.0934	.0918	.0901	.0885	.0869	.0853	.0838	.0823
−1.2	.1151	.1131	.1112	.10936	.1075	.1056	.1038	.1020	.1003	.0985
−1.1	.1357	.1335	.1314	.1292	.1271	.1251	.1230	.1210	.1190	.1170
−1.0	.1587	.1562	.1539	.1515	.14691	.1469	.1446	.1423	.1401	.1379
−0.9	.1841	.1814	.1788	.1762	.1736	.1711	.1685	.1660	.1635	.1611
−0.8	.2119	.2090	.2061	.2033	.2005	.1977	.1949	.1922	.1894	.1867
−0.7	.2420	.2389	.2358	.2327	.2296	.2266	.2236	.2206	.2177	.2148
−0.6	.2743	.2709	.2676	.2648	.2611	.2578	.2546	.2514	.2483	.2451
−0.5	.3085	.3050	.3015	.2981	.2946	.2912	.2877	.2843	.2810	.2776
−0.4	.3446	.3409	.3372	.3336	.3300	.3264	.3228	.3192	.3156	.3121
−0.3	.3821	.3783	.3745	.3707	.3669	.3632	.3594	.3557	.3520	.3483
−0.2	.4207	.4168	.4129	.4090	.4052	.4013	.3974	.3936	.3897	.3859
−0.1	.4602	.4562	.4522	.4483	.4443	.4404	.4364	.4325	.4286	.4247
−0.0	.5000	.4960	.4920	.4880	.4840	.4801	.4761	.4721	.4681	.4641

(*Continued*)

Table A (*Continued*)

Z	.00	.01	.02	.03	.04	.05	.06	.07	.08	.09
+0.0	.5000	.5040	.5080	.5120	.5160	.5199	.5239	.5279	.5319	.5359
+0.1	.5398	.5438	.5478	.5517	.5557	.5596	.5636	.5675	.5714	.5753
+0.2	.5793	.5832	.5871	.5910	.5948	.5987	.6026	.6064	.6103	.6141
+0.3	.6179	.6217	.6255	.6293	.6331	.6368	.6406	.6443	.6480	.6517
+0.4	.6554	.6591	.6628	.6664	.6700	.6736	.6772	.6808	.6844	.6879
+0.5	.6915	.6950	.6985	.7019	.7054	.7088	.7123	.7157	.7190	.7224
+0.6	.7257	.7291	.7324	.7357	.7389	.7422	.7454	.7486	.7517	.7549
+0.7	.7580	.7611	.7642	.7673	.7704	.7734	.7764	.7794	.7823	.7852
+0.8	.7881	.7910	.7939	.7967	.7995	.8023	.8051	.8078	.8106	.8133
+0.9	.8159	.8186	.8212	.8238	.8264	.8289	.8315	.8340	.8365	.8389
+1.0	.8413	.8438	.8461	.8485	.8508	.8531	.8554	.8577	.8599	.8621
+1.1	.8643	.8665	.8686	.8708	.8729	.8749	.8770	.8790	.8810	.8830
+1.2	.8849	.8869	.8888	.8907	.8925	.8944	.8962	.8980	.8997	.9015
+1.3	.9032	.9049	.9066	.9082	.9099	.9115	.9131	.9147	.9162	.9177
+1.4	.9192	.9207	.9222	.9236	.9251	.9265	.9279	.9292	.9306	.9319
+1.5	.9332	.9345	.9357	.9370	.9382	.9394	.9406	.9418	.9429	.9441
+1.6	.9452	.9463	.9474	.9484	.9495	.9505	.9515	.9525	.9535	.9545
+1.7	.9554	.9564	.9573	.9582	.9591	.9599	.9608	.9616	.9625	.9633
+1.8	.9641	.9649	.9656	.9664	.9671	.9678	.9686	.9693	.9699	.9706
+1.9	.9713	.9719	.9726	.9732	.9738	.9744	.9750	.9756	.9761	.9767
+2.0	.9772	.9778	.9783	.9788	.9793	.9798	.9803	.9808	.9812	.9817
+2.1	.9821	.9826	.9830	.9834	.9838	.9842	.9846	.9850	.9854	.9857
+2.2	.9861	.9864	.9868	.9871	.9875	.9878	.9881	.9884	.9887	.9890
+2.3	.9893	.9896	.9898	.9901	.9904	.9906	.9909	.9911	.9913	.9916
+2.4	.9918	.9920	.9922	.9925	.9927	.9929	.9931	.9932	.9934	.9936
+2.5	.9938	.9940	.9941	.9943	.9945	.9946	.9948	.9949	.9951	.9952
+2.6	.9953	.9955	.9956	.9957	.9959	.9960	.9961	.9962	.9963	.9964
+2.7	.9965	.9966	.9967	.9968	.9969	.9970	.9971	.9972	.9973	.9974
+2.8	.9974	.9975	.9976	.9977	.9977	.9978	.9979	.9979	.9980	.9981
+2.9	.9981	.9982	.9982	.9983	.9984	.9984	.9985	.9985	.9986	.9986
+3.0	.9987	.9987	.9987	.9988	.9988	.9989	.9989	.9989	.9990	.9990
+3.1	.9990	.9991	.9991	.9991	.9992	.9992	.9992	.9992	.9993	.9993
+3.2	.9993	.9993	.9994	.9994	.9994	.9994	.9994	.9995	.9995	.9995
+3.3	.9995	.9995	.9995	.9996	.9996	.9996	.9996	.9996	.9996	.9997
+3.4	.9997	.9997	.9997	.9997	.9997	.9997	.9997	.9997	.9997	.9998
+3.5	.9998	.9998	.9998	.9998	.9998	.9998	.9998	.9998	.9998	.9998

Reproduced with permission from I. W. Burr, *Engineering Statistics and Quality Control,* McGraw-Hill, New York, 1953, pp. 404, 405.

Table B Poisson Distribution. Probabilities of c or Less, Given c′, Appear in Body of Table Multiplied by 1000.

c_0 or np_0 \ c	0	1	2	3	4	5	6	7	8	9	10	11	12	13	14	15	16	17	18	19	20	21
0.02	980	1,000																				
0.04	961	999	1,000																			
0.06	942	998	1,000																			
0.08	923	997	1,000																			
0.10	905	995	1,000																			
0.15	861	990	999	1,000																		
0.20	819	982	999	1,000																		
0.25	779	974	998	1,000																		
0.30	741	963	996	1,000																		
0.35	705	951	994	1,000																		
0.40	670	938	992	999	1,000																	
0.45	638	925	989	999	1,000																	
0.50	607	910	986	998	1,000																	
0.55	577	894	982	998	1,000																	
0.60	549	878	977	997	1,000																	
0.65	522	861	972	996	999	1,000																
0.70	497	844	966	994	999	1,000																
0.75	472	827	959	993	999	1,000																
0.80	449	809	953	991	999	1,000																
0.85	427	791	945	989	998	1,000																
0.90	407	772	937	987	998	1,000																
0.95	387	754	929	984	997	1,000																
1.00	368	736	920	981	996	999	1,000															
1.1	333	699	900	974	995	999	1,000															

1.2	301	663	879	966	992	998	1,000										
1.3	273	627	857	957	989	998	1,000										
1.4	247	592	833	946	986	997	999	1,000									
1.5	223	558	809	934	981	996	999	1,000									
1.6	202	525	783	921	976	994	999	1,000									
1.7	183	493	757	907	970	992	998	1,000									
1.8	165	463	731	891	964	990	997	999	1,000								
1.9	150	434	704	875	956	987	997	999	1,000								
2.0	135	406	677	857	947	983	995	999	1,000								
2.2	111	355	623	819	928	975	993	998	1,000								
2.4	091	308	570	779	904	964	988	997	999	1,000							
2.6	074	267	518	736	877	951	983	995	999	1,000							
2.8	061	231	469	692	848	935	976	992	998	999	1,000						
3.0	050	199	423	647	815	916	966	988	996	999	1,000						
3.2	041	171	380	603	781	895	955	983	994	998	1,000						
3.4	033	147	340	558	744	871	942	977	992	997	999	1,000					
3.6	027	126	303	515	706	844	927	969	988	996	999	1,000					
3.8	022	107	269	473	668	816	909	960	984	994	998	999	1,000				
4.0	018	092	238	433	629	785	889	949	979	992	997	999	1,000				
4.2	015	078	210	395	590	753	867	936	972	989	996	999	1,000				
4.4	012	066	185	359	551	720	844	921	964	985	994	998	999	1,000			
4.6	010	056	163	326	513	686	818	905	955	980	992	997	999	1,000			
4.8	008	048	143	294	476	651	791	887	944	975	990	996	999	1,000			
5.0	007	040	125	265	440	616	762	867	932	968	986	995	998	999	1,000		
5.2	006	034	109	238	406	581	732	845	918	960	982	993	997	999	1,000		
5.4	005	029	095	213	373	546	702	822	903	951	977	990	996	999	1,000		
5.6	004	024	082	191	342	512	670	797	886	941	972	988	995	998	999	1,000	
5.8	003	021	072	170	313	478	638	771	867	929	965	984	993	997	999	1,000	
6.0	002	017	062	151	285	446	606	744	847	916	957	980	991	996	999	999	1,000

(Continued)

Table B (*Continued*)

c_0 or np_0 \ c	0	1	2	3	4	5	6	7	8	9	10	11	12	13	14	15	16	17	18	19	20	21
6.2	002	015	054	134	259	414	574	716	826	902	949	975	989	995	998	999	1,000					
6.4	002	012	046	119	235	384	542	687	803	886	939	969	986	994	997	999	1,000					
6.6	001	010	040	105	213	355	511	658	780	869	927	963	982	992	997	999	999	1,000				
6.8	001	009	034	093	192	327	480	628	755	850	915	955	978	990	996	998	999	1,000				
7.0	001	007	030	082	173	301	450	599	729	830	901	947	973	987	994	998	999	1,000				
7.2	001	006	025	072	156	276	420	569	703	810	887	937	967	984	993	997	999	999	1,000			
7.4	001	005	022	063	140	253	392	539	676	788	871	926	961	980	991	996	998	999	1,000			
7.6	001	004	019	055	125	231	365	510	648	765	854	915	954	976	989	995	998	999	1,000			
7.8	000	004	016	048	112	210	338	481	620	741	835	902	945	971	986	993	997	999	1,000			
8.0	000	003	014	042	100	191	313	453	593	717	816	888	936	966	983	992	996	998	999	1,000		
8.5	000	002	009	030	074	150	256	386	523	653	763	849	909	949	973	986	993	997	999	999	1,000	
9.0	000	001	006	021	055	116	207	324	456	587	706	803	876	926	959	978	989	995	998	999	1,000	
9.5	000	001	004	015	040	089	165	269	392	522	645	752	836	898	940	967	982	991	996	998	999	1,000
10.0	000	000	003	010	029	067	130	220	333	458	583	697	792	864	917	951	973	986	993	997	998	999
10.5	000	000	002	007	021	050	102	179	279	397	521	639	742	825	888	932	960	978	988	994	997	999
11.0	000	000	001	005	015	038	079	143	232	341	460	579	689	781	854	907	944	968	982	991	995	998
11.5	000	000	001	003	011	028	060	114	191	289	402	520	633	733	815	878	924	954	974	986	992	996
12.0	000	000	001	002	008	020	046	090	155	242	347	462	576	682	772	844	899	937	963	979	988	994
12.5	000	000	000	002	005	015	035	070	125	201	297	406	519	628	725	806	869	916	948	969	983	991
13.0	000	000	000	001	004	011	026	054	100	166	252	353	463	573	675	764	835	890	930	957	975	986
13.5	000	000	000	001	003	008	019	041	079	135	211	304	409	518	623	718	798	861	908	942	965	980
14.0	000	000	000	000	002	006	014	032	062	109	176	260	358	464	570	669	756	827	883	923	952	971

14.5	000	000	000	000	001	004	010	024	048	088	145	220	311	413	518	619	711	790	853	901	936	960
15.0	000	000	000	000	001	003	008	018	037	070	118	185	268	363	466	568	664	749	819	875	917	947
16					000	001	004	010	022	043	077	127	193	275	368	467	566	659	742	812	868	911
17					000	001	002	005	013	026	049	085	135	201	281	371	468	564	655	736	805	861
18					000	000	001	003	007	015	030	055	092	143	208	287	375	469	562	651	731	799
19					000	000	001	002	004	009	018	035	061	098	150	215	292	378	469	561	647	725
20					000	000	000	001	002	005	011	021	039	066	105	157	221	297	381	470	559	644
21					000	000	000	000	001	003	006	013	025	043	072	111	163	227	302	384	471	558
22					000	000	000	000	001	002	004	008	015	028	048	077	117	169	232	306	387	472
23					000	000	000	000	000	001	002	004	009	017	031	052	082	123	175	238	310	389
24					000	000	000	000	000	000	001	003	005	011	020	034	056	087	128	180	243	314
25					000	000	000	000	000	000	001	001	003	006	012	022	038	060	092	134	185	247

Table B (*Continued*)

c_0 or np_0 \ c	22	23	24	25	26	27	28	29	30	31	32	33	34	35	36	37	38	39	40	41	42	43
10.0	1,000																					
10.5	999	1,000																				
11.0	999	1,000																				
11.5	998	999	1,000																			
12.0	997	999	999	1,000																		
12.5	995	998	999	999	1,000																	
13.0	992	996	998	999	1,000																	
13.5	989	994	997	998	999	1,000																
14.0	983	991	995	997	999	999	1,000															
14.5	976	986	992	996	998	999	999	1,000														
15.0	967	981	989	994	997	998	999	1,000														
16	942	963	978	987	993	996	998	999	999	1,000												
17	905	937	959	975	985	991	995	997	999	999	1,000											
18	855	899	932	955	972	983	990	994	997	998	999	1,000										
19	793	849	893	927	951	969	980	988	993	996	998	999	999	1,000								
20	721	787	843	888	922	948	966	978	987	992	995	997	999	999	1,000							
21	640	716	782	838	883	917	944	963	976	985	991	994	997	998	999	999	1,000					
22	556	637	712	777	832	877	913	940	959	973	983	989	994	996	998	999	999	1,000				
23	472	555	635	708	772	827	873	908	936	956	971	981	988	993	996	997	999	999	1000			
24	392	473	554	632	704	768	823	868	904	932	953	969	979	987	992	995	997	998	999	999	1,000	
25	318	394	473	553	629	700	763	818	863	900	929	950	966	978	985	991	994	997	998	999	999	1,000

Reproduced with permission from I. W. Burr, *Engineering Statistics and Quality Control,* McGraw-Hill, New York, 1953, pp. 417–421.

Table C Control Chart Constants for Averages $\bar{x}$, Standard Deviations s, and Ranges R, from Normal Populations. Factors for computing Central Lines and Three Sigma Control Limits.

Sample Size n	Factors for Control limits for $\bar{x}$			Factors for Standard deviations, s			Factors for control limits for s			Factors for ranges R		Factors for control limits for R			
n	A	A_2	A_3	c_4	c_5	B_3	B_4	B_5	B_6	d_2	d_3	D_1	D_2	D_3	D_4
2	2.121	1.880	2.659	.798	.603	0	3.267	0	2.606	1.128	.853	0	3.686	0	3.267
3	1.732	1.023	1.954	.886	.463	0	2.568	0	2.276	1.693	.888	0	4.358	0	2.575
4	1.500	.729	1.628	.921	.389	0	2.266	0	2.088	2.059	.880	0	4.698	0	2.282
5	1.342	.577	1.427	.940	.341	0	2.089	0	1.964	2.326	.864	0	4.918	0	2.115
6	1.225	.483	1.287	.952	.308	.030	1.970	.029	1.874	2.534	.848	0	5.078	0	2.004
7	1.134	.419	1.182	.959	.282	.118	1.882	.113	1.806	2.704	.833	.205	5.203	.076	1.924
8	1.061	.373	1.099	.965	.262	.185	1.815	.179	1.751	2.847	.820	.387	5.307	.136	1.864
9	1.000	.337	1.032	.969	.246	.239	1.761	.232	1.707	2.970	.808	.546	5.394	.184	1.816
10	.949	.308	.975	.973	.232	.284	1.716	.276	1.669	3.078	.797	.687	5.469	.223	1.777
11	.905	.285	.927	.975	.221	.321	1.679	.313	1.637	3.173	.787	.812	5.534	.256	1.744
12	.866	.266	.886	.978	.211	.354	1.646	.346	1.610	3.258	.778	.924	5.592	.284	1.716
13	.832	.249	.850	.979	.202	.382	1.618	.374	1.585	3.336	.770	1.026	5.646	.308	1.692
14	.802	.235	.817	.981	.194	.406	1.594	.399	1.563	3.407	.762	1.121	5.693	.329	1.671
15	.775	.223	.789	.982	.187	.428	1.572	.421	1.544	3.472	.755	1.207	5.737	.348	1.652
16	.750	.212	.763	.983	.181	.448	1.552	.440	1.526	3.532	.749	1.285	5.779	.364	1.636
17	.728	.203	.739	.985	.175	.466	1.534	.458	1.511	3.588	.743	1.359	5.817	.379	1.621
18	.707	.194	.718	.985	.170	.482	1.518	.475	1.496	3.640	.738	1.426	5.854	.392	1.608
19	.688	.187	.698	.986	.165	.497	1.503	.490	1.483	3.689	.733	1.490	5.888	.404	1.596
20	.671	.180	.680	.987	.161	.510	1.490	.504	1.470	3.735	.729	1.548	5.922	.414	1.586
21	.655	.173	.663	.988	.157	.523	1.477	.516	1.459	3.778	.724	1.606	5.950	.425	1.575

(*Continued*)

Table C (*Continued*)

Sample Size n	Factors for Control limits for $\bar{x}$			Factors for Standard deviations, s			Factors for control limits for s			Factors for ranges R		Factors for control limits for R			
n	A	A_2	A_3	c_4	c_5	B_3	B_4	B_5	B_6	d_2	d_3	D_1	D_2	D_3	D_4
22	.640	.167	.647	.988	.153	.534	1.466	.528	1.448	3.819	.720	1.659	5.979	.434	1.566
23	.626	.162	.633	.989	.150	.545	1.455	.539	1.438	3.858	.716	1.710	6.006	.443	1.557
24	.612	.157	.619	.989	.147	.555	1.445	.549	1.429	3.895	.712	1.759	6.031	.452	1.548
25	.600	.153	.606	.990	.144	.565	1.435	.559	1.420	3.931	.709	1.804	6.058	.459	1.541

Formulas for Control Charts for Variables, $\bar{x}$, s, R

Purpose of Chart		Central Line	3-sigma Control Limits
No Standard Given-used for analyzing past data for control. ($\bar{x}$, $\bar{R}$, $\bar{s}$ are average values for data being analyzed.)	Chart for Averages, $\bar{x}$	$\bar{x}$	$\bar{x} \pm A_2\bar{R}$, or $\bar{x} \pm A_3\bar{s}$
	Ranges, R	$\bar{R}$	$D_3\bar{R}$, $D_4\bar{R}$
	Std. Devs., s	$\bar{s}$	$B_3\bar{s}$, $B_4\bar{s}$
Standards Given Used for controlling quality with respect to standards given μ, σ. $R_0 = d_2\sigma$ for n.	Averages, $\bar{x}$	μ	$\mu \pm A\sigma$
	Ranges, R	$d_2\sigma$, or R_0	$D_1\sigma$, $D_2\sigma$ D_3R_0, D_4R_0
	Std. devs., s	$c_4\sigma$	$B_5\sigma$, $B_6\sigma$

$E(s) = c_4\sigma$, $\sigma_s = c_5\sigma$, $E(R) = d_2\sigma$, $\sigma_R = d_3\sigma$.

Table D Random Numbers

1368	9621	9151	2066	1208	2664	9822	6599	6911	5112
5953	5936	2541	4011	0408	3593	3679	1378	5936	2651
7226	9466	9553	7671	8599	2119	5337	5953	6355	6889
8883	3454	6773	8207	5576	6386	7487	0190	0867	1298
7022	5281	1168	4099	8069	8721	8353	9952	8006	9045
4576	1853	7884	2451	3488	1286	4842	7719	5795	3953
8715	1416	7028	4616	3470	9938	5703	0196	3465	0034
4011	0408	2224	7626	0643	1149	8834	6429	8691	0143
1400	3694	4482	3608	1238	8221	5129	6105	5314	8385
6370	1884	0820	4854	9161	6509	7123	4070	6759	6113
4522	5749	8084	3932	7678	3549	0051	6761	6952	7041
7195	6234	6426	7148	9945	0358	3242	0519	6550	1327
0054	0810	2937	2040	2299	4198	0846	3937	3986	1019
5166	5433	0381	9686	5670	5129	2103	1125	3404	8785
1247	3793	7415	7819	1783	0506	4878	7673	9840	6629
8529	7842	7203	1844	8619	7404	4215	9969	6948	5643
8973	3440	4366	9242	2151	0244	0922	5887	4883	1177
9307	2959	5904	9012	4951	3695	4529	7197	7179	3239
2923	4276	9467	9868	2257	1925	3382	7244	1781	8037
6372	2808	1238	8098	5509	4617	4099	6705	2386	2830
6922	1807	4900	5306	0411	1828	8634	2331	7247	3230
9862	8336	6453	0545	6127	2741	5967	8447	3017	5709
3371	1530	5104	3076	5506	3101	4143	5845	2095	6127
6712	9402	9588	7019	9248	9192	4223	6555	7947	2474
3071	8782	7157	5941	8830	8563	2252	8109	5880	9912
4022	9734	7852	9096	0051	7387	7056	9331	1317	7833
9682	8892	3577	0326	5306	0050	8517	4376	0788	5443
6705	2175	9904	3743	1902	5393	3032	8432	0612	7972
1872	8292	2366	8603	4288	6809	4357	1072	6822	5611
2559	7534	2281	7351	2064	0611	9613	2000	0327	6145
4399	3751	9783	5399	5175	8894	0296	9483	0400	2272
6074	8827	2195	2532	7680	4288	6807	3101	6850	6410
5155	7186	4722	6721	0838	3632	5355	9369	2006	7681
3193	2800	6184	7891	9838	6123	9397	4019	8389	9508
8610	1880	7423	3384	4625	6653	2900	6290	9286	2396
4778	8818	2992	6300	4239	9595	4384	0611	7687	2088
3987	1619	4164	2542	4042	7799	9084	0278	8422	4330
2977	0248	2793	3351	4922	8878	5703	7421	2054	4391
1312	2919	8220	7285	5902	7882	1403	5354	9913	7109
3890	7193	7799	9190	3275	7840	1872	6232	5295	3148
0793	3468	8762	2492	5854	8430	8472	2264	9279	2128

(*Continued*)

Table D (*Continued*)

2139	4552	3444	6462	2524	8601	3372	1848	1472	9667
8277	9153	2880	9053	6880	4284	5044	8931	0861	1517
2236	4778	6639	0862	9509	2141	0208	1450	1222	5281
8837	7686	1771	3374	2894	7314	6856	0440	3766	6047
6605	6380	4599	3333	0713	8401	7146	8940	2629	2006
8399	8175	3525	1646	4019	8390	4344	8975	4489	3423
8053	3046	9102	4515	2944	9763	3003	3408	1199	2791
9837	9378	3237	7016	7593	5958	0068	3114	0456	6840
2557	6395	9496	1884	0612	8102	4402	5498	0422	3335
2671	4690	1550	2262	2597	8034	0785	2978	4409	0237
9111	0250	3275	7519	9740	4577	2064	0286	3398	1348
0391	6035	9230	4999	3332	0608	6113	0391	5789	9926
2475	2144	1886	2079	3004	9686	5669	4367	9306	2595
5336	5845	2095	6446	5694	3641	1085	8705	5416	9066
6808	0423	0155	1652	7897	4335	3567	7109	9690	3739
8525	0577	8940	9451	6726	0876	3818	7607	8854	3566
0398	0741	8787	3043	5063	0617	1770	5048	7721	7032
3623	9636	3638	1406	5731	3978	8068	7238	9715	3363
0739	2644	4917	8866	3632	5399	5175	7422	2476	2607
6713	3041	8133	8749	8835	6745	3597	3476	3816	3455
7775	9315	0432	8327	0861	1515	2297	3375	3713	9174
8599	2122	6842	9202	0810	2936	1514	2090	3067	3574
7955	3759	5254	1126	5553	4713	9605	7909	1658	5490
4766	0070	7260	6033	7997	0109	5993	7592	5436	1727
5165	1670	2534	8811	8231	3721	7947	5719	2640	1394
9111	0513	2751	8256	2931	7783	1281	6531	7259	6993
1667	1084	7889	8963	7018	8617	6381	0723	4926	4551
2145	4587	8585	2412	5431	4667	1942	7238	9613	2212
2739	5528	1481	7528	9368	1823	6979	2547	7268	2467
8769	5480	9160	5354	9700	1362	2774	7980	9157	8788
6531	9435	3422	2474	1475	0159	3414	5224	8399	5820
2937	4134	7120	2206	5084	9473	3958	7320	9878	8609
1581	3285	3727	8924	6204	0797	0882	5945	9375	9153
6268	1045	7076	1436	4165	0143	0293	4190	7171	7932
4293	0523	8625	1961	1039	2856	4889	4358	1492	3804
6936	4213	3212	7229	1230	0019	5998	9206	6753	3762
5334	7641	3258	3769	1362	2771	6124	9813	7915	8960
9373	1158	4418	8826	5665	5896	0358	4717	8232	4859
6968	9428	8950	5346	1741	2348	8143	5377	7695	0685
4229	0587	8794	4009	9691	4579	3302	7673	9629	5246
3807	7785	7097	5701	6639	0723	4819	0900	2713	7650

(*Continued*)

Table D *(Continued)*

4891	8829	1642	2155	0796	0466	2946	2970	9143	6590
1055	2968	7911	7479	8199	9735	8271	5339	7058	2964
2983	2345	0568	4125	0894	8302	0506	6761	7706	4310
4026	3129	2968	8053	2797	4022	9838	9611	0975	2437
4075	0260	4256	0337	2355	9371	2954	6021	5783	2827
8488	5450	1327	7358	2034	8060	1788	6913	6123	9405
1976	1749	5742	4098	5887	4567	6064	2777	7830	5668
2793	4701	9466	9554	8294	2160	7486	1557	4769	2781
0916	6272	6825	7188	9611	1181	2301	5516	5451	6832
5961	1149	7946	1950	2010	0600	5655	0796	0569	4365
3222	4189	1891	8172	8731	4769	2782	1325	4238	9279
1176	7834	4600	9992	9449	5824	5344	1008	6678	1921
2369	8971	2314	4806	5071	8908	8274	4936	3357	4441
0041	4329	9265	0352	4764	9070	7527	7791	1094	2008
0803	8302	6814	2422	6351	0637	0514	0246	1845	8594
9965	7804	3930	8803	0268	1426	3130	3613	3947	8086
0011	2387	3148	7559	4216	2946	2865	6333	1916	2259
1767	9871	3914	5790	5287	7915	8959	1346	5482	9251

Reproduced with permission from D. B. Owen, *Handbook of Statistical Tables*, Addison-Wesley, Reading, Mass., 1962, pp. 519, 520.

Table E Single Sample Tests for $\sigma = \sigma_1$ versus $\sigma = \sigma_2 > \sigma_1$, with Risks $\alpha = \beta$

	Ratio of σ_2/σ_1 for $\alpha = \beta$				Multiplier for σ_1^2 to get K $\alpha = \beta$			
(1) Sample size	(2) .10	(3) .05	(4) .02	(5) .01	(6) .10	(7) .05	(8) .02	(9) .01
2	13.1	31.3	92.6	206.	2.71	3.84	5.41	6.64
3	4.67	7.63	13.9	21.4	2.30	3.00	3.91	4.60
4	3.27	4.71	7.29	9.93	2.08	2.60	3.28	3.78
5	2.70	3.65	5.22	6.69	1.94	2.37	2.92	3.32
6	2.40	3.11	4.22	5.22	1.85	2.21	2.68	3.02
7	2.20	2.76	3.64	4.39	1.77	2.10	2.51	2.80
8	2.06	2.55	3.26	3.86	1.72	2.01	2.37	2.64
9	1.96	2.38	2.99	3.49	1.67	1.94	2.27	2.51
10	1.88	2.26	2.79	3.22	1.63	1.88	2.19	2.41
11	1.81	2.16	2.63	3.01	1.60	1.83	2.12	2.32
12	1.76	2.07	2.50	2.85	1.57	1.79	2.06	2.25
13	1.72	2.01	2.40	2.71	1.55	1.75	2.00	2.18
14	1.68	1.95	2.31	2.60	1.52	1.72	1.96	2.13
15	1.64	1.90	2.24	2.50	1.50	1.69	1.92	2.08
16	1.61	1.86	2.17	2.42	1.49	1.67	1.88	2.04
17	1.59	1.82	2.12	2.35	1.47	1.64	1.85	2.00
18	1.57	1.78	2.07	2.28	1.46	1.62	1.82	1.97
19	1.55	1.75	2.02	2.23	1.44	1.60	1.80	1.93
20	1.53	1.73	1.98	2.18	1.43	1.59	1.77	1.90
21	1.51	1.70	1.95	2.13	1.42	1.57	1.75	1.88
22	1.50	1.68	1.91	2.09	1.41	1.56	1.73	1.85
23	1.48	1.66	1.88	2.05	1.40	1.54	1.71	1.83
24	1.47	1.64	1.86	2.02	1.39	1.53	1.69	1.81
25	1.46	1.62	1.83	1.99	1.38	1.52	1.68	1.79
26	1.44	1.61	1.81	1.96	1.38	1.51	1.66	1.77
27	1.43	1.59	1.79	1.93	1.37	1.50	1.65	1.76
28	1.42	1.58	1.77	1.91	1.36	1.49	1.63	1.74
29	1.41	1.56	1.75	1.89	1.35	1.48	1.62	1.72
30	1.41	1.55	1.73	1.87	1.35	1.47	1.61	1.71
31	1.40	1.54	1.72	1.84	1.34	1.46	1.60	1.70
40	1.34	1.46	1.60	1.71	1.30	1.40	1.52	1.60
50	1.30	1.40	1.52	1.61	1.27	1.35	1.46	1.53
60	1.27	1.36	1.46	1.54	1.24	1.32	1.41	1.48
70	1.25	1.33	1.42	1.49	1.22	1.30	1.38	1.44

(Continued)

Table E (*Continued*)

	Ratio of σ_2/σ_1 for $\alpha=\beta$				Multiplier for σ_1^2 to get K $\alpha=\beta$			
(1)	(2)	(3)	(4)	(5)	(6)	(7)	(8)	(9)
Sample size	.10	.05	.02	.01	.10	.05	.02	.01
80	1.23	1.30	1.39	1.45	1.21	1.28	1.36	1.41
90	1.21	1.28	1.36	1.42	1.20	1.26	1.33	1.38
100	1.20	1.26	1.34	1.39	1.19	1.24	1.31	1.36

For n, seek entry in columns (2)–(5), $\leq \sigma_2/\sigma_1$, giving n. Then for this n and $\alpha=\beta$, find in columns (6)–(9), the multiplier for σ_1^2 to give K. Then $s^2 \leq K$, accept $\sigma=\sigma_1$, $s^2 > K$, reject $\sigma=\sigma_1$, and conclude $\sigma > \sigma_1$.

Reproduced with permission form I. W. Burr, *Applied Statistical Methods*, Academic Press, New York, 1974, p. 448.

ANSWERS TO ODD-NUMBERED PROBLEMS

Numerical answers rounded off at end of calculation after having carried more precision during calculation. Discussions mostly omitted here.

Chapter 2

2.1 $\bar{x} = 69.3$, $s = 2.66$, $R = 6$.

2.3 $\bar{x} = 140.4$, $s = 3.21$, $R = 7$ .001 in.

2.5 $\bar{x} = 40.3$, $s = 8.39$, $R = 15$ .0001 in.

2.7 $\bar{x} = 2.503825$, $s = .000479$, $R = .0010$ g/cm^3.

2.9

Class	0–9	10–19	20–29	30–39	40–49	
	4	19	10	23	11	
Class	50–59	60–69	70–79	80–89	90–99	.0001 in.
	14	3	3	2	1	

2.11 $\bar{x} = 36.28$, $s = 19.75$ .0001 in.

2.13 $\bar{x} = 2.50428$, $s = .001283$ g/cm^3.

2.15 $\bar{x} = 52.86$, $s = 1.003$ lb.

Chapter 3

3.1 $\sigma_d = 5$.

3.3 P(2g) = .9604, P(1g) = .0392, P(0g) = .0004.

3.5 P(3g) = .778, 688, P(2g) = .203, 136, P(1g) = .017, 664, P(0g) = .000, 512.

3.7 $P(d=0)=.625$, $P(d=1)=.375$.
3.9 $P(d=0)=10/28$, $P(d=1)=15/28$, $P(d=2)=3/28$.
3.11 210, 210, 5040.

Chapter 4
4.1 .663, .121.
4.3 .677, .271, .594.
4.5 (a) .590, (b) hypergeometric, (c) binomial.
4.7 .819, .163, .017.
4.9 $\sigma_c=1.10$, $P(c \geq 4.5)=P(5 \text{ or more})=.008$.
4.11 $\sigma_c=1.41$, $P(c \geq 6.23)=P(7 \text{ or more})=.005$.
4.13 .4096, .4096, .1536, .0256, .0016.
4.15 10/21, 10/21, 1/21.

Chapter 6
6.1 $n\bar{p}=5.5$, UCL = 12.1, control perfect. $np'=5$, UCL = 11.4, compatible
6.3 $\bar{p}=.113$, UCL = .247. Three in succession out, one other close. $\bar{p}=.0289$, UCL = .100. Tow points out of control.
6.5 $np'=2$, UCL = 6.2. In control relative to $p'=.02$. Yes.
6.7 Use $\bar{p}=.1064$, $3\text{limits}_p = .1064 \pm 3\sqrt{.1064(.8936)/\bar{n}}$ for n's used. Cracks might well be related from iron. Still investigate points out.
6.9 Prefer np chart to avoid division by 39. $n\bar{p}=1.72$, UCL = 5.6. Probably to insure delivery of 36 unbroken articles.
6.11 $\bar{c}=405.7$, $\text{limits}_c=345.3$, 466.1. Many points out. Probably repetitive defects.
6.13 $\bar{c}=7.6$, $UCL_c=15.9$, homogeneous, $c'=8$, $UCL_c=16.5$. In control relative to $c'=8$.
6.15 Alignment: $\bar{c}=9.46$, $\text{limits}_c=.2$, 18.7. In control.
6.17 $\bar{u}=1.97$, $\text{limits}_u=1.59$, 2.35. One high point, 12 low in 18. For present data $\bar{u}=1.6$, limits_u 1.26, 1.94. Several indications of assignable causes. Unless causes found and eliminated, might use 1.6.

Chapter 7
7.1 $\bar{\bar{x}}=3.694$, $\text{limits}_{\bar{x}}=3.25$, 4.14, in control. $\bar{R}=.77$, $UCL_R=1.63$, in control. $\text{Limits}_x=2.70$, 4.69, justified by control. Must lower process average. Distribution runs above specification limit with one very high x.
7.3 $\bar{\bar{x}}=.2818$, $\text{limits}_{\bar{x}}=.222$, .341. $\bar{R}=.0317$, $UCL_R=.104$. Evidence of assignable causes on both charts. Not meeting specification limit. Seek better control, but raise $\bar{\bar{x}}$ at once. (d) Set up two equations, $\bar{x}$ and R in terms of x_1, x_2: .26, .25.

	$\bar{\bar{x}}$	$\text{Limits}_{\bar{x}}$	$\bar{R}$	UCL_R	$\bar{x}$ chart	R chart	x's
7.5	2.37	−.4, +5.1	4.8	10.2	two out	3 near limit	many out
	2.65	+.2, +5.1	4.25	9.0	in control	two high	many out
	2.68	+2.0, +3.4	1.20	2.5	one high	two high	OK

7.7 $\bar{\bar{x}} = -2.10$, limits$_{\bar{x}} = -5.5, +1.3$. $\bar{R} = 5.85$, $UCL_R = 12.4$. $\bar{\bar{x}} = -.17$, limits$_{\bar{x}} = -1.6, +1.2$, $\bar{R} = 2.45$, $UCL_R = 5.2$. All four charts show lack of control. For second period a preliminary estimate of σ is 1.05; x limits $-3.3, +3.0$. All right if control is improved.

7.9 $\bar{\bar{x}} = 78.28$, limits$_{\bar{x}} = 67.2, 89.3$, in control. $\bar{R} = 10.81$, $UCL_R = 27.8$, in control. Can set x limits: 59.1, 97.4, which are far outside of specifications. Need a fundamental change in process.

7.11 Must emphasize that it is just as easy to put $\bar{x}$'s between limits for $\bar{x}$ as it is to run $\bar{x}$'s between x limits. The natural spread of x's is much greater that for $\bar{x}$'s, hence $\bar{x}$ limits must be *well inside* of x specifications.

Chapter 8

8.3 (a) $\bar{R} = 1.7$, $UCL_R = 3.6$, in control. $\hat{\sigma} = .73$ $A_2\bar{R} = .98$ $3\hat{\sigma} = 2.19$ used for safe process averages. (c) Tool wearing in control. First run may have been let run too long.

8.5 (a) $\bar{x} = 484.8$, $\bar{R} = 28.2$, $UCL_R = 92.0$, R's in control. 2σ limits for x's: 434.8, 534.8, x's in control. (b) Because of control can estimate proportion of heats outside (35%).

8.9 $U = 9.2$, control lines: 4.11, 6.35, 8.59. Out of control.

Chapter 9

9.1 $p'_{95} = .017$, $p'_{10} = .084$.

9.3 Check points:

(a) $p' = .02$, $Pa = .783$, $AOQ = .0157$, $ASN = 80.0$, $ATI = 280$
$p' = .05$, $Pa = .238$, $AOQ = .0119$, $ASN = 80.0$, $ATI = 781$

(b) $p' = .02$, $Pa = .842$, $AOQ = .0168$, $ASN = 77.6$, $ATI = 224$
$p' = .05$, $Pa = .268$, $AOQ = .0134$, $ASN = 73.1$, $ATI = 755$

9.5 Check points:

$p' = .02$, $Pa = .819$, $AOQ = .0164$, $ASN = 74.5$, $ATI = 411$
$p' = .05$, $Pa = .299$, $AOQ = .0150$, $ASN = 97.0$, $ATI = 1418$

9.7 $Ac = 7$; $n = 392$ and $n = 3920$.

Chapter 10

10.1 (a) $n = 32$, $Ac = 0$; (b) $n = 50$, $Ac = 0$; (c) $n = 13$, $Ac = 0$; (d) $n = 32$, $Ac = 0$; (e) $n_1 = n_2 = 80$, $Ac_1 = 0$, $Re_1 = 3$, $Ac_2 = 3$, $Re_2 = 4$; (f) $n = 200$, $Ac = 0$; (g) $n_1 = n_2 = 50$, $Ac_1 = 11$, $Re_1 = 16$, $Ac_2 = 26$, $Re_2 = 27$; (h) $n_1 = n_2 = 200$, $Ac_1 = 0$, $Re_1 = 3$, $Ac_2 = 3$, $Re_2 = 4$;

(i) $n_i = 80$, Ac ∗ 0 0 1 2 3 4
Re 2 3 3 4 4 5 5

(i) $n_i = 80$, Ac ∗ ∗ 0 0 1 1 2;
Re 2 2 2 3 3 3 3

(k) $n_1 = n_2 = 20$, $Ac_1 = 0$, $Re_1 = 4$, $Ac_2 = 3$, $Re_2 = 6$;
(l) $n = 32$, $Ac = 2$, $Re = 5$.

10.3 I a = .062 at $p' = .04$. Compatible.
10.5 C SP-1: i = 96, CSP-2: i = 128. Explain i and f.
10.7 i = 14 or 15. Discussion.

Chapter 11

11.1 Test seven fuses, finding $\bar{x}$. $\bar{x} \leq K$ accept, otherwise reject. K = 112.1, 112.5 or 112.9, by choice made. Sketch.
11.3 Weigh contents of four packages and find $\bar{x}$. $\bar{x} \geq K$ accept, otherwise reject. K about 501.5. Sketch.
11.5 Measure 11 gaskets and find $\bar{x}$. $\bar{x}$ inside .1044, .1076 in. accept, otherwise reject. Sketch.
11.7 Use n = 9 and find s^2. $s^2 \leq 4.91$ accept. Sketch and experiment.
11.9 Use n = 22, and find s^2 (in $(.00001 \text{ in.})^2$). $s^2 \leq 5.64$ accept, otherwise reject.
11.11 Use n = 147, k = 2.31. $\bar{x} + ks \leq 150$ sec accept, otherwise reject.
11.15 Code K. n = 35, M = 1.87%. Estimate percent P_U above U by $Q_U = (U - \bar{x})/s$. $p_U \leq M$ accept.

Chapter 12

12.1

Three dice total	3	4	5	6	7	8	9	10
Expected d, 108	.5	1.5	3.0	5.0	7.5	10.5	12.5	13.5
Three-dice total	11	12	13	14	15	16	17	18
Expected d, 108	13.5	12.5	10.5	7.5	5.0	3.0	1.5	.5
	108							

12.3 $\hat{\sigma}_{OD} = .0001376$, $\hat{\sigma}_{ID} = .0001720$. $\hat{\mu}_W = .00063$, $\hat{\sigma}_W = .0002202$ (all in inches), $z = -2.41$, $p' = .0080$.
12.5 $\mu = .1987$, $\sigma = .001, 855$ in. Needed independence. 3σ limits .19314, .20426 in.
12.7 For $z = x + y$, $\mu = 200$, $\sigma = .467$, $3\sigma = 1.40$ ohms. Will meet 200 ± 1.5 ohms.
12.9 For w, $\mu = 175.7$, $\sigma = 6.08$ lb. 3.9% low.

Chapter13

13.1 (b) r = .9964, m = 1.12261, b = 1.794, $s_{y \cdot x} = 17.52$.
(c) t = 35.2, significant. (d) $\hat{y} = 166.6$.
13.3 (b) r = .39745, m = 1.9101, b = 6.05, $s_{y \cdot x} = 17.88$. $\hat{y} = 9.87$ at x = 2.
13.5 t = −41.0, significant. $s_{y \cdot x} = .168$. $\hat{y} = 6.18, 4.86$.
13.7 (a) $\bar{x} = 112.714$, $\bar{y} = 105.347$, $s_y = 3.172$, m = .89714, b = 4.2, r = .81634, $s_{y \cdot x} = 1.852$. (b) t = 9.69 significant. (c) $\hat{y} = 102.0$, 111.0.

chapter 14

14.1 P(1) = .1, P(2) = .09, P(3) = .081, P(4) = .0729, $\mu_n = 10$, $\sigma_n = 9.5$.
14.3 Table 14.1 gives 22 tested with zero failures.
14.5 .981, .972, .963.
14.7 h(x) = 1/2000, E(d failures) = 10h(x)1200 = 6.
14.9 .9998.

Index